Sustainable Engineering Forum 2018

Topical Conference at the 2018 AIChE Spring Meeting and 14th Global Congress on Process Safety

Orlando, Florida, USA
22 - 25 April 2018

ISBN: 978-1-5108-6436-8

Printed from e-media with permission by:

Curran Associates, Inc.
57 Morehouse Lane
Red Hook, NY 12571

Some format issues inherent in the e-media version may also appear in this print version.

Printed by Curran Associates, Inc. (2018)

For permission requests, please contact AIChE
at the address below.

AIChE
120 Wall Street, FL 23
New York, NY 10005-4020

Phone: (800) 242-4363
Fax: (203) 775-5177

www.aiche.org

Additional copies of this publication are available from:

Curran Associates, Inc.
57 Morehouse Lane
Red Hook, NY 12571 USA
Phone: 845-758-0400
Fax: 845-758-2633
Email: curran@proceedings.com
Web: www.proceedings.com

TABLE OF CONTENTS

Some abstracts in this book refer to images that were not made available by the publisher

513988 Digitalization Strategies in an Age of Rapid Change

Abstract:

As with many large chemical producers, Dow has launched a wide ranging strategy towards achieving significant value through implementation of developing technologies that will speed decision making, make work safer, and improve reliability. A common thread in all these initiatives is the unification of IT and analytics into the digitalization of the work place. Digitalization, in its most classic form, simply means turning information into a computer-readable format that can be easily rendered in a variety of formats. A digitalization strategy, therefore, means planning for the use of that information from inception, through storage, to end-user, all sitting upon an appropriate digital thread. In manufacturing in general, applications are as limitless as humans can find practical use, but in the chemical industry there are particular areas that Dow has identified as key for extracting the value we seek. We will discuss Dowâ€™s broad initiatives in two areas: Business / Supply Chain, and Manufacturing / Engineering, as well as featuring the formation of two key capabilities that will support this strategy â€“ Dowâ€™s Chicago-based Digital Marketplace supporting our commercial strategy, and the newly opened Houston-based Digital Operations Center supporting our M&E strategy.

513481 Digital Twins for Predicting Early Onset of Failures Flow Valves

Abstract:

Unplanned shutdowns in petrochemical refineries due to equipment failures cause significant financial losses. Across our operations, Shell has been collecting real-time data for decades. At present, more than 10 million operational variables per minute are collected, streamed, archived, integrated with operational control systems. There is enormous potential to exploit this data further. With predictive analytics and machine learning algorithms, unexpected failures and unnecessary maintenance can be avoided, saving assets millions of dollars per annum in optimized maintenance and deferment avoidance.

This paper presents the results of two different proof-of-concept study carried out in Shell Pernis and Shell Martinez manufacturing sites, to use unlabelled historical process control data to develop a digital twin algorithm which predicts failures. Experiments for the use case were performed on multiple control valves. The target was to verify, whether the machine learning methods are capable of distinguishing between normally behaving valves from abnormally behaving valves. The difference between the prediction and the measured values, i.e. the error rate, should be as low as possible for the normal valves while being as large as possible for the abnormal valves.

Multiple solutions based on artificial neural networks and statistical approaches were implemented to model the normal behavior of the monitored systems in Pernis. Mismatches with predictions of the modelled systems were then used to predict failures in advance, where the prediction horizon reached more than a month for some use cases. 4-layer GRUs (Gated Recurrent Unit) with tanh activation functions and an input sequence length of 4 samples provided the best results. GRUs were 7% faster to train than LSTMs (Long Short-Term Memory) while reducing the prediction error by 15%. Furthermore, the approach allows for predicting failure with high accuracy, characterising failing signals with deviations that are 5x larger than the deviations during normal operation. This indicates that machine learning models can predict failures in petrochemical refineries for the studied use case, without the need for industry-specific knowledge, if the model is trained with data representing fault- free operation.

In case of Martinez, first, a traditional mass balance with meters upstream and downstream of the target flow element was utilized. A secondary accuracy check occurred by developing a novel, machine learning based predictive deterioration model that relied on first principles and statistical multivariate regression to augment and validate traditional mass balancing. This model successfully verified that the mass balance approach provided acceptable meter accuracies in this instance and allowed engineers to track predicted flow meter performance vs. actual metering. Going forward, these models will enable equipment deterioration analysis and be the catalytic step change towards predictive maintenance.

511696 Accelerating Time to Value By Integrating Smart Technology and Business Operations

Abstract:

As smart manufacturing technologies have proliferated in factories, business operations have struggled to keep pace with the new requirements these technologies present. With advanced manufacturing systems come demands on sourcing, maintenance, scheduling, information management, human resource and financial functions. Automation offers higher levels of throughput and process consistency. But it can come at a cost. Raw materials may need to be delivered to tighter tolerances. Information on pedigree may need to be provided to smart analytics. Reduced operator touch time might be replaced by increased technical skill to troubleshoot equipment or monitor quality through real time charting. Few suppliers of advanced technologies are equipped to understand these critical business demands, let alone incorporate new requirements into their designs. The result is a protracted period of debug and redesign after equipment is delivered and accepted by factories, characterized by higher scrap rates, increase labor cost and missed schedules. This presentation will offer a different approach to developing advanced manufacturing systems that achieves rapid return on investment by including business outcomes in the development. Examples will be provided that show how integration of business operations and system design achieved better performance in less time.

513432 Challenges for Big Data Use of Production Process Data

Abstract:

There are a number of initiatives for the future of manufacturing such as Industry 4.0(1), Digital Transformation Initiative(2), and Smart Manufacturing Leadership Coalition(3). These bring together a variety of concepts to create a seamless connection between the management of various data sources, advanced computer simulations/analysis, and adaptive user/machine/organization interactions. All of these include process data as part of their corporate integration, but there are a number of issues with this type of data that will need to be address if it is going to be made transparent to the user. In particular, this is the incorporation of knowledge within Big Data(4). This presentation will look at some of these issues and consider examples where these issues impact the understanding of the data.

In particular, the presentation will consider issues of the compression of production data, examples about various types of errors in the measurements, the time delays within a process, and the high correlation of process operation. The knowledge applied to these issues includes the details about the processes that ranges from the physical information about the equipment to the materials and conditions used to the theoretical understanding of the system. The knowledge needed also includes the type of sensors that are used for the measurements and the management and sampling of the data.

In addition, there is a need to understand the various intended use of the data and the appropriate analysis for that intention. The computations should be different if the intent is to predict future action compared to an intent of understanding why a certain defect occurred. This is a difference that is often ignored when analyzing data.

The examples are derived from actual production data to provide actual insight to how these issues have been traditional handled and provide some examples of how advanced analytics has helped with these issues. The intention is to encourage further development of working with production data that deals with these issues and that could be truly fulfill some of the vision of data transparency and advance analytics for manufacturingâ€™s next evolution.

1. Kagermann H, Wahlster W, Helbig J. Recommendations for implementing the strategic initiative INDUSTRIE 4.0. Final report of the Industrie 4.0 Working Group. Acatech- National Academy of Science and Engineering; 2013.
2. World_Economic_Forum. Digital Transformation Initiative Chemistry and Advanced Materials Industry. Geneva, Switzerland: World Economic Forum; 2017.
3. Davis J, Edgar T, Porter J, Bernaden J, Sarli M. Smart manufacturing, manufacturing intelligence and demand-dynamic performance. Computers & Chemical Engineering. 2012;47(Supplement C):145-56.
4. Wuest T, Weimer D, Irgens C, Thoben K-D. Machine learning in manufacturing: advantages, challenges, and applications. Production & Manufacturing Research. 2016;4(1):23-45.

513906 Risk Analysis for Count Data

Abstract:

The methodology for performing risk analysis calculation in the presence of process and measurement variation has been well established in the literature [1-4]. However current methods are limited to continuous distribution, specifically normal distribution. Even though a majority of quality measurements made within Dow are continuous in nature, there are still discrete measurements such as counts of defects, grade level assignments to name a few. In addition to this, the advances made in the field of metrology have also allowed higher resolution measurements be made on the process or products. One such capability is the ability to count low-level defects continuously on product, and such count data typically do not fit a normal distribution. The work here is to extend the risk analysis on low level defect count data which can be best described using the Poisson distribution.

[1] Burdick, R. K.; Borror, C. M.; and Montgomery, D. C. (2003). A Review of Methods for Measurement Systems Capability Analysis. *Journal of Quality Technology*, 35 (4), pp. 342â€"354.

[2] Mader, D. P.; Prins, J.; and Lampe, R. E. (1999). The Economic Impact of Measurement Error. *Quality Engineering* 11(4), pp. 563â€"574.

[3] Mottonen, Matti; Belt, Pekka; Harkonen, Janne; Haapasalo, Harri; Kess, Pekka; Manufacturing Process Capability and Specification Limits, *The Open Industrial and Manufacturing Engineering Journal*, 2008, 1, 29-36.

[4] Basnet, Chuda; Case, Kenneth E. (1992). A Review of Methods for Measurement Systems Capability Analysis. *Quality Engineering*, 4 (3), pp. 383â €"397.

513401 Dealing with Small Data in Biopharmaceutical Batch Process Monitoring: A Machine-Learning Approach

Abstract:

In recent years, biotechnology-based products have been gaining increasing visibility and success in treating chronic diseases, such as arthritis, diabetes and cancer. As per the report [1] published in 2005, approximately 27% of the new medicines in active development are in fact biotechnology-based products. Another report published in 2013 by the United States Food and Drug Administration (US FDA) states that a growing percentage, now over 40%, of all pharmaceutical industry research are now biopharmaceutical rather than classical drugs.

To efficiently monitor and control biopharmaceutical processes, multivariate statistical techniques are commonly deployed for batch process monitoring (BPM). A BPM framework uses multivariate statistical models (e.g., principal component analysis (PCA) and partial least squares (PLS)) to capture common-cause variations in the batch [2]. The control charts (e.g., Hotelling T2 and the squared prediction error (SPE) statistics) and control limits are then used to determine whether a new batch demonstrates normal operating behavior or not. An alarm is raised if a batch is statistically different from the normal operations.

Despite over two decades of research in BPM, the biopharmaceutical BPM framework suffers from a unique challenge the Low-N problem (or small-data problem). The Low-N problem represents a scenario where a product has a limited production history, denoted here by N. It is common for companies to have only one or two runs for a new drug product at a manufacturing facility. While a new product may require a limited number of runs to meet clinical or early commercial demand, it also creates a Low-N scenario for the product. In terms of BPM, a Low-N scenario poses several challenges. First, under the Low-N, it is nontrivial to capture the common-cause variations in its entirety. Second, the predictive capabilities of PCA and PLS are less accurate under the Low-N scenario. Further, under the Low-N scenario, model over-fitting becomes much harder to avoid and the effects of outliers are much more pronounced.

The Low-N problem is a longstanding, industry-wide problem in biopharmaceutical manufacturing that challenges the theoretical foundations and practical applicability of the existing BPM platform. We propose an approach to transition from a Low-N scenario to a Large-N scenario by generating an arbitrarily large number of in silico batch data sets. The proposed method is a combination of hardware exploitation and algorithm development. To this effect, we propose a block-learning method for a Bayesian non-parametric model of a batch process, and then use probabilistic programming to generate an arbitrarily large number of dynamic in silico campaign data sets. The proposed solution not only alleviates the monitoring issues associated with a Low-N scenario, it is also compatible with the industrial BPM framework. To the best of authors knowledge, this is the first method that describes a systematic approach to address the small data problem using the tools for big data. The efficacy of the proposed solution is elucidated on an industrial biopharmaceutical process.

References

[1]. Hamilton G. "The Biotechnology Market Outlook: Growth Opportunities and Effective Strategies for Licensing and Collaboration". Dublin: Research and Markets. 2005.

[2]. P. Nomikos and J. F. MacGregor, "Monitoring batch processes using multiway principal component analysis, AIChE Journal, vol. 40, no. 8, pp. 1361–1375, 1994.

513487 Analysing Big Data in Dairy Processing, By Throwing Most of It Away

Abstract:

Dairy processing in New Zealand has a desire to reduce out-of-specification production of value-added milk powders. While the process is tightly controlled, and such events are rare, the detection of some faults takes up to three days post-production, and leads to significant downgrades given very high production rates. Grading tests indicate if powders are fit for release, but are not clear measurements useful for data analysis. It is hard to identify the plant operating conditions that lead to failure, and furthermore, the underlying physical mechanisms are not always well understood.

The examination of multiple industrial plants spread over geographically separate sites, all producing similar products requires careful management of significant amounts of data. Ideally this "Big Data" approach would underpin models describing the process, which could be used to predict key functional properties of the milk powder, and deliver early warning signals before product was to slip out of specification. Extensive work was done in the very difficult task of collecting and aligning several years of process and quality data, from plants around New Zealand, representing a range of designs, ages, geographical locations, process control schemes, and data storage types.

However, as recent publications have shown, and this study reinforced, the creation of the dataset is typically a much larger job than the actual modelling and analysis, which at times can seem as trivial as sending the data to a regression function. It is becoming increasingly evident that advanced technical knowledge is not simply waiting to be unlocked by the "Big Data Revolution". Our early approaches led to a range of models that suffered from poor predictability and were of little practical use. It appeared to be a combination of several factors, where measurements to discern the key physical phenomena were not available (or possible), and that there was not enough data in the main regions of interest, due to the rarity of failure.

Whilst adding new instruments is difficult and expensive, substantial improvements in predictability were made, not by collecting even more data, but by "throwing much of it away". Changes in plant operations, equipment, and products mean that only `small data' models can be applied to subsets of the data. It was necessary to go back to first principles and apply detailed operational knowledge to find ways to improve the time alignment of process data and quality samples. Statistical resampling of data sets was applied to give more weight to rare occurrences, as well as using categorical modelling to find these occurrences.

This approach, and the need for fundamental engineering information and careful examination of the physical process was reinforced by investigation of a different big data set for cream cheese production. The data was analysed to give an overview of the behaviour of the plant, but progress in modelling and prediction required significant work and low-level calculations to dissect the data. Rather than multivariate data analysis, this required actual engineering knowledge to understand how the plant is automatically controlled, how the operators physically conducted manual tasks, and the vagaries of how the different instruments measure and are recorded.

This paper will describe two examples of how "Big Data" - that is the collection and preparation of the data - is not a solution to our problems, rather than a tool which still needs careful investigation and engineering knowledge to understand engineering problems. It is the unfortunate situation that often the biggest data still does not contain enough information, either because key information is not measured, or because there is not enough data in the regions we are really interested in. In this case, models that are intended to predict certain failures need to have substantial data that exhibits these failures. We describe this problem and how we were able to address it.

513821 Designing Carbon-Based Materials for Efficient Hydrocarbon Synthesis

Abstract:

Electrochemical reduction of CO_2 (CO_2RR) using renewable electricity has emerged as an attractive approach to synthesize hydrocarbons and mitigate atmospheric CO_2 concentration. Carbon-based materials have recently attracted a particular attention as electrocatalysts for CO_2RR due to their extraordinary properties such as low cost and flexibility in tuning the morphology and electronic structure. Wide range of different N-doped carbon materials have been experimentally tested and reported for CO_2RR. The performance map for CO_2RR over reported N-doped carbon-based materials presents diverse activities and in most cases the major product of the CO_2 reduction is found to be CO. Based on a simple thermochemical analysis Wu et al previously suggested pyridinic-N as the active sites.[1] Using density functional theory (DFT) calculations we study a wide range of different possible active motifs included in the defective carbon materials to rationalize the associated trends in their intrinsic activities. We study both CO_2RR and hydrogen evolution reaction (HER) using thermochemical analysis. We identify several active motifs that are expected to exhibit a comparable or even better activity to the state-of-the-art gold catalyst, and several configurations are suggested to be selective for CO_2RR over HER. This study provides insights in the nature of active and selective motifs for CO_2RR and rationalizes a number of reports in the literature.[2] We also show that in the presence of metal impurities further reduced products can be synthesized with significantly improved activity and selectivity.[3]

References:

(1) Wu, J.; Yadav, R. M.; Liu, M.; Sharma, P. P.; Tiwary, C. S.; Ma, L.; Zou, X.; Zhou, X.-D.; Yakobson, B. I.; Lou, J.; Ajayan, P. M. *ACS Nano* **2015**, *9* (5), 5364â€“5371.

(2) Siahrostami, S.; Jiang, K.; Karamad, M.; Chan, K.; Wang, H.; Nørskov, J. *ACS Sustain. Chem. Eng.* **2017**, acssuschemeng.7b03031.

(3) Jiang, K.; **Siahrostami, S.**; Li, Y.; Lu, Z.; Gardener, J.; Lattimer, J; Stokes, C.; Hill, W.; Bell, D.; Chan, K.; Nørskov, J.K.; Yi Cui, Y.; Wang, H. *Chem* **2017** https://doi.org/10.1016/j.chempr.2017.09.014.

513277 Leveraging Separations to Improve Sustainability of Chemical Processes

Abstract:

Typically, separations technologies for chemical processes are designed to ensure that the desired product(s) meet technical requirements such as purity and that the overall process economics are optimized. We can improve the sustainability of the chemical products by also considering separations that improve energy efficiency, reduce waste and avoid toxic solvents, entrainers, and adsorbents. This presentation will use real world examples to review developments in this area and outline strategies to select and design separations technologies for more sustainable chemical products.

513173 Soot Generation of Olefin and Alkane Flares

Abstract:

Soot Generation of olefin and alkane Flares

Anan Wang[1], Helen H. Lou[1*], Daniel Chen [1], XianChangLi [2], Christopher B. Martin[3]

(1)Dan F. Smith Department of Chemical Engineering, Lamar University, Beaumont, TX, (2)Mechanical, Lamar University, Beaumont, TX, (3)Chemistry and Biochemistry, Lamar University, Beaumont,

Abstract

Flares are used to combust unwanted natural gas or hydrocarbon by-products of petroleum refining and chemical manufacturing. Recently, flare operations have been scrutinized due to their potential environmental impacts. Industrial flares may generate black carbon i.e. soot. However, the prediction of soot generation from flares is tricky due to fundamental combustion chemistry.

In this work, detail comparison of soot precursor species and soot yields between olefin and alkane flares were conducted based on the combustion chemistry. Optimal settings of controlled variables, including flow rate of steam/air assist and flow rate of make-up fuel, were identified in order to minimize soot yields.

513109 Modular Coal Gasification Systems: A Fit for Future of Power Generation?

Abstract:

This paper explores the addition of gasifiers to coal-fired power plants to provide a flexible technology option for the accommodation of renewable power that is only available intermittently, and is the primary source of base-load coal fired plants being required to cycle. Modular gasifiers could enable leap-ahead strategies for power plant operations, and for emerging economies, by providing an integrated modularized coal utilization technologies that can define a power plant of the future.

Coal gasification is very amenable to the modular concept, whereby coal is converted into syngas or hydrogen for the production of electricity, fuels, and chemicals in small modular systems. These modular units provide a flexible technology option for using domestically available resources while meeting future environmental emission standards. In addition to efficiently producing electric power, a wide range of transportation fuels and chemicals can be produced from the cleaned syngas, thereby providing the flexibility needed to capitalize on the changing economic market. As a result, gasification provides a flexible technology option for using domestically available resources while meeting future environmental emission standards. Poly-generation plants that produce multiple products are uniquely possible with gasification technologies.

Meanwhile, coal-fired power generating units in the U.S. continue to switch from high-capacity-factor operation to various modes of flexible operation. Cycling and load following are not new. However, over the past decade, and in particular the past few years, a perfect storm of factors has gathered to require even baseload coal plants to prepare for increased flexible operation.

Integration of the gasifier with the coal fired units is possible because of the many advances we have made in modeling, materials research, and advanced manufacturing. New developments in Computational Fluid Dynamics (CFD) modeling should be used to create new gasifier designs. These new designs for coal gasification systems which incorporate modular concepts will allow the gasifier to be easily added, expanded, or replaced as necessary. New materials that are being designed can be coupled with 3-D manufacturing can be used only where needed, for example in high temperature regions of the gasifier, to minimize costs.

Case Studies and Lessons from Process Safety Culture Improvement Initiatives

Steve Arendt, PE
ABS Group
16855 Northchase Drive
Houston, TX 77060
sarendt@absconsulting.com

Rick Curtis
ABS Group
16855 Northchase Drive
Houston, TX 77060
rcurtis@absconsulting.com

Prepared for Presentation at
American Institute of Chemical Engineers
2018 Spring Meeting and 14th Global Congress on Process Safety
Orlando, Florida
April 22 – 25, 2018

Case Studies and Lessons from Process Safety Culture Improvement Initiatives

Steve Arendt, PE
ABS Group
16855 Northchase Drive
Houston, TX 77060
sarendt@absconsulting.com

Rick Curtis
ABS Group
16855 Northchase Drive
Houston, TX 77060
rcurtis@absconsulting.com

Keywords: Culture, process safety, organizational culture, culture improvement

Abstract

Many have recognized the importance of organizational culture in major accident prevention. As a result, companies have undertaken long-term efforts to assess, improve and maintain a good process safety/HSE culture. This paper will summarize lessons from over 40 safety culture improvement initiatives and some especially challenging issues faced by companies wanting to improve process safety culture.

1 Introduction

Companies have been taught many times that organizational factors have been important contributors to major incidents. Some of those organizational characteristics have to do with not having a proper safety culture, failing to exhibit strong leadership to support the culture, and not creating the consistent operational discipline at all organizational levels. One theme common to all three of these aspects has been the failure of companies to learn from experience – either from their own or from others. The following are some characteristics of a "process safety learning-disabled company:"

- Dysfunctional safety culture
- People mask information and "kill the messengers"
- Employees reluctant to question authority – fear of retribution
- Procedures are not followed without accountability
- Unclear or insufficient safety and/or production information
- Complacency, low trust, silo mentality during execution of key tasks
- Misplaced safety ownership

- Invisible or ineffective leadership
- Superficial causal analysis of problems
- Broken equipment or problems with procedures either don't get fixed, or stay fixed
- No company memory

If a company wants to sustain continuous improvement in process safety and HSE, or wants to reach for operational excellence, it must address improvement of culture throughout the organization beginning from the CEO to the field operations as well as maintenance. This paper compares the strengths and weaknesses of safety culture frameworks. The American Institute of Chemical Engineer's (AIChE) Center for Chemical Process Safety (CCPS) has made process safety culture an official process safety management element when it published *"Risk Based Process Safety Guidelines" in 2007.* Embedded in that element chapter are the key work practices to establish a management system to nurture process safety culture and 12 essential features of a good culture. Over the past ten years, other organizations in several industries have proposed various frameworks to understand and assess safety culture. During that same period, ABS Group has performed over 40 culture evaluations of organizations involving hundreds of facilities, and tens of thousands of employees and contractors, across all global regions and ethnic cultures.

The evaluations have determined weaknesses of safety culture and linked them as causal factors to actual company safety performance. These weaknesses represent the current symptoms of the cultural 'disease' for these organizations. In several cases, ABS Group was able to determine the historical 'disease pathways' in which conditions that influence the safety culture were allowed to degrade over time. However, we have also witnessed how companies have adopted practical measures to prevent, mitigate, and recover from 'safety culture disease'.

This paper summarizes practical means which the senior executives, plant managers, engineers, and operating and maintenance staff can adopt to avoid 'safety culture disease'. We summarize the lessons from a number of case study evaluations, characterize common culture degradation patterns among these different companies and provide suggestions for vaccinating companies against future process safety culture disease.

2 Culture, Leadership and Operational Discipline

Process safety culture has been a key organizational factor in several major process safety incidents. Leadership has been a focal point of several industry groups promoting performance improvement successes as described in the CCPS Risk Based Process Safety (RBPS) culture and conduct of operations (ConOps) elements[1,2] and its Vision 220 initiative. The UK Process Safety Leadership Group's report following the Buncefield incident also alluded to leadership direction as an important factor for process safety improvements.[3] Subsequent tools provided by the Organization for Economic Cooperation and Development (OECD) and the United Kingdom Petroleum Industry Association (UKPIA).[4,5] In the U.S., the American Petroleum Institute (API) and the American Fuel & Petrochemical Manufacturers (AFPM) have been focusing on process safety leadership as one of its key assessment areas. [6] Finally, lack of operating discipline, as a product of a Conduct of Operations program, has been a sore point for many companies whose incidents causal factors increasingly involve 'people who did not follow the procedures.'[7]

With consideration for the numerous definitions for culture, the author's definition of culture is – ***"Culture is the tendency in all of us – and our organization - to want to do the right thing in the right way at the right time, ALL the time – even when/if no one is looking."*** Leadership is an essential feature of a good culture. Operational discipline (or the lack thereof) is a behavioral result of your culture and leadership.

3 Frameworks for Understanding Safety Culture

The CCPS introduced "culture" as an official safety management system (SMS) element for the first time when it published its *RBPS* guidelines.[1] A CCPS safety culture working group and ABS Consulting evaluated major organizational accidents and prepared a safety culture awareness tool, which has been widely distributed via the CCPS's web page. Subsequently, process safety culture was defined as an element in the RBPS Guidelines that created a culture management practice and outlined the 12 essential features of a strong process safety culture. These 12 essential features are given in Table 1.

Table 1. Essential Features of a Strong Process Safety Culture

1	Establish process safety as a core value
2	Provide strong leadership
3	Establish and enforce high standards of performance
4	Document the process safety culture emphasis and approach
5	Maintain a sense of vulnerability
6	Empower individuals to successfully fulfill their process safety responsibilities
7	Defer to expertise
8	Ensure open and effective communications
9	Establish a questioning/learning environment
10	Foster mutual trust
11	Provide timely response to process safety issues and concerns
12	Provide continuous monitoring of performance

Other organizations and authors stated below have also established their own structures for understanding safety culture:

- James Reasons, Department of Psychology, University of Manchester), numerous publications including *Achieving a safe culture; theory and practice*, 1998
- HRO (High Reliability Organization)
- UK HSE (Health and Safety Executive)
- PSLG (Process Safety Leadership Group)
- IAEA Safety Culture Characteristics Nuclear Industry)

- IAEA Generalized Safety Culture Characteristics
- NRC Traits of a Strong Safety Culture
- INPO's Principles for a Strong Nuclear Safety Culture
- BSEE Safety Culture Aspects
- CNEB - split into four positives and four negatives (defenses needed)
- Pipeline Hazardous Materials Safety Administration (PHMSA)
- Energy Institute

As one could expect, as more of these frameworks are created by various groups, with respect to which industry they come from, the more the frameworks tend to cover the same aspects for improving safety culture. Figure 1 compares the CCPS safety culture features with the other frameworks to determine which framework appears to be most complete.

Our belief is that, while the CCPS culture feature framework is the most complete one. Figure 1 illustrates how several of the frameworks compare in coverage to the CCPS-RBPS 12 essential features where a green cell indicates similar content in the various frameworks. Ultimately, it will not matter which framework one would follow. The main intent by following any of the frameworks is ensuring excellence by adhering to the cultural improvement aspects stated within a framework. When this does not happen and a poor culture persists, there are some lessons that the authors have learned regarding why there is unfavorable culture and what needs to be done for improvements.

- If an organization has a poor culture, marked by mistrust or requires major improvements, the worst thing that the organization can do is to only begin “talking” about it at the top management level.
- The “top” management needs to first start “behaving” in a manner substantiating the importance of a good safety culture in order to address the culture weaknesses. After that, the management’s talk can build on improvements steps from the bottom up.
- If the organization conducts surveys, they should be done anonymously and voluntarily. The company should commit to sharing the results of the survey early.
- Any continuing education or training should extend to all of the workforce including contractors.
- Organizations must strive to build ownership.

	CCPS - BBPS	INPO	IAEA	BSEE	NRC	CNEB	UK HSE	HRO	PHMSA
1	Establish process safety as a core value	■	■			■		■	■
2	Provide strong leadership	■	■	■	■	■	■		
3	Establish and enforce high standards of performance	■	■	■	■	■			■
4	Document the process safety culture emphasis andapproach		■	■			■		■
5	Maintain a sense of vulnerability					■		■	
6	Empower individuals to successfully fulfill their process safety responsibilities		■		■		■		■
7	Defer to expertise		■				■	■	
8	Ensure open and effective communications		■	■	■		■		
9	Establish a questioning/learning environment	■	■	■	■				■
10	Foster mutual trust	■	■		■				■
11	Provide timely response to process safety issues and concerns								
12	Provide continuous monitoring of performance	■	■	■			■		

Figure 1. Cross-reference of Safety Culture Frameworks to CCPS Features

Once the framework that is to be used in the process safety culture assessment, the assessment tools can begin, using an appropriate assessment methodology.

4 How to Measure Culture and Leadership

Process safety culture is hard to measure and more difficult to change. There are few direct indicators of process safety culture, and because of its nature, it cannot be evaluated very frequently. Leadership and operational discipline are essential attributes of sustaining a healthy process safety culture. So, how does a company know if better culture or operational discipline is needed? What evidence would lead the company to believe that a better operational discipline and a ConOps element are needed?

There are several tools that can be used to assess the process safety culture. The authors typically use a combination of all these means, anchored by some variation while conducting a culture survey. Typical ways to get a handle on process safety culture are:

Employee surveys – Surveys are the most frequently used method. Typically, a company will prepare an anonymous survey (20-70 questions, shorter is better) for management, hourly employees and contractors. The content of the survey has historically been focused more on occupational safety issues, but recently they have been adapted to addressing process safety issues. Survey questions are developed to gauge how employees "feel" about important process safety-related matters. Table 2 lists some best practices for developing and conducting process safety culture surveys. Some difficulties encountered with surveys are: (1) that they should be voluntary, which may lead to insufficient participation to achieve statistical validity, and (2) surveys cannot be repeated very often or else, workers will become accustomed to it and can tend to provide feedback which companies "what you want to hear", rather than what they are really feeling. While survey data alone is important and typically forms the foundation for the process safety culture assessment, it (like all assessment tools) contains inherent weaknesses and should be supported with other culture assessment tools.

Interviews – Another way to elicit process safety culture insights is through targeted representative interviews of company personnel. These interviews may last anywhere from 15 minutes to an hour. A disadvantage of interviews is that they are very time-consuming and resource intensive. Therefore, the results are more difficult to amass together in a consistent framework for analysis.

Site observations – Process safety culture issues that deal with the tendency for employees to not follow procedures, safe work practices, etc., can be identified via workplace observations. These observations can prove to be very effective but, are difficult to conduct in a consistent fashion using a large number of observers. The biggest limitation is that they are difficult to do without the person being observed knowing that they are being watched or evaluated. If they know they are being observed then the observer is unlikely to get the "real" information about how the worker behaves "without anyone looking." Observations of physical conditions (e.g., equipment or "housekeeping") often reveal culture insights. Still, even if the worker corrects his/her behavior because he/she is being observed, it would serve as a reiteration of the expected safety practice to the worker.

Process safety leading indicator metrics – More companies are using leading indicators of process safety as a window into understanding their process safety culture. For example, the rate of reporting of near-misses, the rate of close-out of action items, and the completeness of training compliances can be used to gauge a company's leadership in process safety. Metrics are good because they can be refreshed frequently but, they are usually very "indirect" measurements of process safety issues.

Analysis of Other Process safety Performance Data – Other facility data can shed light on the process safety culture. The assessment team may also briefly review previous incidents and audits as well as hazard assessment tools such as a process hazard analysis. These are reviewed not as a compliance audit exercise, but instead, the culture assessor will look for depth and breadth of the documents to provide insight as to, for example, whether managers, employees, and contractors take these important activities seriously.

Table 2. Important Considerations for Doing a Safety Culture Survey

1. Surveys should be engaging, but not too short (can't achieve sufficient inquiry into measurement objectives) or too long (people get bored and just "tick the box"), i.e., ideally 30-50 questions or a 15-20 minute survey.
2. Survey questions should be simple.
3. Questions should be based on designed measurement objectives. Don't just categorize the questions by "topic" (e.g., training, reporting). Link the survey to the company "values".
4. Develop questions that will stand the test of time. Be willing to augment previous surveys with some "new" topics. Best to make questions and answers either all "positive" or all "negative".
5. Use a Likert psychometric scale such as Strongly disagree, Disagree, Neutral, Agree, Strongly agree
6. Include everyone, even contractors.
7. As surveys are voluntary, agree on communicating the results and actions to employees within a reasonable period of time.
8. Be careful with demographic questions because they can make some groups so small that it invalidates anonymity.
9. Conduct surveys electronically and augment with polling workshops as needed.
10. Have at least one essay answer question in the end.
11. Surveys must be preceded by a good rollout communication. Make the surveys available for at least a couple of weeks. Do monitor the number of surveys completed and provide reminders but, not too aggressively.

Another important aspect of survey and interview question development is to determine the focal point and "measurement direction" for each question. It is important to frame questions about what a respondent sees others doing - subordinates, peers, direct supervisors, or management/company" – rather than asking all questions about 'how they perform their duties themselves.' Figure 2 illustrates the various measurement directions available.

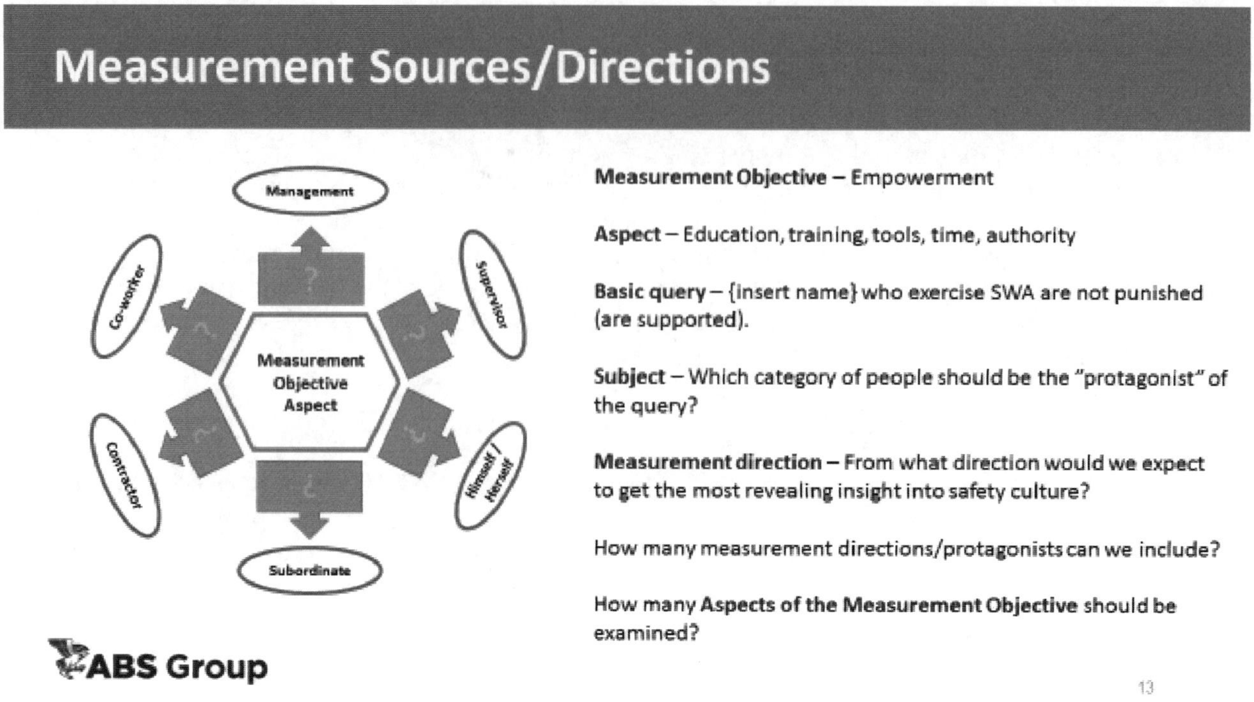

Figure 2. Measurement Sources and Directions

Table 3 is a partial list of the kinds of questions that may be asked in a process safety culture survey. A delicate balance must be struck between soliciting data to adequate address each of the elements in the framework and being as brief as possible.

Table 3. Typical Process Safety Culture Survey Questions

- Plant management is sincerely interested in feedback on process safety from the employees.
- My co-workers routinely report all process safety issues - including near misses - to our supervisor.
- Process safety is #1 at this company – we do not cut corners in order to maintain or improve production
- When an incident occurs, management focuses on correcting the issue rather than on "blaming someone."
- Our supervisor never asks us to perform an unsafe operation or task.
- My coworkers never "work around" a safety problem, rather than report it
- Overall, I am satisfied with the process safety training we receive
- Safety systems are not left disabled, but are restored as rapidly as possible

4.1 Process Safety Performance Assurance Review

ABS Group has devised an approach for connecting process safety culture survey results to process safety outcomes. Figure 3 is an example of overall process safety survey results alone for a facility where the figure illustrates the positive responses across all 12 essential features.

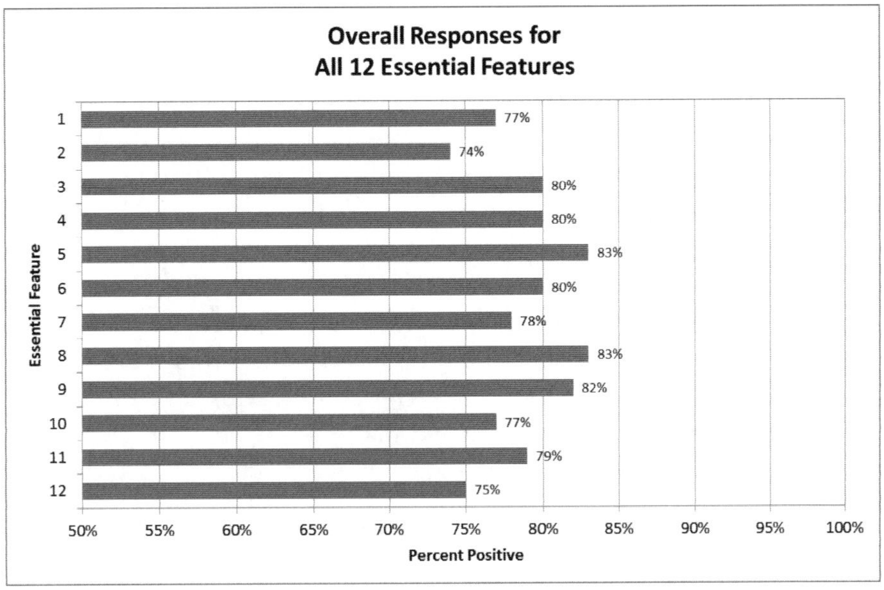

Figure 3. Example of Process Safety Survey Data Alone

The process safety Performance Assurance Review (PAR) approach (Figure 4) includes data from all culture assessment data sources (e.g., survey, interviews, performance) and maps them in to the 12 essential features of a good process safety culture. The PAR approach categorizes the results from a review of recent and historical process safety performance at a plant (e.g., current PSM or HSE audit results) and maps these results to the same 12 essential features. The weighted outcome of the mapping of contributions of all culture data collected to the 12 essential features are totaled and the most significant process safety culture issues are identified for the plant or company.

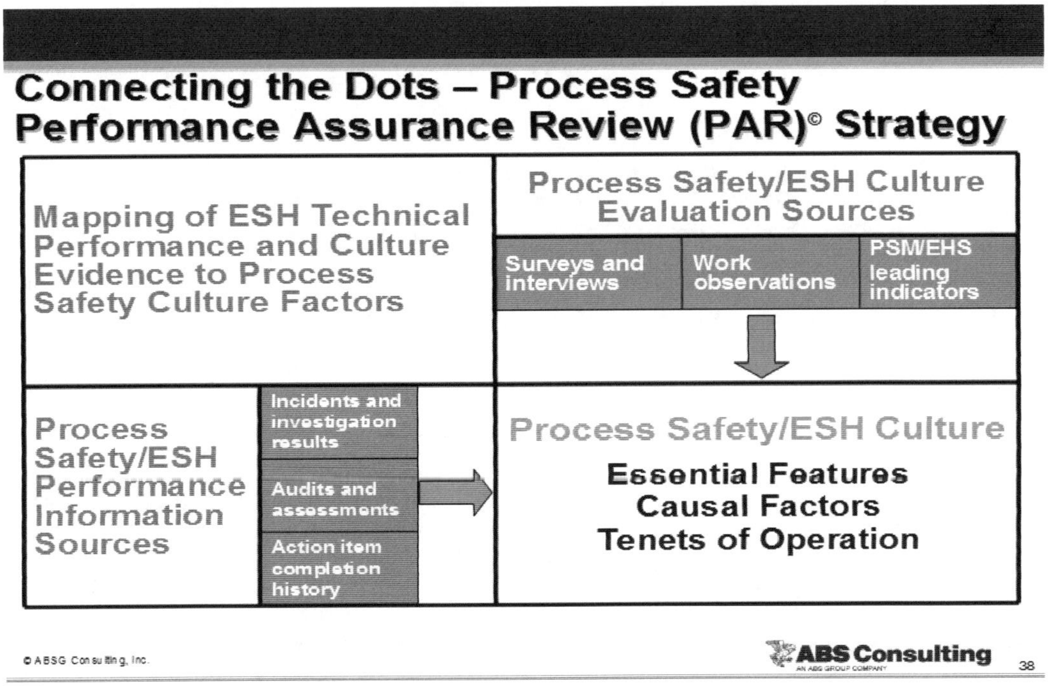

Figure 4. Process Safety Performance Review Culture Evaluation Approach

4.2 Categorizing the Process Safety Performance Review Results

Technical findings are given varying degrees of "risk significance" (Figure 5) based upon, (a) the type of finding (e.g., imminent integrity issue vs. improper training records) and (b) the number of items contributing to the finding (e.g., 93 problem situations found for one category of finding vs. 8 situations found for another). Each of these categories is mapped to the 12 essential features of a good process safety culture. This mapping shows the relationship between the existence of the apparent underlying culture problem and how it may have results in the evident process safety outcome. For example, a plant that has, (1) known imminent apparent safety hazard situations that have not been fixed in a long time, (2) a large number of action items not completed, (3) large deficient equipment and inspection backlogs, and (4) poor operating discipline indicated by a lack of following safety work practices, all indicate that the plant has the culture disease symptom of "Normalization of Deviance – Lack of Setting and Follow Process Safety Standards and Rules".

Figure 5. Process Safety Performance Deficiency Categories

- Containment integrity issue allowed to exist
- Unsafe work practice
- Action item not completed or late
- Safety hazard situation is allowed to exist
- Inadequate inspection, testing, maintenance
- Inadequate management practice/ system / procedure
- Inadequate hazard, risk, root cause analysis
- Inadequate monitoring or auditing
- Inadequate training
- Inadequate recordkeeping and documentation
- Inadequate communication or signage/placards

↑ **Increasing Risk Significance**

4.3 Connecting Process Safety Performance to Culture Results

The survey results and the technical results are both weighted and qualitatively mapped to the same 12 essential features. The weighted contributions from survey results that also have contributions from the technical results are summed to give the overall score for the essential feature. Features that have high positive survey results but do not have high scores for other process safety evidence (not survey data) are reduced in significance. Features that have high positive survey scores that also have strong process safety evidence scores are enhanced in importance.

The process safety PAR approach seeks to find technical evidence consisting of process safety outcomes (e.g., findings and improvement areas) that seem to support the opinions and feelings of the respondents from the process safety culture survey. The results of this comparison showed which process safety culture problems indicated by "opinions" in the survey were "proven out" by the company's recent historical performance in the field. This way, a company can have greater confidence that the potential culture issues uncovered from the survey are, in fact, real – and supported by visible evidence in the field.

4.4 Understanding Safety Culture Degradation

Safety culture evaluations, whether using the approach outlined above or other methods, determine the existing status of the entity's culture, at a point of time. As such, this is a picture of the current "state" or condition of the culture. Figure 8 is a typical process safety culture scorecard from a recent study.

Figure 6. Typical Process Safety Culture Scorecard

Typically, these culture states evolve over a fairly long period of time. To use a health care metaphor, the degradation of safety culture is similar to a person getting sick, i.e., acquiring 'safety culture disease'. Rarely is there information available from a company about how they acquired the particular safety culture disease and the specific pathway on how they got "culture-sick". Normally, this information is only available from major investigations where organizational factors were identified as causal factors and the investigation team invested significant time to go back in history to chart the various events that happened and conditions that accumulated over a long time to create the dysfunctional organizational culture or disease that contributed to the incident. Figure 9 is a degradation model that we have found useful in charting the flow of safety culture dysfunction in companies over time.

Typically, a failure will occur at the foundation with notably unrealized process safety or HSE values or peer leadership. That opens the door for the company to be susceptible to failure at the next level. A failure at the 2nd level – say, a lack of sense of vulnerability can lead to weakness at the 3rd level. That could results in people not having a questioning/learning attitude about non-normal conditions, which could them lead to a failure at the 4th level. This relationship is most notably apparent as a lack of responsiveness to safety concerns (e.g., safety systems left disable for long periods). The sequence of these maladies, if they last long enough, can make a company more susceptible to a major incident, where all the holes in their Swiss cheese lineup.[9]

Supporting evidence for this thought model is at best anecdotal. However, what the authors have found is that, although there is not a certain absolute culture degradation model, there is a typical (almost predictable) trend of process safety culture degradation. This degradation tends to transition upwards from the "foundational failures" of core value and leadership failures to sequential failures within the other three "notional layers" shown in Figure 8. If this is true, then safety culture protection features can be designed to address the foundation levels of the culture model.

Having information on how a company slowly declined to the point where its safety culture was poor is important information to have. Devoid of that information, some improvement measures can be prescribed for a 'disease safety culture'. Nevertheless, in order to prevent future similar 'disease', it is better to know what the specific safety culture disease pathway(s) were so that specific safety culture preventative 'medicines, vaccinations, or vitamins' can be applied to strengthen a company against 'safety culture disease' in the future. Lacking that specific disease pathology information, sometimes the best remedy is to avoid future 'safety culture disease' and remain healthy is to takes "broad-spectrum safety culture antibiotics" or pursue safety culture "exercises" to try to maintain a healthy safety culture. The next section describes some remedies that we have seen work to improve safety culture weaknesses and strengthen overall process safety performance.

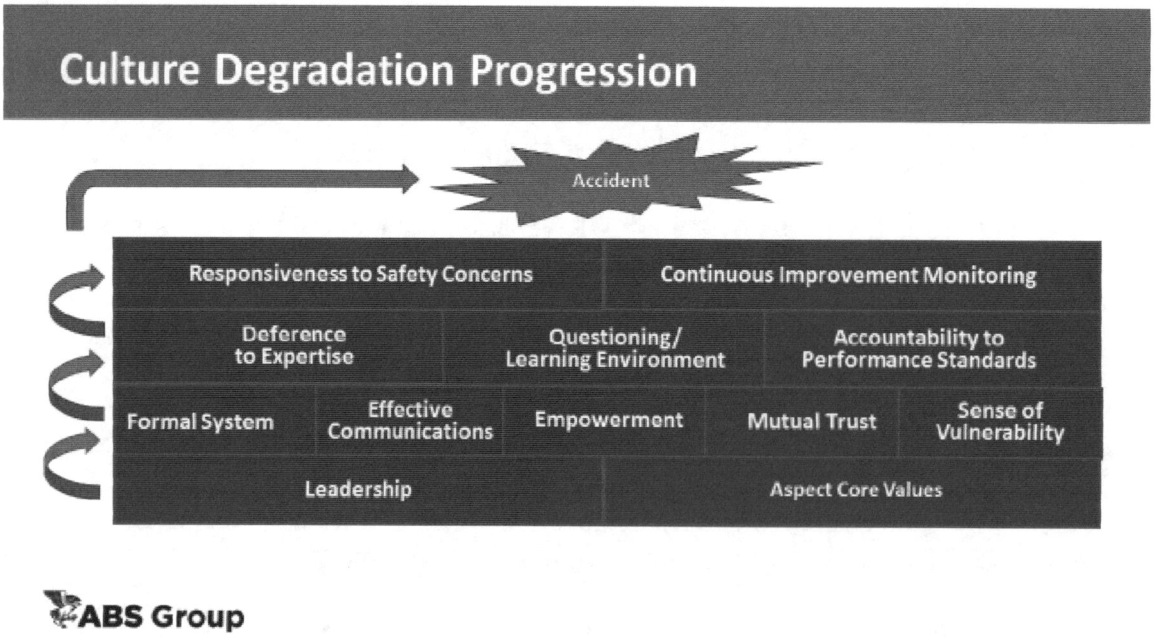

Figure 7. General Process Safety Culture Degradation Flow

5 Lessons Learned from a Progression of Industry Data

There is a lot to be learned from analyzing safety culture data from the various layers of the organization. For example, a single facility may have a process safety culture scorecard (as a result of all information obtained in the assessment) that may look like Figure 6.

Strengths	Moderate	Weaknesses
1. Establish safety as a core value	2. Provide strong leadership	3. Establish and enforce high standards of performance
5. Maintain a sense of vulnerability	4. Formalize the safety culture emphasis/approach	6. Empower individuals to successfully fulfill their safety responsibilities
	7. Defer to expertise	8. Ensure open and effective communications
	9. Establish a questioning/learning environment	10. Foster mutual trust
	11. Provide timely response to safety issues and concerns	
	12. Provide continuous monitoring of performance	

Figure 8. Example Culture Scorecard for a Single Facility

Which can also be represented as in Figure 7 with S (green) representing a strength for a given essential feature, M (yellow) for moderate, and W (pink) for weakness:

	Essential Feature											
Facility	1	2	3	4	5	6	7	8	9	10	11	12
Facility A	S	M	W	M	S	W	M	W	M	W	M	M

Figure 9. Matrix Format for Culture Scorecard

The company may have four different facilities that are included in the assessment as shown in Figure 8.

Facility	Essential Feature 1	2	3	4	5	6	7	8	9	10	11	12
Facility A	S	M	W	M	S	W	M	W	M	W	M	M
Facility B	W	M	M	S	W	M	M	W	S	W	S	M
Facility C	W	M	M	S	S	M	M	M	M	W	W	S
Facility D	S	M	W	M	M	S	W	S	W	S	W	M
Company Average	M	M	W	M	M	M	M	W	M	W	W	M

Figure 10. Matrix Scorecard for Multiple Facilities in a Company

We could then assemble the average scores for many different companies as in Figure 9.

Facility	Essential Feature 1	2	3	4	5	6	7	8	9	10	11	12
Company A	S	M	W	M	S	W	M	W	M	W	M	M
Company B	W	M	M	S	W	M	M	W	S	W	S	M
Company C	W	M	M	S	S	M	M	M	M	W	W	S
Company D	S	M	W	M	M	S	W	S	W	S	W	M
Company E	W	S	W	S	W	S	W	S	M	S	M	M
Company F	M	M	M	M	M	S	W	S	W	S	S	W
Company G	W	W	M	M	M	S	M	S	M	S	M	W
Company H	M	M	S	S	S	S	S	S	S	S	S	M
Company I	M	W	W	S	S	M	M	S	M	S	M	M
Company J	W	W	W	W	W	W	W	W	W	W	W	W
Company K	W	W	W	W	W	M	W	W	M	M	M	W
Company L	S	M	W	M	W	M	M	M	W	M	M	M
Company M	S	W	M	S	M	M	M	M	S	M	M	M

Figure 11. Process Safety Performance Deficiency Categories

This matrix of culture assessment scores represents actual PAR data for companies in very diverse industries, such as upstream and downstream oil and gas, pharmaceuticals, and petrochemicals. It is apparent that some of the companies have a stronger culture than others. In this format it is difficult to do arithmetic functions on the data, so we can assign as S = 9, M = 3, and W = 1. That allows us to show the "digitized" data in Figure 10.

Company	Essential Feature 1	2	3	4	5	6	7	8	9	10	11	12	Average
Company A	9	3	1	3	9	1	3	1	3	1	3	3	3.33
Company B	1	3	3	9	1	3	3	1	9	1	9	3	3.83
Company C	1	3	3	9	9	3	3	3	3	1	1	9	4

Company D	9	3	1	3	3	9	1	9	1	9	1	3	4.33
Company E	1	9	1	9	1	9	1	9	3	9	3	3	4.83
Company F	3	3	3	3	3	9	1	9	1	9	9	1	4.5
Company G	1	1	3	3	3	9	3	9	3	9	3	1	4
Company H	3	3	9	9	9	9	9	9	9	9	9	3	7.5
Company I	3	1	1	9	9	3	3	9	3	9	3	3	4.67
Company J	1	1	1	1	1	1	1	1	1	1	1	1	1
Company K	1	1	1	1	1	3	1	1	3	3	3	1	1.67
Company L	9	3	1	3	1	3	3	3	1	3	3	3	3
Company M	9	1	3	9	3	3	3	3	9	3	3	3	4.33
Average	3.9	2.7	2.4	5.5	4.1	5.0	2.7	5.2	3.8	5.2	3.9	2.8	

Figure 12. Digitized Data from Multiple PAR Studies

With the data in in this digitized format we can look at each of the essential features across all of the companies and sort out which of the essential features tend to be strengths and which ones tend to be weaknesses over a very large set of data.

EF #	Abbreviated Essential Feature Name	Average	Delta
3	Establish and enforce high standards of performance	2.4	
2	Provide strong leadership	2.7	0.3
7	Defer to expertise	2.7	0
12	Provide continuous monitoring of performance	2.8	0.1
9	Establish a questioning/learning environment	3.8	1.0
1	Establish safety as a core value	3.9	0.1
11	Provide timely response to safety issues and concerns	3.9	0
5	Maintain a sense of vulnerability	4.1	0.2
6	Empower individuals to successfully fulfill their safety responsibilities	5.0	0.9
8	Ensure open and effective communications	5.2	0.2
10	Foster mutual trust	5.2	0
4	Formalize the safety culture emphasis/approach	5.5	0.3

Figure 13. Essential Features by Strength, Moderate, and Weakness

By analyzing the difference (delta) between each of the two adjacent essential features there appears to be a Pareto-type of distinction between essential features 2 and 9, and 5 and 6. Coincidentally, the PAR data for a diverse set of safety culture assessment data, there appears to be 3 distinct strata for cultural strengths, weaknesses, and moderate cultural aspects of the combined data, each of which contain 4 essential features.

While this multi-company comparison should not be used to make conclusions about potential process safety culture strengths and weakness at a single facility or company, it does provide some insight into common areas of cultural strengths and weakness across a broad spectrum of industries.

We can focus on the four essential features shown as cultural weakness across industry and observe that if emphasis were placed on the associated attributes, broad-ranged strengthening would likely occur. The four weakest essential features with a brief summary of their attributes are given in Table 4.

Table 4. Attributes of the Weakest Essential Features

EF#	Title	Key Attributes
2	*Provide strong leadership*	• Manage by walking around • Fund important safety related projects • Provide adequate staffing • Walk the talk
3	*Establish and enforce high standards of performance*	• Follow policies and procedures, do what you say you are going to do • Normalization of deviance is sought out and eliminated • Zero tolerance for willful violations • Accurate and current documents (e.g., process safety information)
7	*Defer to expertise*	• Subject matter experts are trained, developed, and recognized, including process safety expertise • Organization turns to the experts rather than rank • Leaders ensure that there is always a network of experts available • Experts relied on in safety-related decision making
12	*Provide continuous monitoring of performance*	• Key performance indicators are set, periodically reviewed/revised and communicated; leading and lagging indicators are included • Performance against the KPIs is routinely measured and communicated • Action plan is implemented when performance falls outside of the KPIs

6 How to Improve Process Safety/HSE Culture

Usually, culture improvement activities takes many months if not years to have a deeply rooted impact. Table 5 lists the general steps we have found to be effective for improving safety culture at a facility.

Table 5. Employee Engagement Culture Improvement Work Process

- Personalize to each person's job
- Connect to their "ownership basis"; create belief in results
- Brainstorm how they got to where they are
- List current and future conditions that could affect culture
- Select one or two weaknesses/strengths
- Discuss how they are relevant to daily jobs
- Provide culture improvement idea menu
- Ask them to create/commit personal daily/weekly plan
- Solicit ideas for department/facility/company plan
- Monitor culture improvement – plan, personal growth, and metrics
- Implement this approach with top leaders first, then First-Line Supervisors, then workers/ contractors, and only then go up the rest of the organization

6.1 Culture Workshop Ideas

The engagement workshop type of training, whether for executives or hourly workers, is more effective is the delegates are fully engaged and thinking about culture change. Their opinions matter and it is important that organizations discover their employee's ideas for culture change. Interactive facilitated workshops provide an effective means for dialogue, soliciting ideas, engagement, and learning. The workshops are most effective if done using small groups that consider some of the following issues:

1. What can you do to assess the culture in your company and/or your facility?
2. What can you do to identify which of the process safety culture features will have the greatest impact in your company? Which is the weakest in your company?
3. Pick one of the culture areas and develop an action plan to address the following issues at your company/site.
    - Develop a culture improvement plan for your company/site. Be as specific as possible.
    - Decide which culture elements you are going to address
    - Decide who in management is needed to support your efforts
    - Develop a plan for the next month, the next six months, next year, and next three years.
    - Determine who owns the culture change initiatives at different levels of the organization and how the process will be maintained evergreen.
    - Determine what metrics will be used to monitor progress.

6.2 Identify Culture Improvement Options

The following steps should be addressed in culture workshops at all levels of the plant or company.

1. <u>Identify possible culture weakness root causes.</u>
    - Brainstorm what has happened in the past that could be in workforce memory.
    - Look at culture survey and interview results by feature. What is going on in the past year or recently that could affect this feature?
2. <u>Identify existing activities that could be used to help correct the cultural issues.</u>
    - What things are going on now in the plant or company that could improve or adversely affect this issue?
        - Capital expansion or operations contraction.
        - Employee welfare.
        - Workforce changes due to attrition.
        - Implementation of new safety programs (e.g., Behavior Based Safety).
3. <u>Look at the PSM/HSE audit finding resolution actions.</u>
    - See how they could be done in a culture-smart way.
    - Avoid initiative overload and unfunded mandates that could adversely affect the progress of culture change initiatives.
4. <u>Identify additional actions to fix the culture issues.</u>
    - Utilize the list of potential remedial actions in the next section as a starting point.

- Have small groups during culture awareness training to develop company and personal culture action plans using this list.

5. Develop an overall, integrated list of activities to help deal with the culture challenges.
    - Work through the plan as you would any process safety or HSE action plan.
    - Monitor implementation for progress.
    - Develop culture improvement metrics.

6.3 Remedies for Culture Weaknesses

Following the above approach will require that the company eventually address what actions need to be undertaken to improve culture. The organization will have to establish and follow an interactive process that engages all levels of the plant and company. At each occasion early on, the organization will seek personnel opinions about the root causes of culture weaknesses, factors that contributed to the problems, and conditions that the company faces in the near-term as it pursues culture improvements. Normally, a plant or company faces one or more culture issues at the same time. These issues should be addressed together to avoid adverse interactions of improvement actions and to increase efficiency. Table 6 gives an example of the possible remedies that have been used to address weaknesses where process safety is not a core value.

Table 6. Process Safety/HSE Is NOT a Core Value

- Establish a process safety policy.
- Communicate the business case for process safety to senior executives; communicate the process safety policy regularly.
- Hold training courses and workshops on the company policy.
- Reinforce policy via company newsletters and meetings.
- Ask the workforce for feedback on whether they consider process safety to be a core value and whether management also holds process safety as a core value.
- Demonstrate that process safety is a core value by making decisions and taking visible action in support of the policy.
- Celebrate workforce actions that demonstrate adherence to process safety being a core value.
- Encourage senior executives to participate in process safety functions, committees and symposia for benchmarking with other company executives.
- Widely share the CCPS Process Safety Beacon newsletter, which provides useful industry incident information.
- Share significant incidents from similar facilities and relate the broader consequences of a process safety incident at all levels of the organization.
- Help people at all levels in the organization to understand the importance of process safety.
- Appeal to the business case for process safety, the humanitarian case for process safety, and the legal case for process safety.
- Find a process safety champion high in the organization that is willing to try to influence other senior leadership.
- Provide safety culture ideas and input for annual reports and other visionary documents that are periodically published.

We have lists of remedies for other culture weaknesses that are too long to include in this paper, many of which are documented in the CCPS RBPS guidelines book. For purposes of an example, some key weaknesses have been documented in this paper.

7 Conclusions

An increasing number of companies are undertaking organizational culture improvement programs. For example, Cal-OSHA has promulgated changes to its PSM regulation that requires refining companies to conduct Process Safety Culture Assessments and implement a long-term culture improvement/assurance program. Over the past twelve years, we have conducted over 40 major culture evaluations around the world, involving more than 250 facilities and interactions with upwards of 100,000 personnel. As a result, we have compiled a number of lessons and identified ways to ensure value from investing in a culture improvement program.

Table 7. Ten Ways to Ensure Value from a Culture Evaluation

1.	Use an industry practice framework for measurement objectives to focus data gathering and analysis (e.g., CCPS 12 Essential Features of a Good Culture) to build your measurement tools.
2.	Employ anonymous, voluntary data gathering methods whether it be survey, focus group, workshop polling or observations.
3.	Use thorough representative sampling of assets, invite all work groups, from top management to different organizational levels including contractors.
4.	Use multiple measurement methods such as surveys (online and small group voting cards), interviews (individual preferred over focus group), field observations and recent past performance data.
5.	Do the assessment in a way that will make it easier to measure culture change on a quarterly or annual basis and to allow for efficient updating of collected information.
6.	When analyzing results and determining improvement actions, focus on weaknesses revealed by measurement objectives rather than just "fixing" the answers to worst response questions individually.
7.	Employ a good communications plan to maximize authentic employee participation throughout the initiative and to improve process safety and HSE ownership and leadership.
8.	Implement culture improvement actions in all levels of the organization using a variety of methods such as education, training, coaching and mentoring. Measure progress quarterly and re-direct as needed
9.	Develop a written management practice to integrate with the organization's process safety and HSE management system to guide future culture sustainment journey. Assessment is only the first step towards improvement.
10.	Follow through. Say what you are going to do and do what you say you are going to do.

In addition, many companies have overall safety policies or visions that embody some sort of "pursuit of zero accidents". Include process safety in this vision, namely, develop a vision for "Perfect Process Safety."

- A culture based on proper ownership of process safety and HSE empowered by visionary leadership.
- Risk-informed sensitivity that guides priorities (e.g., funding resources, change management)
- Effective, fit-for-purpose management systems.
- Process Safety practices are embraced and followed with good operational discipline at all levels of the organization.
- Learning from all sources – internal, external and outside industry group.
- Well-formed or visible process safety leading and lagging metrics at every level that drive intended behaviors
- Goals and actual performance that reflects continuous improvement

"Perfect Process Safety" is a worthy and valuable stretch goal for all. It can only be effectively pursued by highly reliable organizations that embody effective learning patterns and have a growing culture. Sustainable process safety cannot not allow learning to just evaporate or "retire". Companies that focus on improving process safety culture, leadership and operational discipline have the greatest potential to succeed in process safety performance improvement.

References

[1] Guidelines for Risk Based Process Safety, Written by ABS Consulting, AIChE Center for Chemical Process Safety, John Wiley & Sons, New York, NY, April 2007.

[2] A Management Approach for Creating a Strong Process Safety Culture – 20 Elements for Success, AIChE Center for Chemical Process Safety, New York, 2007.

[3] Principles of Process Safety Leadership, Process Safety Leadership Group, UKHSE, July 2009

[4] Successful Leadership Self-Assessment Module, UK Petroleum Industry Association Ltd., November 2010

[5] Corporate Governance for Process Safety - Guidance for Senior Leaders in High Hazard Industries, OECD Environment, Health and Safety, Chemical Accidents Programme, June 2012

[6] Process Safety Assessment Program, API/AFPM, January 2012

[7] Final Safety Culture Policy Statement, Bureau of Safety and Environmental Enforcement, May 9, 2013

[8] Guidelines for Conduct of Operations and Operational Discipline, Written by ABS Consulting, American Institute of Chemical Engineers, Center for Chemical Process Safety, John Wiley & Sons, New York, NY, March 2011.

[9] Swiss cheese model is widely used in literatures, but was originally postulated by Dante Orlandella and James T. Reason of the University of Manchester

Contact Energy - Process Safety 'Our Story'.

How leadership and staff engagement ensured that "Safe to Run" was a fundamental culture change supported by system and process changes

Brendan Bleasdale
Engineering Authority
Contact Energy
Wellington
New Zealand

What process safety means for us

For Contact Energy, safety is top of mind in everything we do, and process safety is an integral part of our safety ethos. We recognise that having good measures, systems, procedures and policies in place to help us proactively manage the risks that our major hazards represent is just one aspect of safety management. At its heart, process safety is about engagement and creating a culture that empowers people to play a meaningful role in identifying potential safety problems and coming up with ways to solve them.

Our goal at Contact is to have a generative safety culture, where safety is part of our DNA and where we provide a safe environment for our people to learn.

If something does go wrong, instead of demanding answers to why it went wrong, we ask: Where are our defences weak? What can we learn? How can we improve? And, how can we build the capacity to fail safely?

Where our process safety journey started

Back in 2013, Contact Energy was concerned about the number of process safety events happening globally. While our industry has an excellent record in occupational safety, the increase in significant asset failures worldwide is a leading indicator for deteriorating process safety. We had just introduced a system, based on API754[1], for identifying and recording Process Safety Incidents and were becoming increasingly aware, from our own leading indicators, of the need to improve.

There were other drivers as well, including evolving international awareness of, and guidance around process safety[2], legislative changes and stronger business imperatives.

Who is Contact Energy?

Contact Energy is one of New Zealand's largest companies. Our generation portfolio includes geothermal, thermal, hydro and natural gas storage. We supply electricity, natural gas and LPG (bottled and reticulated) to customers across New Zealand. Our technical portfolio includes three upper-tier Major Hazard Facilities (MHFs), five lower-tier MHFs, an upper-tier petroleum production installation, and three hydro dams with a "High" Potential Impact Classification

How we went about improving process safety

Rather than reinventing the wheel, our organization took the approach of building on our existing process safety improvement activities and on the learning and knowledge of others.

Our company took a look in our own backyard and asked ourselves: Are we doing enough? Are we focused on the right things? Do we have the balance right between what we're doing in occupational safety and in process safety? Do we have the right culture? And, how's our process safety incident record?

We identified that we needed to be able to systematically demonstrate that we could answer three fundamental questions:

Do we understand what can go wrong?

Do we know what our systems are for preventing this?

Do we have information to assure us our systems are working effectively?

Creating common understanding

It was important to develop common understanding, across the organisation, of what was needed to ensure the integrity of our operations.

To provide a structure for assessing and monitoring the performance of our risk control areas and processes we produced a Process Safety Management Framework (adapted from a framework initially developed by Scottish Power and based on multiple inputs, including the Energy Institute framework[3]).

Contact Process Safety Management Framework

PLANT		PROCESS			PEOPLE	RECOVERY
Technical Risk Management	**Safety Critical Systems**	**Alarm & Instrument Management**	**Maintenance Management**	**Operations Management**	**Staff Competence**	**Emergency Preparedness**
Management of Change	Critical Systems	Process Control	Work Prioritisation / Planning / Scheduling	Start Up / Shut Down Procedures	Competency Assurance	Emergency Planning, Arrangements, & Equipment
Plant Status Review / Technical Risk Register	Emergency Supplies	Critical Instrumentation	Work Identification / Routine Plant Inspections	Routine Plant Checks	Leadership	Crisis Management
Strategic Spares	Fire Systems	Protective Systems / Devices	Work Execution	Routine Testing	Training Needs Analysis	Environmental Protection Systems
Engineering Standards	HV Electrical Systems	Alarm Systems	Personal Risk Assessment	Shift Log / Handover	Communications	
Technical Knowledge Management	Main Protection Systems	Operational Technology Security		Safety Rules & Personal Risk Assessment	Staff Knowledge Management	
Design Review	Environmental Systems			Operating Limits / Envelopes	Contractor Competency	
Asset Investment				Operational Risk Assessment		
Civil Asset Inspections				Third Party Activities		
Pressure System Safety						
Operations & Compliance Audit	External Audit		Integrated Internal Audit	Incident Reporting, Investigation & Action Tracking		Controlled Documents (Policies, Procedures, Standards)

Table 1 – Contact Energy Framework for Process Safety Management (PSM)

Assessing our defences

With our Process Safety Management Framework in place, our company knew where our defences needed to be strongest. This allowed us to undertake a process safety 'health check' to measure the effectiveness of our process safety controls and barriers.
This included:

- a gap analysis to assess our process safety management culture, leadership and processes against industry good practice and guidelines, and to identify our process safety gaps.

- a hazard risk analysis to formally identify our Major Accident Hazards (MAHs), the control and mitigation barriers we use to manage them, and to establish whether our MAHs were adequately controlled.

Visualising our Major Accident Hazards

Bowtie diagrams were identified as a very effective way of allowing staff at all levels to visualise and assess our MAHs. Figure 1 below is a simplified illustration of how these bow ties were structured. Each bowtie workshop included a wide cross section of site staff and also had our Chief Executive or one of his Leadership Team participating. Their attendance signalled to staff the importance of Process Safety but also exposed the full senior company leadership to our major hazards and what needs to be in place to control them. The workshops generated considerable discussion on what could cause a failure event (threats), and what we had (or should have) in place (controls/ barriers) to prevent the event.

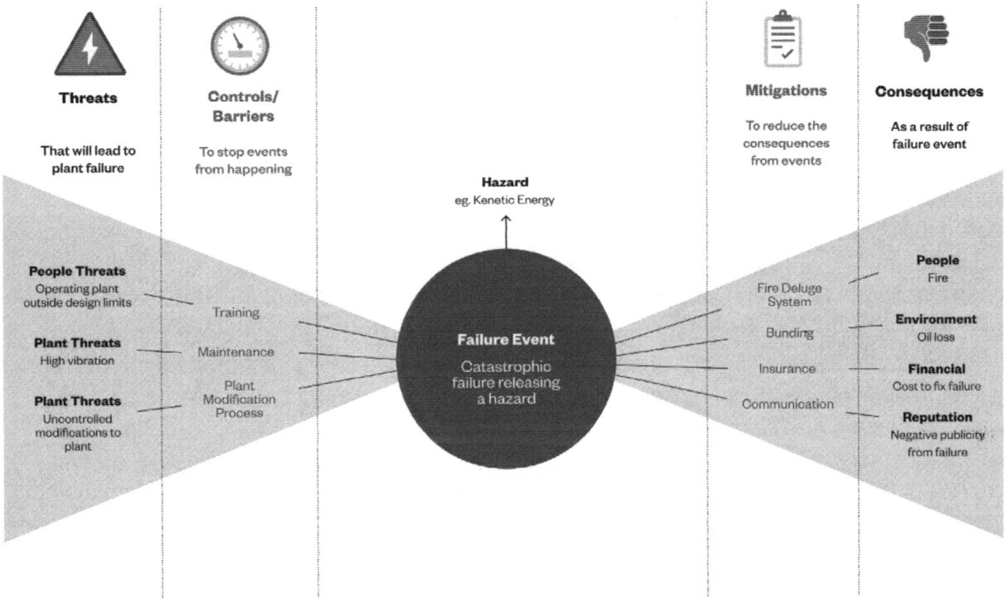

Figure 1 – High Level Bow Tie

Prioritising what needed to be done

We prioritised our improvement opportunities using weighted criteria which included priority/benefit, workforce "pull" for the change, change impact, flow-on work, resource needed, and cost.

Presenting the case for change

In 2014, we developed the business case, which brought together a number of parallel activities including the simplification of our HSE Management System, process safety improvements and asset integrity work. It also included the prioritised recommendations from the gap analysis.

The business benefits were compelling. In addition to having a demonstrably safer operation, with a reduction in operation and maintenance costs, we were also expecting improved engagement, portfolio risk, plant availability, business efficiency and insurance cost management.

Bringing it together

Knowing what needed to be done, and with our Process Safety Management Framework in place, it was time to pull everything together and decide how to bring process safety to life for people.

We knew that improving, sustaining and in some cases replacing our safety critical processes and systems for the future meant some big changes, so we looked to some fundamental principles of change to focus our approach:

We knew that successful change:

- has to be supported
- should happen within people, not to people
- needs to be integrated into everyday business
- needs effective communication
- must have full executive and management commitment
- needs the right focus and culture
- has to be supported by good measurement.

1. Change has to be supported

We integrated our process safety, HSE simplification and asset integrity initiatives into a single programme called OSIP (Operational Safety Improvement Programme). The programme was designed to support the business with getting our safety initiatives happening much faster. The programme took a change management approach to ensure alignment with our aspiration of progressing to a generative safety culture.

The 30-month programme consisted of 20 projects, managed under four focus areas and nine workstreams based on the high-level risk control areas from our Process Safety Management Framework. The scope involved changes to business processes, the use and integration of new technology and the development of capability, skills and competency. A resourced project framework was put in place so that we could maximise the technical input from our front-line subject matter experts without burdening them with the associated project management activities.

2. Making it meaningful

Although the business benefits of process safety were compelling, we needed to make it meaningful for people. It had to be engaging if we wanted it to remain top of mind for people. We did this in three ways:

- We talked about outcomes, rather than outputs to the extent that we gave the project a name that summed up, for us, what process safety is all about, which is keeping Contact 'Safe to Run':

  Safe to Run represents a fundamental shift in thinking and has helped us position process safety as a way of thinking. We now have people asking if we're safe to run; whereas in the past they'd want to know what was stopping them from restarting the plant.

  - We brought to life the idea of leaving nothing to chance by creating a character of the same name. Chance represents the 'gremlin in the system', the stuff we can't always see, but left unchecked, can cause havoc. We use him to remind people that process safety needs our constant attention. In everything we do, we challenge people to give Chance, no chance.

Figure 2 - Chance

- We helped our leaders with their safety leadership.
  Through an externally facilitated Leader to Leader programme we brought together our top 50 operational leaders and exposed them to the tools they could use to be effective safety leaders. Through tools such as "my 50%", "creative tension", and "courageous conversations" our leaders were helped to understand their safety leadership style, taken out of their comfort zone, and encouraged to use "small experiments" to improve. This programme had a transformational effect on our leaders and is now being rolled out to front-line supervisors
- We helped people see where they make a real difference.
  We engaged our people on the importance of always having a heightened sense of awareness of safety and of communicating effectively, particularly when involved in hazardous operations. We emphasised the important role they play in developing our process safety culture by ensuring they have the competencies they need to do the job, with a focus on skills which are safety critical. We encouraged our people to understand the processes, procedures and instructions that apply to the activities they carry out and to question anything that didn't seem right or was hard to follow.

3. Change should happen within people, not to people

Process safety is more than simply having good measures, systems, procedures and policies in place to manage the risks that our major hazards represent. At its heart is engagement – getting the right people involved at the right time, gaining their buy-in, creating the right culture, and having the people who use the processes involved in designing and implementing the improvements.

So, it was important that we had engaged people from awareness of the need to improve, right through to designing and measuring the improvements. We wanted everyone sharing their knowledge and ideas. We achieved this by:

- using our collective intelligence and enabling as many people as possible to be part of the improvements
- identifying 100+ subject matter experts from across the business to lead the workstreams and projects and provide guidance through steering committees
- trusting our people to come up with solutions.

4. Change needs to be integrated into everyday business

To be successful we knew we needed to take an integrated approach to how we viewed safety across Contact. We did this by positioning safety as 'how we do business round here' and applying process safety together with other safety processes as part of overall operational safety, and underpinning this with a generative safety culture.

5. Change needs effective communication

It was important to set the scene for the continuous improvement journey ahead of us. We wanted to be clear about what OSIP and Safe to Run were all about and why we were focusing on process safety. We supported the launch with a video and a number of key foundational resources:

Figure 3 – Our communication Materials

As the programme progressed we looked for simple and creative ways to drive engagement and help our leaders explain the 'why', like in a video where one of our Asset Managers talks about why our new Maintenance Management process is so important.

Figure 4 – Utilising staff to help the message

We talk about 'smashing complexity' a lot at Contact, and communicating things simply is central to this. We use simple, visual communications to help people understand what the big picture looks like, before diving into the technical detail; an approach that helps provide context and drives good healthy discussion. Figure 5 below is an example of how we communicated the simplification of our maintenance management process.

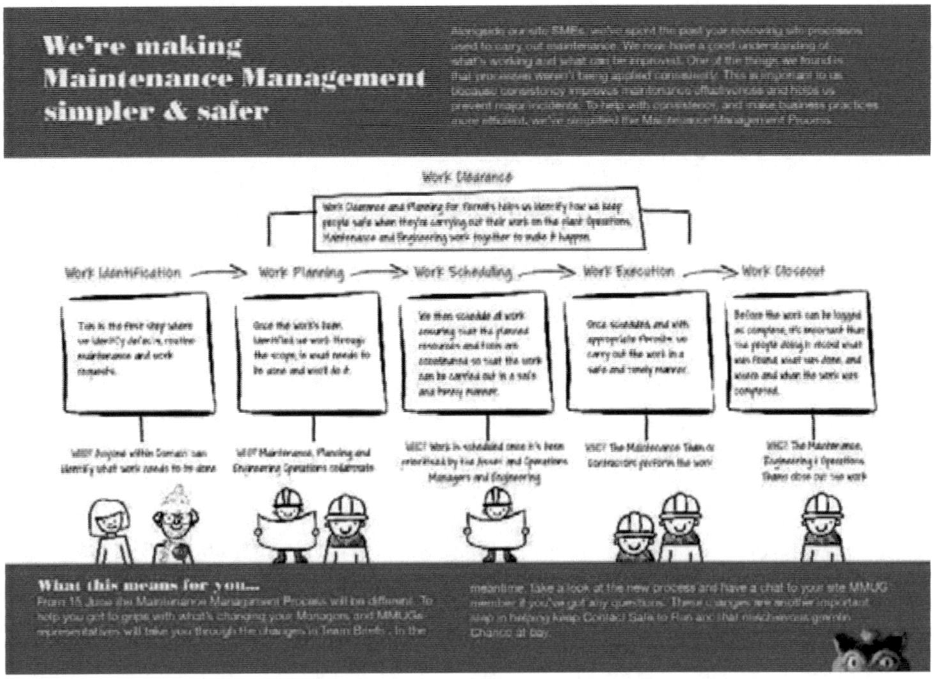

Figure 5 – Maintenance Management Simplification Poster

6. Change must have full executive and management commitment

We engaged our leadership team early through a series of briefings and presentations. We also brought our operations leaders together to talk about how, as a team, we'd generate success. Our CEO, Dennis Barnes appears in communications to reinforce the importance of the work we're doing, and we established process safety and the improvement programme as standing agenda items at senior management meetings

7. Change needs the right culture and focus

When OSIP was set up, Contact was already well underway with developing an underlying safety culture that supported people to try new things and to build the capacity to fail safely. It was important to build on this to create a strong culture around process safety.

- We aim for perfection – but are practical about it. This is about the importance of making small, incremental steps where we can, and not waiting until we have everything absolutely perfect and considering cultural aspects, such as how people 'think' and 'feel' rather than just the 'how' and 'what' of compliance.

- We balance compliance with culture. This is based on an approach which emphasises the importance of considering cultural aspects, such as how people 'think' and 'feel', rather than just the 'how' and 'what' of compliance.

8. Change has to be supported by good measurement

Although we were confident that the programme would help us answer the first two process safety questions...

1. Do we understand what can go wrong?
2. Do we know what our systems are for preventing this?

...we weren't able to answer the third question with any confidence:

3. Do we have information to assure us our systems are working effectively?

This was because we didn't have a clear view of how all of our controls and assets were performing. So, in parallel with our other process safety initiatives, we developed a process safety dashboard which gives us visibility of the health and strength of the barriers/controls we have in place to keep us safe from major hazards.

It's made up of a comprehensive set of leading and lagging process safety performance indicators, each linked to a risk control area.

Leading indicators measure the effectiveness of the barriers in place to prevent incidents (eg maintenance plans). Each leading indicator has a defined target, a tolerance band, and information on good practice. The dashboard provides near real-time information on performance against target by extracting data daily from our systems (including SAP maintenance management, alarm management monitoring software, Management of Change software and Incident Management). The information is presented graphically, contains point in time and trend information, and allows drill down to individual indicators and the data feeding in to them.

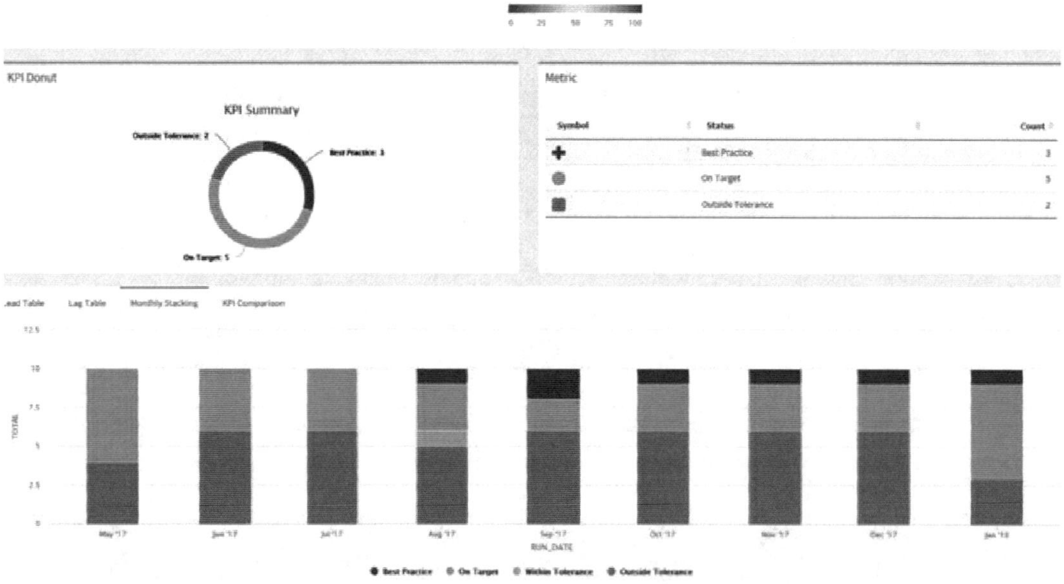

Figure 6- Example dashboard information

Transition into an embedded Continuous Improvement culture

The OSIP programme was of fixed duration and that to sustain momentum and embed improvements and careful planning was needed to maintain momentum and transition what we were doing into "business as usual" when the programme ended. This was achieved by establishing the concept of continuous improvement, and setting up a simple infrastructure to support and encourage ongoing improvements. The infrastructure includes:

- A small central team of business analysts, facilitators and "disruptors"
- Site based staff trained in delivery of our safety leadership tools and in facilitating Continuous Improvement teams who identify and deliver improvements
- Maintaining the subject matter expert groups (such as the Maintenance Management User Group and the Management of Change User Group) established during the OSIP Programme, giving them ownership of the processes they use, and charging them with refining and improving.

What we've learned so far

We've learned lots on our journey towards improving our process safety. Most importantly that it isn't just about having processes and procedures in place. Here are some of the key themes that emerged for us:

It starts from the top

Process safety needs to be of corporate importance. Executive and management buy-in and involvement is key.

It's about people

While processes and procedures are important, it's people that make the real difference. It was important to us that our people were part of the improvement and that we supported leaders in empowering our people to drive the change.

At its heart is engagement

From getting buy-in, to getting design input, implementing change relies on good engagement. That's why we made sure people knew where we were headed and why, and where they made the real difference. Through the programme we've created a culture that encourages engagement and where people feel valued.

It has to be integrated

Process safety can't be a bolt-on. It has to be integrated with other safety disciplines. In our case the programme aligned with our aspiration of progressing to a generative safety culture.

Make it simple

Processes and procedures need to be simple to use and easy to find. We've found that simple language and catchy metaphors go a long way towards helping people understand why they do what they do every day. Simplifying entire processes down to a single page has helped lift people above the technical detail, and provided an opportunity for good healthy discussion, and decluttering has helped support our people better manage risk.

Visualise it for people

Bowtie diagrams have been a great way to help explain our risks because they visualise them in a single, easy to understand picture.

Shaped like a bow tie, the diagram shows a clear differentiation between proactive and reactive risk management. The power of the Bowtie method is that it provides an overview of multiple plausible scenarios. In short, it provides a simple, visual explanation of a risk that would be much more difficult to explain otherwise.

Don't wait 'til its perfect

As Mark Twain once said 'continuous improvement is better than delayed perfection', and it's a principle that's been core to the success of our programme, where we aim for perfection, but are practical about it. We've found the approach of our people designing, developing and testing new processes, and then refining them, is better than waiting until we've got everything 100% right.

You need the right information

Having the right information is vital. We started with a 'health check' so we could review the effectiveness of our process safety approach in demonstrating our MAH management performance. And, we implemented a process safety dashboard as a priority knowing that you 'can't fix what you can't see'. Without the right information, we wouldn't have been able to prioritise our initiatives, or know how well our systems were working.

Where to next for us

Process safety has been a catalyst for broader improvements across a number of areas including culture, leadership, breaking down silos, realising process efficiencies, developing new ways of working and transitioning into a culture of 'continuous improvement'.

"Safe to Run" is an ongoing journey of improvement for Contact Energy, and we will continue to utilise the commitment and expertise of all of our staff to ensure that we leave nothing to Chance. Our drive for process safety excellence will continue to underpin safe, reliable and efficient operation of our assets.

References:

1. Process Safety Performance Indicators for the Refining and Petrochemical Industries. ANSI/API Recommended Practice 754, April 2016
2. High level framework for process safety management. 1st Edition December 2010. Energy Institute, London. ISBN 978 0 85293 584 2
3. Developing process safety indicators. A step-by-step guide for chemical and major hazard industries. UK Health and Safety Executive. ISBN 978 0 7176 6180 0

Recommendations and the Broken KPI
Optimizing PHA Recommendation Implementation

Ronjit Mukherjee
Risk Alive Analytics Inc.
#300, 926 5th Avenue
Calgary, AB, Canada, T2P 0N7
rmukherjee@acm.ca

Allison De Man
Risk Alive Analytics Inc.
#300, 926 5th Avenue
Calgary, AB, Canada, T2P 0N7
ademan@acm.ca

Prepared for Presentation at
American Institute of Chemical Engineers
2018 Spring Meeting and 14th Global Congress on Process Safety
Orlando, Florida
April 22 – 25, 2018

Recommendations and the Broken KPI Optimizing PHA Recommendation Implementation

Ronjit Mukherjee
Risk Alive Analytics Inc.
#300, 926 5th Avenue
Calgary, AB, Canada, T2P 0N7
rmukherjee@acm.ca

Allison De Man
Risk Alive Analytics Inc.

Keywords: Manage Risks, Process Safety, Hazard Identification and Risk Analysis, Management Review and Continuous Improvement, Process Safety Competency

Abstract

Process company metrics and CCPS Industry Surveys show that the Percentage of Process Hazard Analysis (PHA) Recommendations Completed can be considered a process safety "Leading KPI" [1]. This type of metric focuses companies on the quantity of PHA recommendations implemented, and not necessarily the risk reduction impact or benefit of the recommendations. As PHAs often identify tens to hundreds of recommendations, the philosophy to implement as many recommendations as possible can lead to unexpected risk exposure, either by not implementing risk critical recommendations, or by not implementing them in a reasonable time frame. This can also lead to inefficient financial expenditures and resource utilization.

Process industries can improve on these current practices and focus on implementing the most risk reducing recommendations, instead of implementing "as many as possible".

Three key elements which will assist in this change of philosophy from quantity to quality, and in turn ensure workers are more aware of minimizing risk exposure in time, are:

1) Determine a key performance indicator which represents the risk reduction benefit of each recommendation.
2) Determine the optimal order of implementation to maximize risk reduction, while including consideration of minimizing the number of recommendations required.
3) Improve risk exposure and work efficiency by comparing actual implementation to optimal implementation, and close gaps by analyzing the difference.

By determining the risk reducing benefit of recommendations, simulations can be run which represent what would happen to the risk exposure of potential consequences if the most risk reducing recommendation were implemented. For example, "If Recommendation X was implemented, how would the risk reducing benefit of other recommendations change? How would the risk profile change? What would happen to the risk profile if Recommendation Y were implemented?" Using an iterative methodology, we can determine the optimal order of implementation, and provide an implementation sequence target to strive toward.

1 Introduction

Process Hazard Analysis (PHA) studies are used, and in some countries required by regulatory bodies, to identify process hazards and operational issues for facilities. These studies enable management, engineers and site personnel to make decisions that reduce risk previously identified in the PHA.

A critical output from most PHA studies is the list of recommendations. These recommendations often are in the form of future safeguards or process design changes that are intended to reduce the risk of consequences. Process facility teams are often expected to implement these recommendations, or address PHA findings and resolve them in a timely manner, soon after the PHA is completed. However, with constraints of time, budget, and resources, and often with company directives or key performance indicators (KPI) which focus on implementing as many recommendations as possible [1], there is a chance for companies to implement recommendations in such a way that can lead to unanticipated risk exposure.

Furthermore, in discussions with site personnel or Process Safety Management (PSM) personnel, 3 key issues are often identified with the current methodology of implementation recommendations:

- Industry typically uses the KPI "# of Recommendations Implemented", which incentivizes quantity and not quality of benefit of actions.
- "Complete all recommendations associated with very high risk scenarios first." This approach does not consider how risk changes over time as recommendations are implemented.
- Each layer of an organization has differing understanding of PHA, Risks and Process Safety.

By improving the methodology with which recommendations are prioritized to be implemented, the opportunity for unanticipated risk exposure can be minimized. By taking into account other variables such as the cost, time, and effort it takes to implement a recommendation, the financial and resource expenditures can also be optimized, while having a primary focus on reducing risk.

This paper presents two case studies which demonstrate how integration of PHAs with recommendation optimization analytics can assist in minimizing risk exposure, while potentially optimizing financial opportunity costs. In these case studies, recommendation optimization analytics were performed after PHA recommendations were completely or partially implemented, and the difference between the actual and optimal implementation order was analyzed.

The PHAs analyzed were Hazard and Operability (HAZOP) studies performed on two existing compressor stations. The Center for Chemical Process Safety's (CCPS) publication on Layers of Protection Analysis (LOPA) was used for the case study when risk reduction credits were applied to existing safeguards and recommendations to be implemented in the future [2].

2 Recommendation Risk Reduction

2.1 Calculating the Risk Reduction provided by Recommendations

A current method often use to assist in prioritizing recommendations is to generate the dataset below (Figure 1.), an export from PHA software tools. In this dataset summary, personnel can utilize the number of scenarios in which a recommendation was applied to, and the maximum residual risk (risk after existing safeguards) associated with a recommendation in order to drive the prioritization process. However, this information is an over summarized dataset that may be misleading because:

- Personnel have to make several assumptions to determine which recommendation is the most important. For example, although multiple recommendations are shown to be associated with high risk scenarios, personnel are unable to quickly determine how many high risk scenarios each recommendation is related to.
- Multiple recommendations may be associated with a high risk ranking. However, without observing the severity and likelihoods of the scenarios, it may be difficult to differentiate the benefit of the recommendations.
- This dataset is a static snapshot in time. Once a recommendation is implemented, the risk profile of scenarios may change and the benefit of remaining recommendations may change.

Recommendation	Max Risk ▼	Place(s) Used in HAZOP
4. Recommendation 4	Very High	Consequences: 3.3.2.1
8. Recommendation 8	Very High	Consequences: 10.1.5.1, 10.1.5.2, 10.1.5.3, 10.1.5.4, 10.1.5.5, 10.1.6.1, 10.1.6.2, 10.1.6.3, 10.1.6.4, 10.1.6.5, 10.3.1.1, 10.3.1.2, 10.3.1.3, 10.3.1.4, 10.3.1.5
9. Recommendation 9	High	Consequences: 10.1.7.1, 10.2.5.1
10. Recommendation 10	High	Consequences: 10.1.7.1, 10.2.5.1
11. Recommendation 11	High	Consequences: 10.1.7.1
2. Recommendation 2	High	Consequences: 1.1.2.2, 1.1.2.3, 1.1.2.4, 1.1.2.5, 1.1.2.6, 1.9.2.1, 1.9.2.2, 1.9.2.3, 1.9.2.4, 1.9.2.5, 10.1.1.2, 10.1.1.3, 10.1.1.4, 10.1.1.5, 10.1.1.6
5. Recommendation 5	High	Consequences: 10.1.3.2, 10.1.3.3, 10.1.3.4, 10.1.3.5, 10.1.3.6
1. Recommendation 1	Medium	Consequences: 1.2.1.1, 1.2.1.2, 1.2.1.3, 1.2.1.4, 1.2.1.5, 1.2.1.6, 10.1.2.1, 10.1.2.2, 10.1.2.3, 10.1.2.4, 10.1.2.5, 10.2.1.1, 10.2.1.2, 10.2.1.3, 10.2.1.4, 10.2.1.5
3. Recommendation 3	Medium	Consequences: 3.3.1.1, 3.3.1.2, 3.3.1.3, 3.3.1.4, 3.3.1.5
6. Recommendation 6	Medium	Consequences: 10.1.4.1, 10.1.4.2, 10.1.4.3, 10.1.4.4, 10.1.4.5, 10.2.3.1, 10.2.3.2, 10.2.3.3, 10.2.3.4, 10.2.3.5

Figure 1. Sample PHA Recommendation HAZOP Export

In order to minimize these assumptions and more objectively determine the actual risk reduction (benefit) provided by recommendations, the following need to be considered:

- The current frequency of the consequence and/or the initiating event of the cause.
- The probability of failure of demand (PFD) of the recommendation of focus and any existing safeguards on related consequences.

- The severity level and associated tolerable likelihood of each cause/consequence pair the recommendation is associated with. These tolerable likelihoods are usually identified on the Corporate Risk Matrix, and are based on a company's risk tolerance color zones.

On the presented cases, the following assumptions were made when calculating the risk reduction provided by each recommendation. This was based on the Risk Matrix shown in Figure 2:

- Initiating Event Frequencies (IEF) of causes were assigned based on the likelihoods provided by the PHA Team. (Table 1.)
- A PFD of 0.1 was assigned to each safeguard/recommendation given 1 "credit" of risk reduction in the PHA. Safeguards/recommendations given 2 "credits" were assigned a PFD of 0.01[2].
- Tolerable Frequencies were determined based on the below assumption for each severity. (Table 2.)

Companies can use specific IEFs and PFDs from their internal equipment databases for the risk reduction calculations to improve the precision of the calculations. Risk Reducing Safeguards and recommendations reviewed in this report are assumed to be specific, independent, dependable and auditable.

Table 1. Frequency Correlation for Risk Matrix Likelihoods

Likelihood	Initiating Event Likelihood (/yr)
1	0.0001
2	0.001
3	0.01
4	0.1
5	1

Table 2. Target Tolerable Frequency Likelihoods for Risk Matrix Severities

Severity Code	Tolerable Frequency (/yr)
1	1
2	0.1
3	0.01
4	0.001
5	0.0001

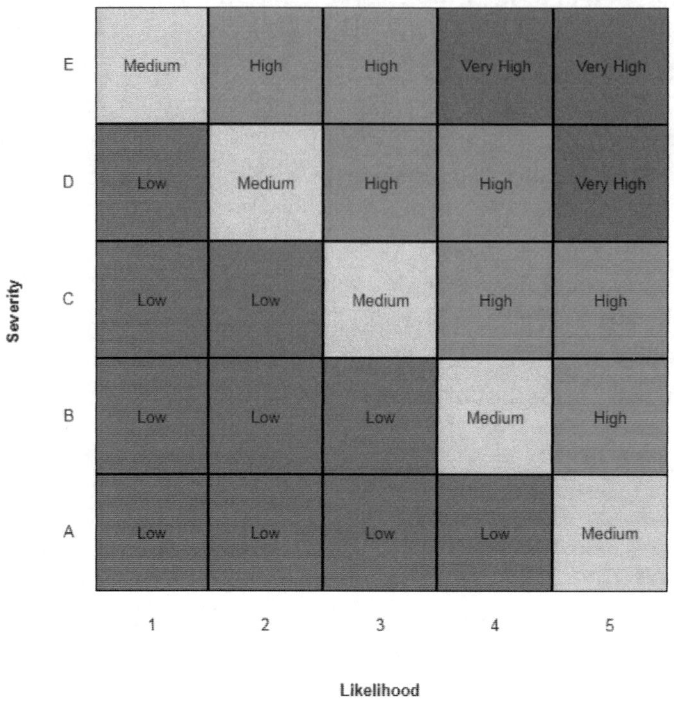

Figure 2. Case Study Risk Matrix

The methodology used to calculate risk reduction provided by recommendations included the following criteria:

- Safeguards and recommendations provide only reduction of the likelihood of a consequence from occurring. The severities of the worst credible consequences identified in the PHA were assumed to be constant.
- The assumed tolerable frequency of a consequence is indicative of its relative impact. The less tolerable a consequence is considered, the more severe its impact.
- It is assumed that all risk reducing safeguards identified in the PHA are specific, functionally independent from each other and the initiating event, and are dependably effective. If existing safeguards or future recommendations are not as effective as described in the session, the corresponding risk reduction provided for the recommendation may change.
- The assumed frequency of the cause or initiating event, and the PFD of the safeguards documented can be used to determine the current mitigated frequency of the consequence scenario.
- The assumed frequency of the cause or initiating event, the PFD of the safeguards, and the PFD of the recommendations being considered as risk reducing (whether they are future safeguards or design changes), can be used to determine the future frequency of the consequence if the recommendation were to be implemented.
- The risk reduction of a Recommendation is the change in risk associated with implementing that individual recommendation. This means that if multiple recommendations are applied to the same consequence scenario, their collective impact

on reducing the risk is not considered as it is not guaranteed that all recommendations on that scenario will be implemented concurrently.

- Risk reduction across different severities and consequence receptors were aggregated to determine the cumulative risk reduction associated with a recommendation. This means that the more times a recommendation was used in a PHA, the more risk reduction it will provide. These cumulative values can be broken down into different consequence receptors and severities as required.
- These calculated recommendation risk reductions to be provided are snap shots in time that can be continually updated as recommendations are implemented. This will be further elaborated in Section 3.

2.2 Case Study Recommendation Risk Reduction

Below are "snap shots" for the case study which presents the initially determined static risk reduction of each recommendation identified in the PHAs. Figures 3 and 4 below show:

- The y-axis represents the recommendation risk reduction provided, and the x-axis shows which recommendation is being analyzed.
- The data is sorted in descending order of risk reduction provided by recommendations.
- Each bar, which represents risk reduction, is broken down by risk reduction provided above tolerable risk (green), and risk reduction provided below tolerable risk (blue). For example, if a recommendation reduces risk from a very high risk level to a tolerable risk level, the recommendation bar would only show green. However, if a recommendation reduces risk level from tolerable to low, the recommendation bar would only show blue.
- Risk reduction across different consequence receptors are aggregated, however can be broken down into each receptor as required.

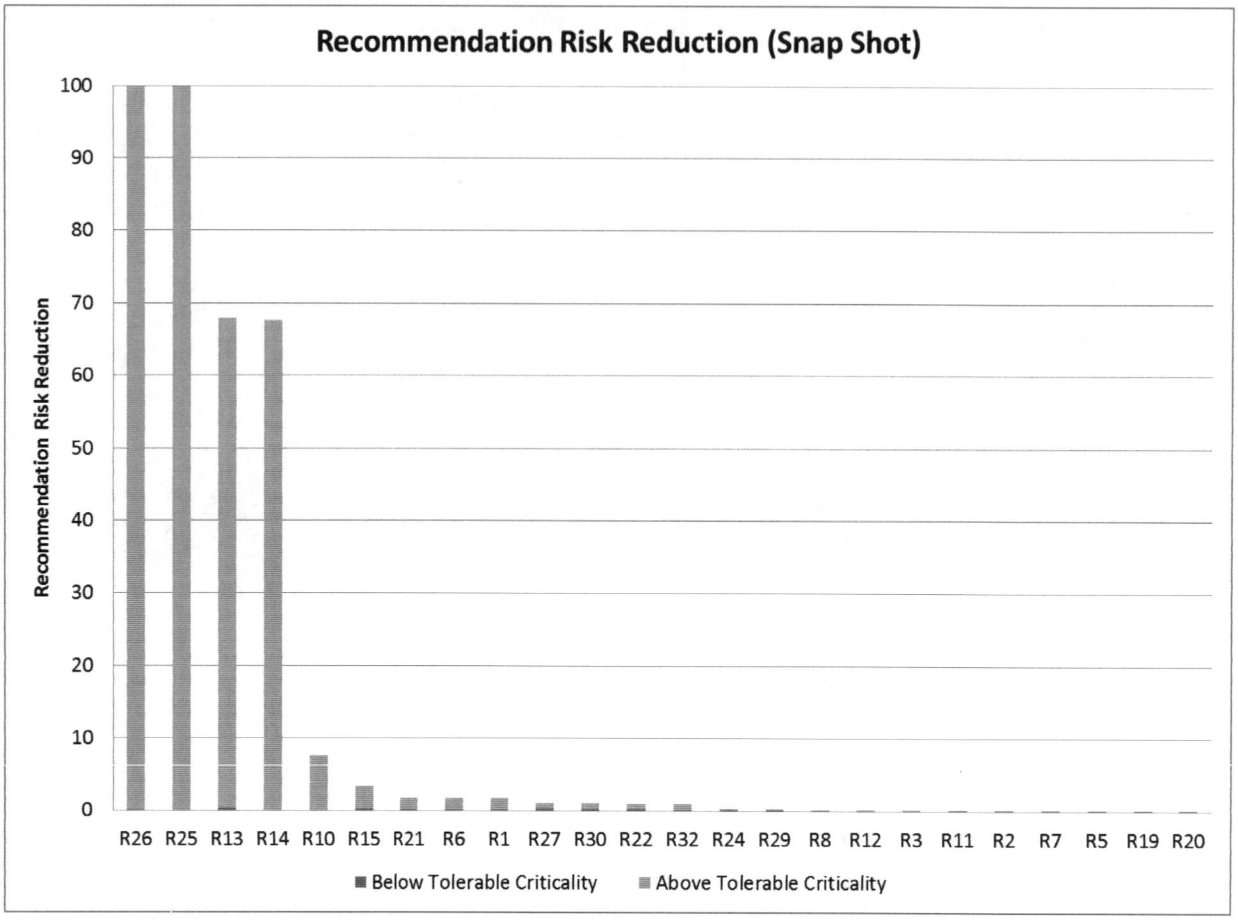

Figure 3. Recommendation Risk Reduction Snap Shot (Case Study # 1)

In Case Study # 1 (Figure 3.):

- There is a notable difference between the relative risk reductions provided of the Top 4 recommendations versus the remaining recommendations.
- Rec #25 and #26 are applied to many of the same consequence scenarios, but are different in risk reduction provided.
- Rec # 13 and # 15 have slight contributions in criticality from below tolerable risk scenarios. From Rec #24 onwards, all recommendations only have below tolerable risk reduction.

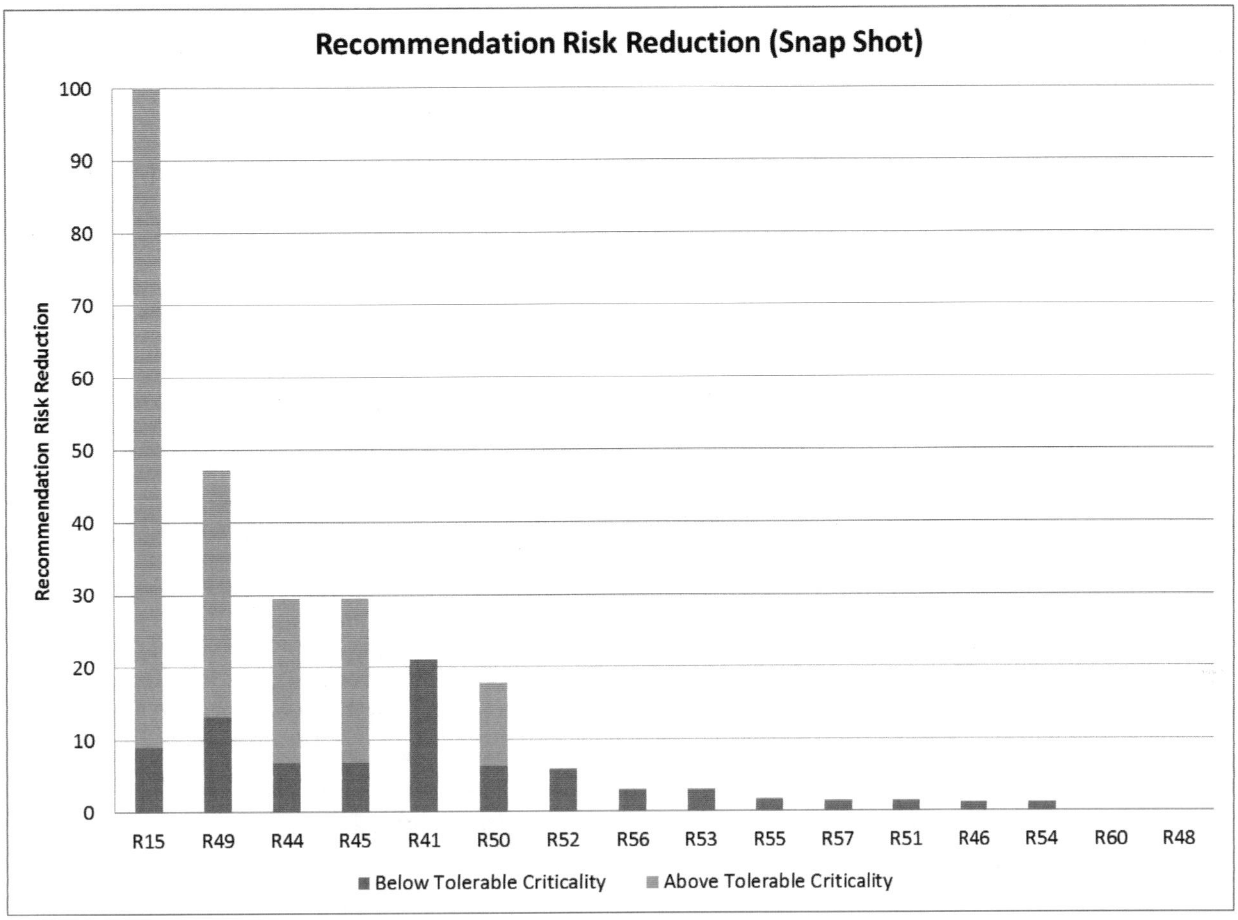

Figure 4. Recommendation Risk Reduction Snap Shot (Case Study # 2)

In Case Study # 2 (Figure 4.):

- Both Rec #44 and # 45 are tied for risk reduction as they are applied to the same consequence scenarios.
- Note that Rec # 41, which only has below tolerable risk reduction, is more important than a recommendation which has above tolerable risk reduction. This is primarily because Rec # 41 was utilized many times more than Rec # 50, adding to its cumulative risk reduction.

3 Implementing Recommendations (Optimal vs Actual)

3.1 Determining the Optimal Order of Implementing Recommendations

As mentioned in the prior sections, simply calculating the risk reduction of the recommendations in a snapshot in time dataset may not be enough to understand how you should sequentially and optimally implement your recommendations.

When decisions are made to implement a recommendation, its impact on the frequency of the consequence and the risk reduction of remaining and unimplemented recommendations needs to be considered. Implementing one recommendation may reduce the importance or the need for sequentially implementing additional recommendations on the same consequence scenario. This iterative methodology will improve the ability of personnel to optimally choose which recommendation should be implemented next, and ensure that risk exposure is minimized. Personnel can also consider financial optimization in scenarios where risk can be reduced to an "as low as reasonably practicable level" (ALARP).

The below Optimal Recommendation Implementation Charts (Figures 5 and 6) identifies the optimal sequential order to implement recommendations to minimize risk exposure through recommendations which provide the most risk reduction. After each recommendation is implemented, calculations are re-run for recommendation risk reduction and consequence risk ranking. In this simulation it is assumed that there are no constraints to implement recommendations such as budget, cost or effort. These parameters can be taken into consideration if provided by those implementing recommendations.

In the charts below:

- Only risk reducing recommendations are listed.
- The left y-axis is associated with the stacked chart area, which shows the number of very high and high risk scenarios that remain as recommendations are sequentially implemented. The height of the red section area represents the number of very high risk scenarios, and the height of the orange area section represents the number of high risk scenarios. Note that when the risk ranking of very high risk scenarios are reduced, it is possible that the number of high risk scenarios increase. This is because the risk level of a very high risk scenario can be reduced to high risk through implementation of a recommendation. Depending on the risk matrix, it is also possible that when risk reducing recommendations are implemented, that the risk ranking will not change even though the likelihood of the scenario is reduced.
- The right y-axis is associated with the bars which represents the recommendation risk reduction. These bars represent the same data as in the prior snapshot charts (Figure 3. and 4.)

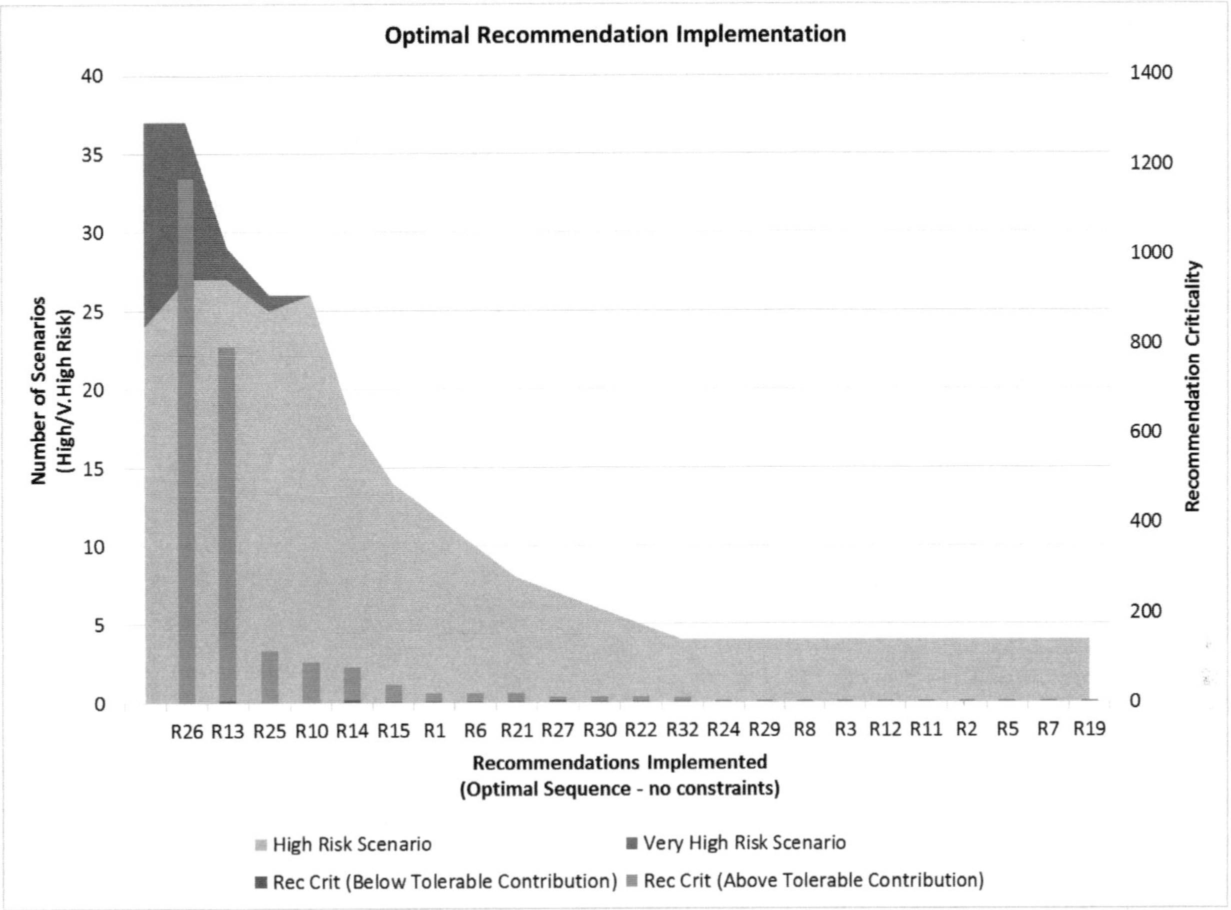

Figure 5. Optimal Recommendation Implementation by Risk Reduction (Case Study # 1)

In Case Study # 1 (Figure 5.):

- In Figure 3. (snapshot risk reduction) Rec #25 and #26 had very similar risk reduction, whereas in Figure 5., Rec # 25 had notably less risk reduction. This is because Rec # 25 and # 26 were applied to several of the same consequence scenarios and when Rec # 26 was implemented and reduced the risk of its respective scenarios, Rec # 25 had less risk reduction associated with it.
- After all risk reducing recommendations were implemented there are still high risk scenarios remaining. This means that it should be verified whether additional risk reducing recommendations are required in order to achieve the minimum tolerable risk level required across all scenarios. This demonstrates that through this analytic exercise, unintended risk exposure due to sequential recommendation implementation can be quickly identified right after a PHA.

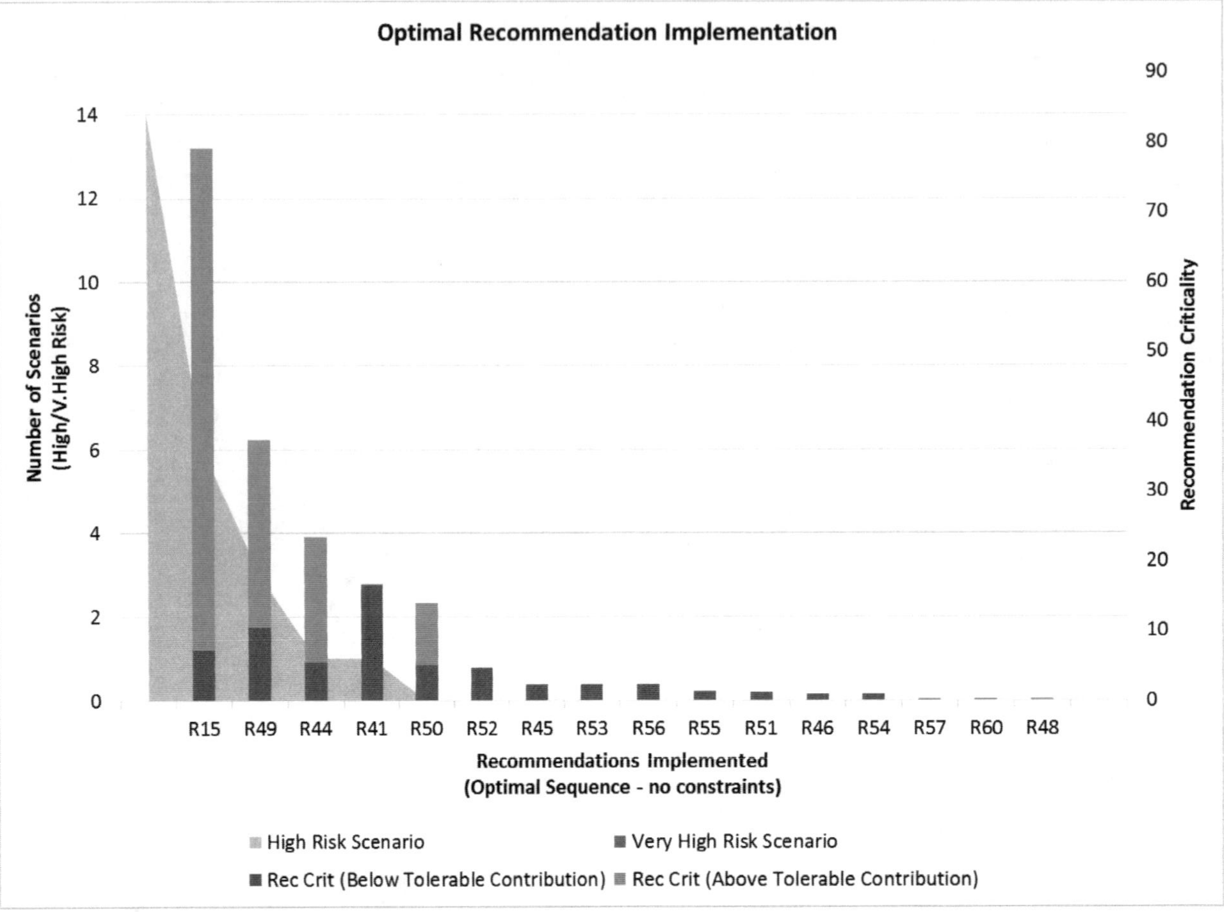

Figure 6. Optimal Recommendation Implementation by Risk Reduction (Case Study # 2)

In Case Study # 2 (Figure 6.):

- Only 4 risk reducing recommendations are required to lower the risk of all high risk scenarios to a tolerable risk level.
- In Figure 4. Rec # 44 and # 45 had the same criticality, however in Figure 6. Rec # 45 had notably less risk reduction. This is because these recommendations were applied to the same scenarios, and when Rec # 44 was implemented and the risk of the associated scenarios were reduced, the risk reduction provided by Rec # 45 decreased. Without this iterative approach, personnel may think Rec # 45 is still as beneficial as Rec # 44, and have potential financial opportunity cost.

Across these two optimal sequences (Figure 5. and Figure 6.) it can be seen that:

- An optimal sequence can help quickly identify whether additional risk reducing recommendations are required in order to achieve tolerable risk level across all scenarios.
- Implementing a risk reducing recommendation on a consequence scenario will have an impact on the risk exposure and risk reduction of the remaining recommendations identified in that scenario.

- Although in many cases the recommendation with higher risk reduction will impact more high risk scenarios, it is possible for a recommendation to provide a higher risk reduction and impact fewer high risk scenarios, simply because it was used more times for risk reduction.

3.2 Gap Analysis of the Optimal vs Actual Order of Implementing Risk Reducing Recommendations

The charts below show how recommendations were actually implemented in Case Study # 1 and # 2 without considering an optimal implementation sequence. These charts follow the same format as Figures 5. and 6., except that all recommendations, regardless of whether they are risk reducing or not, are shown in the graph as per the actual implementation results.

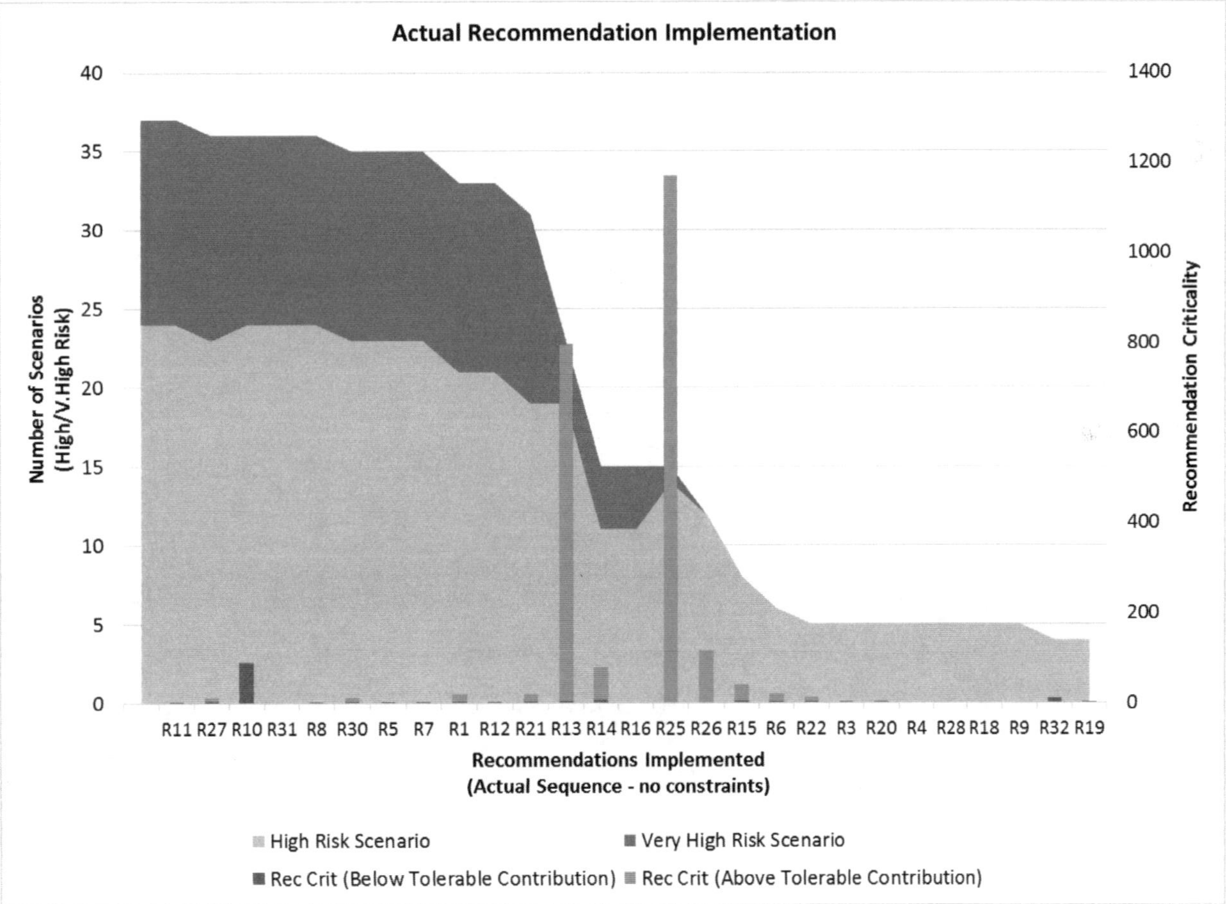

Figure 7. Actual Recommendation Implementation (Case Study # 1)

In Case Study # 1 (Figure 7.):

- It took the team 16 recommendations (~ 6 months from the completion of the HAZOP) to eliminate all very high risk scenarios, whereas optimally (Figure 5.) it would have taken 4 recommendations to achieve a more efficient outcome.
- In discussion with the key Process Safety contact, the recommendation implementation team was focused on implementing as many recommendations as possible without any specific focus on risk reduction. This meant that they were focused on items which were low effort but also provided low risk reduction.
- Although there were some critical risk reducing recommendations which could not be completed immediately, others, such as Rec # 13, an additional operator round for operations, could have been implemented much sooner and provided reduced risk exposure.

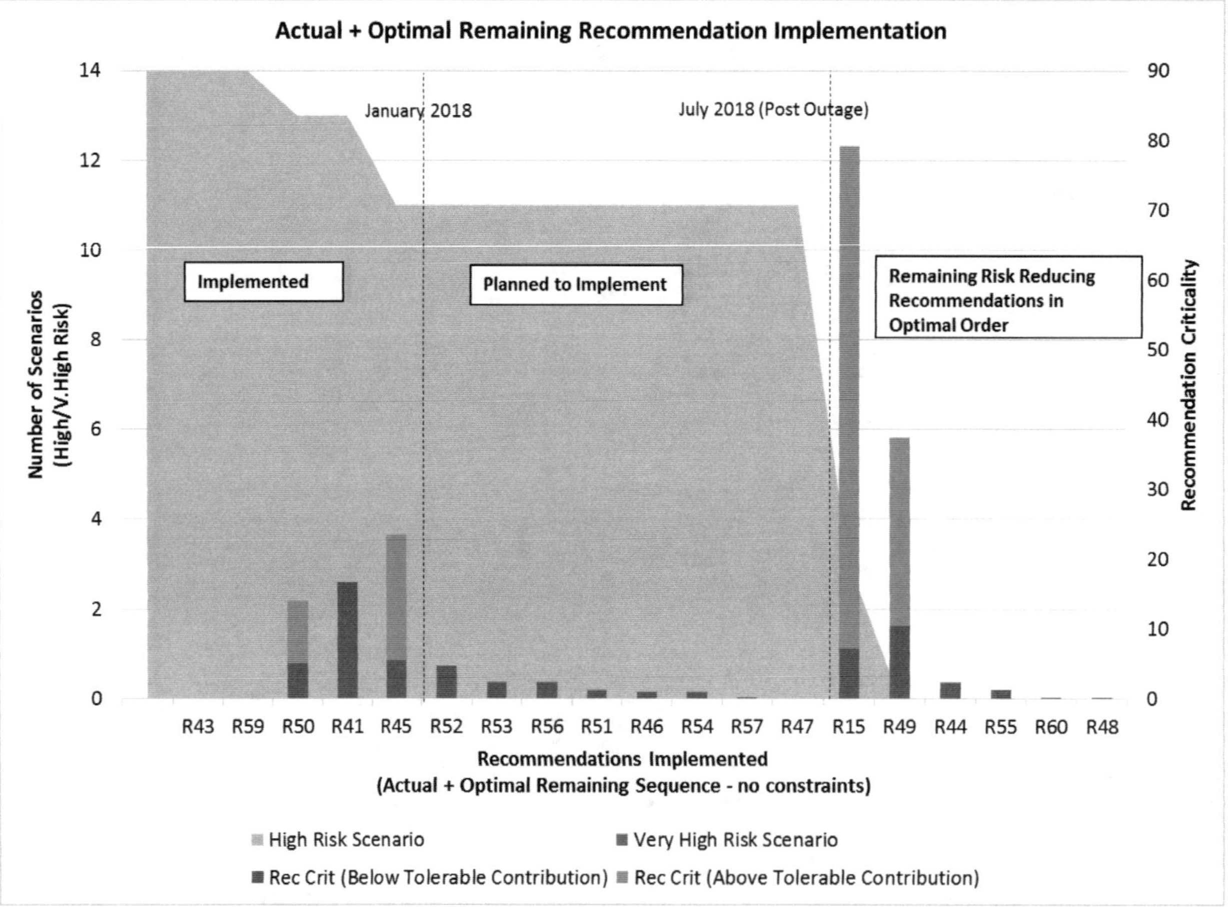

Figure 8. Actual Recommendation Implementation (Case Study # 2)

In Case Study # 2 (Figure 8.), not all PHA recommendations were implemented at the time of the analysis and there was an opportunity to assist in prioritization the remaining recommendations to attempt to achieve lower risk exposure. Some of these critical recommendations were planned to be implemented in the short term, and others recommendations were not identified in any short term planning. In Case Study # 2.:

- Recommendations are broken down into three groups:
  - Implemented
  - Planned to Implement
  - Remaining Recommendations
- Only 4 risk reducing recommendations would have been required to be implemented to reach a tolerable risk level exposure across all scenarios. However, 50% of those recommendations were not in any immediate plans for implementation. This means that there might have been long term unintended risk exposure if this analysis was not performed.
- Many recommendations were planned to be implemented during an outage, all of which were reducing risk levels from tolerable to below tolerable risk. This is a potential area of budget savings depending on the benefit to cost ratio of these recommendations, and the resources that would potentially have been used could have been utilized instead for the remaining high risk reducing recommendations.

The common issues across these two case studies when comparing Optimal to Actual Recommendation Implementation sequence were:

- Personnel appeared to be more focused on completing as many recommendations as possible versus the most critical risk reducing recommendations.
- In both case studies there was a notable difference between the minimum and actual number of recommendations that were implemented before the minimum tolerable risk level would have been reached. This identifies a risk exposure opportunity cost which could have been minimized.
- In both case studies there were several recommendations with low or no risk reduction and that were implemented before more critical risk reducing recommendations. This identifies the financial and resource opportunity cost that could have been reduced either through postponing or not requiring to implement certain recommendations.
- Ideally recommendation optimization simulations can be completed immediately after the PHA and provide the implementation team with a sequential guide for implementing recommendations in order to minimize financial cost, and risk exposure.

4 Conclusions

PHA Recommendation Optimization Analytics can provide a detailed analysis and prioritization assessment of a facility's PHA recommendations.

Current industry's methodology for implementing recommendations have key performance indicators with primary focus on implementing as many recommendations as possible, without necessarily focusing on optimally reducing risk exposure.

Based on Case Study # 1 and # 2, it was found that in the process of implementing recommendations for 2 separate PHAs there were situations where the personnel:

- Unintentionally did not plan to implement recommendations which were associated with providing high risk reduction, and
- Prioritized recommendations which had little or no risk reduction as they were prioritized based on less required effort to implement.

If recommendations have already been partially or completely implemented from a PHA, recommendation optimization sequencer can be used to determine and find any risk reducing gaps in the implementation sequence, which can be adjusted for future or remaining PHA recommendations.

If no recommendations have been implemented, recommendation optimization can be used to improve awareness of risk exposure and to plan on how to implement a sequence of recommendations, taking into consideration items such as time, effort and cost. This approach can minimize risk exposure and financial and resource opportunity cost.

5 References

[1] AICHE. 2011. *Process Safety Leading and Lagging Metrics*. [ONLINE] Available at: https://www.aiche.org/sites/default/files/docs/pages/CCPS_ProcessSafety_Lagging_2011_2-24.pdf. [Accessed 20 October 2017].

[2] Center for Chemical Process Safety (CCPS). *Guidelines for Enabling Conditions and Conditional Modifiers in Layer of Protection Analysis*. Hoboken, NJ, USA: John Wiley & Sons, Inc., 2012

[3] Center for Chemical Process Safety (CCPS). *Layer of Protection Analysis*. New York, New York, USA. American Institute of Chemical Engineers, 2001

[4] Havard Business Review. 2017. *Know the Difference Between Your Data and Your Metrics*. [ONLINE] Available at: https://hbr.org/2013/03/know-the-difference-between-yo. [Accessed 20 October 2017].

514288 Low-Touch Machine Learning Is Fulfilling the Promise of APM (poster)

Abstract:

New methods and cutting-edge technologies are driving asset performance management (APM) well beyond historical capabilities, rapidly increasing its bottom-line value. Technologies such as cloud computing, data science and machine learning are now being integrated with automated methodologies directly into APM solutions.

This wave of integration firmly places advanced analytical techniques into the hands of operators and engineers with previously unimagined scale and ease of use. The incremental progress in APM over the last 20 years pales in comparison to what's now possible through digital transformation.

Low-touch machine learning is the key catalyst to scale APM's potential well beyond existing first principles-based solutions and "armies" of consultant engineers and data scientists. A widespread integration of machine learning in APM will mark a transition from estimated engineering and statistical models towards measuring actual patterns of asset behavior.

Manufacturing facilities staff can now readily extract value from decades of existing design and operations data to better manage and optimize asset performance. This "low-touch" machine learning method continuously embraces changes in asset behavior, empowering real-time APM value creation. Vetted and tested across diverse industries, scalable across multiple assets and powered by cloud and parallel computing, low-touch machine learning ushers in a new era of performance and optimization for all industries.

514063 Big Data Analytics Platforms: The Key to Enabling the Analytic Enterprise (poster)

Abstract:

Analytics and the Industrial Internet of Things are the new hot buzzwords these days, but analyzing data is nothing new, and manufacturers have been generating data off the line for decades. So why all the hype? The last few years have seen the evolution and advancement of several technologies that are converging to allow for unprecedented analysis and insights into business problems. But leveraging these takes more than just smart coders. Analytic insights need to be deployed into production and action has to be taken to truly realize the benefits. Find out why analytic platforms are the key to enabling the Analytic Enterprise.

514064 Industry 4.0 and Driving Better Organizational Decisions (poster)

Abstract:

Industry 4.0 is more than just about the Industrial Internet of Things, it is also about leveraging the Industrial Internet of People (IIOP). This talk will focus on using the tools now available to bring sensors, data, and analytics together to drive better organizational understanding and continuous improvement. Who are the key stake holders, and what do they need to improve to improve their contribution to the way a chemical plant is operated? How do you structure these projects to ensure successful deployments? This talk will focus on practical use cases and examples of Industry 4.0 in action.

514067 Early-Stage Digital Transformation, a Case Study (poster)

Abstract:

Enthought transforms the business of science with artificial intelligence. We are a world-class provider of scientific software solutions and digital transformation services to companies in a range of industries, from chemical manufacturing to oil and gas upstream research and production to aerospace, finance, and others. For more than 15 years Enthought has been building artificial intelligence and machine learning based solutions with science and engineering at the core.

In this talk, we present a case study of digital transformation at a polymer design and manufacturing company looking to improve their data management and increase extraction of the value hidden in their data. We will cover details of the on-going project, including steps taken to digitize previously manual, paper-based operations; changes in workflow and processes to support digitalization; and examples of the practical applications of machine learning techniques. In addition, we will discuss lessons learned, successes and pitfalls in the project. Positive transformations were not limited to the business only but also included new opportunities and personal transformation for employees and a positive change of culture at the company.

514070 Integrated Technical Computing for Process Engineering (poster)

Abstract:

Process engineering leverages data, domain knowledge, and modeling to design, scale, and continuously improve plants, from R&D to pilot to production. The workflow fundamentally involves data analysis and visualization, model development, and process optimization.

Data analysis starts with preprocessing of the dataset, to impute missing values, correct errors, and often impart structure. Descriptive analysis is used to characterize what has been observed, for each individual quantity, as well as compute correlations between quantities. Visualization is used to rapidly develop understanding regarding the relationships between predictors and responses.

Insights gained from data analysis are subsequently leveraged to create a model, to describe the responses as mathematical functions of the predictors. In simple cases, first principles understanding is confirmed, and a governing equation can be written. In cases of intermediate complexity, the basic mechanisms for observed behavior are described by a theoretical framework, and parameter estimation or process identification methods can be used. Complex cases are typified by a multiplicity of superposed physical effects and absence of an economic functional form â€"in these cases, machine learning techniques are required to infer input-output relationships.

Once the model has been created, predictions can be made about the responses that would be recorded if an experiment based on the predictor values was conducted. Since the model enables the parameter space to be explored continuously, optimization methods can be employed to find the combination of predictors that minimizes a specified function of the responses. The optimal conditions so determined inform the selection of the plant operating setpoint. Sensitivity studies are conducted to define the plant operating region around the setpoint, and thus specify the control system requirements.

There are numerous software tools available which address select subsets of the process engineering workflow. MATLAB is one of very few that cover it end-to-end, and is distinguished in that it also offers utilities for distributed computing, big data processing, automatic code generation, and real-time interfacing with other software tools â€"including process simulators.

In this talk we describe MATLAB functionality relevant to each stage of the process engineering workflow, and provide application examples and graphics to bring the concepts to life. Specific topics will include data analysis (tall objects, logical indexing, statistical plotting), model development (principal components, system identification, neural networks), and optimization (linear programming, pattern search, and genetic algorithms).

514071 Tony Starck Needs Jarvis to Run Iron Man. What's the Future in AI for Continuous Process Industries? (Poster)

Abstract:

Why bigdata is such a game changer? We have run plants with automation and scada levels for 30 years, and we encounter a lot of data. We visualize, we set, we adjust, we regulate. With more and more advanced technologies. We make a lot of decisions, impacting the performance of our plants. The bigdata era creates another level of opportunity. By crossing all kinds of information, we open the gate of a new worldâ€¦ of complexity. We understand that there is a gold mine awaiting our discovery, but our biological brain, the main technology we use to dig into it, is now outdated. Our brains will not crunch the bigdata! We need new technology : artificial intelligence.

Today, the most advanced continuous process plants are about to become autonomous. Transforming the decisions of process optimization to dedicated AI. This is coming fast for all of us. What does already exist? How do we implement it for what profit? What can we expect for the future? How can this benefit us?

514072 Approaches to Multiblock Modeling and Dimension Reduction for Combining Disparate Data Types (poster)

Abstract:

Continuous manufacturing, which is being strongly promoted by the FDA, is an innovative approach to pharmaceutical processing that utilizes information from multiple sensors throughout the entire process. Due to the numerous factors that need to be monitored and controlled at the same time, a multi-step approach is necessary for optimizing individual unit operations as well as the overall process. In this presentation we will show some approaches to combining process and spectroscopic data, which can be applied to process monitoring for continuous manufacturing as well as for batch processes.

513463 Manufacturing Analytics - the Journey so Far

Abstract:

The goal of Enterprise Manufacturing Intelligence (EMI) is to make these high value relevant data available in real time to all levels of production management (Operations to Plant and Business Management). Therefore, plants can take advantage of this information to make improvements in raw material usage, asset utilization by reducing unnecessary process wait times and customer satisfaction through improved product quality by reducing product and process variation. This applies to both batch and continuous processes.

At The Dow Chemical Company, we have been actively pursuing EMI since 2012. Since that time we have done 70+ EMI implementations in a variety of environments. This poster talk will discuss the different audiences / kinds of problems our installations have sought to address. We will also discuss some challenges (both technical and cultural) and how we overcame them. Finally we will discuss how these experiences have shaped our view of the future for EMI at Dow.

516856 Applying Predictive Analytics to Develop Actionable Demand Forecasts

Abstract:

The Dow Chemical Company is a leading chemical manufacturer in the world. Dow manufactures plastics, chemicals, and agriculture products. With a presence in over 150 countries, it employs over 50,000 people worldwide. Dow serves to customers in different market segments, with a wide variety of products made for each one. The magnitude of Dow makes demand forecasting a challenge. Dow is using hierarchical statistical demand forecasting techniques to accurately forecast demand for the key products and customers. This poster is about the data and predictive analytics methods and tools used to develop the actionable demand forecasts.

Rare but Conceivable: Determining the Likelihood of Meteors and Other Infrequent Events

Michael S. Schmidt
Engineering Management and Chemical Engineering Departments
Missouri University of Science and Technology (MS&T)
and
Bluefield Process Safety
3101 Cherokee St.
St. Louis, MO 63118
MSchmidt@BluefieldSafety.com

Maria Vega-Westhoff
Bluefield Process Safety

Prepared for Presentation at
American Institute of Chemical Engineers
2018 Spring Meeting and 14th Global Congress on Process Safety
Orlando, Florida
April 22 – 25, 2018

Rare but Conceivable: Determining the Likelihood of Meteors and Other Infrequent Events

Michael S. Schmidt
Engineering Management and Chemical Engineering Departments
Missouri University of Science and Technology (MS&T)
and
Bluefield Process Safety
3101 Cherokee St.
St. Louis, MO 63118
MSchmidt@BluefieldSafety.com

Maria Vega-Westhoff
Bluefield Process Safety

Keywords: External Event Analysis, Facility Siting Data, Risk Assessment

Abstract

In the aftermath of a rare but devastating disaster, there is a tendency to overestimate how frequent such an event may be, hence a tendency to overestimate the risk associated with such an event. Unlike "black swans," which are inconceivable until they occur, events like meteors, hurricanes, and airplane crashes that impact chemical plants are rare, but conceivable. They occur with calculable likelihood and have quantifiable consequences, but most would not fault a PHA or LOPA team for dismissing a meteor strike as "not credible".

This paper considers several "meteors", both natural and man-made, that often become a cause of general concern after such an event occurs: meteors, earthquakes, and tsunamis; tornadoes and hurricanes; and airplane crashes, pipeline ruptures, train derailments, and truck crashes. It describes approaches to estimating the frequency of such events at a specific facility, and provides specific methods given certain assumptions. With these frequencies in hand, the paper goes on to develop a framework for giving meaning to the term "credible".

1 Conceivable *vs.* Credible

Nassim Taleb's Black Swan Theory can explain or excuse all sorts of catastrophic events. When Taleb repurposed the term "black swan" in 2007 [1], he described it this way: "First, it is an *outlier*, as it lies outside the realm of regular expectations, because nothing in the past can convincingly point to its possibility. Second, it carries an extreme impact. Third, in spite of its outlier status, human nature makes us concoct explanations for its occurrence *after* the fact, making it explainable and predictable." In other words, it is not simply rare but inconceivable, which means that it is not credible by any definition of the word. The fact that an explanation can be constructed after the event, in hindsight, does not mean that it was in any way predictable.

As a result, the Black Swan Theory is an interesting construct, but has no practical place in risk assessments. By their very nature, black swans are not something we can assess prospectively. We can only study them retrospectively.

1.1 "Meteors"

Catastrophic meteor strikes are not black swans. While quite rare, rarity is not what qualifies an event as a black swan. Black swans are not impossible, but they are inconceivable. Meteor strikes are not inconceivable. There is ample evidence from the past to convincingly point to their possibility.

Most meteors entering the earth's atmosphere explode in the upper atmosphere, where the solids vaporize and pose no threat to human activity. Asteroids larger than 20 m across (about the size of a barn) can make it close enough to the earth's surface that either the airburst or the impact crater have an effect. The Chelybinsk meteor (Figure 1) that exploded over Russia in 2013 had an estimated diameter of approximately 20 m; the airburst was about 500 kTons, around 30 times greater than that of the Hiroshima bomb.

More recently, a bolide streaked over Michigan. The boom from the exploding meteor registered in Michigan as an earthquake at 2.0 on the Richter Scale. A still taken from the security cameras at a chemical process facility in southern Michigan (Figure 2) tags the event at 8:08 pm EST on January 16, 2018.

Clearly, destructive bolides in the vicinity of chemical facilities are credible.

As metaphors, "meteors" should be far more interesting to risk analysts than "black swans". "Black swans," are inconceivable and so cannot be evaluated for risk. Once they occur, they are no longer inconceivable and hence, no longer "black swans". At that point, they can become part of the spectrum of potential events evaluated by future risk analysts.

"Meteors," on the other hand, are conceivable before they occur. They are rare, but occur at frequency that can be determined. When they are sufficiently rare, risk analysts will assume the catastrophic events they trigger are "not credible".

Photo credit: http://www.pbs.org/wgbh/nova/earth/meteor-strike.html

Figure 1. The Chelybinsk meteor that exploded over Russia on February 15, 2013.

Figure 2. The meteor that exploded over southern Michigan on January 16, 2018.

1.2 Credible scenarios

To allay concerns that imaginations will run wild, most analysts suggest limiting process hazard analyses and quantitative risk assessments to “credible scenarios”. This then invites questions about what is and is not credible.

Exactly what does “credible” mean?

For some, credible means “possible” or “not impossible.” This definition reduces “not credible” to “does not violate the laws of nature.” Evil twins are no longer a soap opera plot device, but credible. Zombies, on the other hand, are still not credible.

For some, credible means something that has happened, as in “if it happened once, it could happen again”. An aircraft once flew into a chemical plant in the southeast United States. As unlikely as that is to happen again, personnel at that plant deem any hazard review that does not consider being struck by an airplane as incomplete.

It is not the intent of this paper to declare what frequency is sufficiently rare that it can be treated as “negligible” or “not credible”. Instead, it is the intent of this paper to give guidance in calculating the frequency of certain “meteors” so that risk analysts can make informed assessments of whether those events are sufficiently rare to treat as not credible, or conversely, determine the frequency to consider “not credible” based on the frequency of events that they do not consider credible.

2 External Causes — Improbable But Not Impossible

Consider nine improbable events that can not only result in catastrophic releases, but are of such a nature that analysts encounter them periodically while conducting risk assessments. The intent of this section is to address those events as potential causes for Layers of Protection Analysis (LOPA) and give guidance to analysts, so they can calculate frequency at which these events may impact their specific facility. This section divides the nine events into three categories: geophysical, climate-related, and transportation-related.

The geophysical events:

- Meteors
- Earthquakes
- Tsunamis

The climate-related events:

- Hurricanes
- Tornados

The transportation-related events:

- Pipeline ruptures
- Train derailments
- Truck crashes
- Airplane crashes

2.1 Geophysical

2.1.1 Meteors

An excellent and well annotated discussion of meteor impacts with earth [2] explains that while asteroids, comets, and meteoroids frequently impact the earth, the vast majority burn up or explode on entry into the earth's atmosphere as airbursts. We call these exploding meteors bolides and typically they have no effect on the surface of the earth. An asteroid must be at least 20 m across before the explosion is strong enough and at a low enough altitude to cause damage or before it can strike the earth and form a crater.

Using a table constructed with an on-line tool developed at Imperial College London [3], we determined that asteroids 20 m across or bigger impact the earth about once every 40 years. Typically, a once-every-40-year event will get the attention of a HazOp team. However, these events usually do not affect the entire world—the affected area is more like 250 km^2 (100 mi^2). The surface of the earth is about $5.2 \cdot 10^8$ km^2 (200 million mi^2), meaning that the likelihood that an asteroid strike will have an impact on a specific plant is low:

$$1 \text{ event/40 years} \cdot 250 \text{ km}^2 / 5.2 \cdot 10^8 \text{ km}^2 = 1.2 \cdot 10^{-8} \text{ events per year.}$$

2.1.2 Earthquakes

The occurrence of earthquakes around the globe follows a straight-forward relationship: as the earthquakes get 10 times stronger, they occur 10 times less often [4, 5]. Table 1 shows the rate of occurrence, as well as the effects of earthquakes by category.

Table 1. Earthquake Frequency

Category	**Moment Magnitude (Mw)**	**Effect**	**Global Occurrences**
Great	≥ 8	Causes serious damage in areas hundreds of miles across	1.3/yr
Major	7 – 7.9	Causes serious damage over larger areas	13/yr
Strong	6 – 6.9	Destructive in areas up to 160 km (100 mi) across	130/ yr
Moderate	5 – 5.9	Causes major damage to poorly constructed building over small regions; slight damage to well-design buildings	1,300/yr
Light	4 – 4.9	Noticeable shaking of indoor items; rattling noises	13,000/yr
Minor	3 – 3.9	Often felt	130,000/yr
Micro	2 – 2.9	Generally not felt but can be detected with instruments	1,300,000/yr

What should be clear from Table 1 is that a facility risk assessment need only consider the top three categories—Great, Major, and Strong. Worldwide, there are less than 150 earthquakes of interest per year. Unlike meteors, however, earthquakes do not occur randomly across the planet. Popular culture has made us all acutely aware of the "Ring of Fire" around the Pacific, as well as other seismic zones. The approach taken with meteors, that they are purely random events with equal likelihood anywhere on the planet, is not appropriate.

Instead, a map showing the frequency of earthquakes is the appropriate approach. Or more correctly, a map showing the frequency of damaging earthquake effects is the appropriate approach. Fortunately, at least for facilities in the United States, the U.S. Geological Survey [6] has prepared such a map (Figure 3). The map shows zones going from "cool," meaning less than 2 destructive earthquakes per 10,000 years, to "hot," which has more than 250 destructive earthquakes per 10,000 years. The upper Midwestern United States and the Gulf Coast are in the coolest earthquake zone. Facilities there can expect $1 \cdot 10^{-4}$ damaging earthquakes per year. Coastal California, central Alaska and the Aleutian Islands, and the northwest corner of Puerto Rico, on the other hand, are in the hottest earthquake zone and facilities there can expect $5 \cdot 10^{-2}$ damaging earthquakes per year.

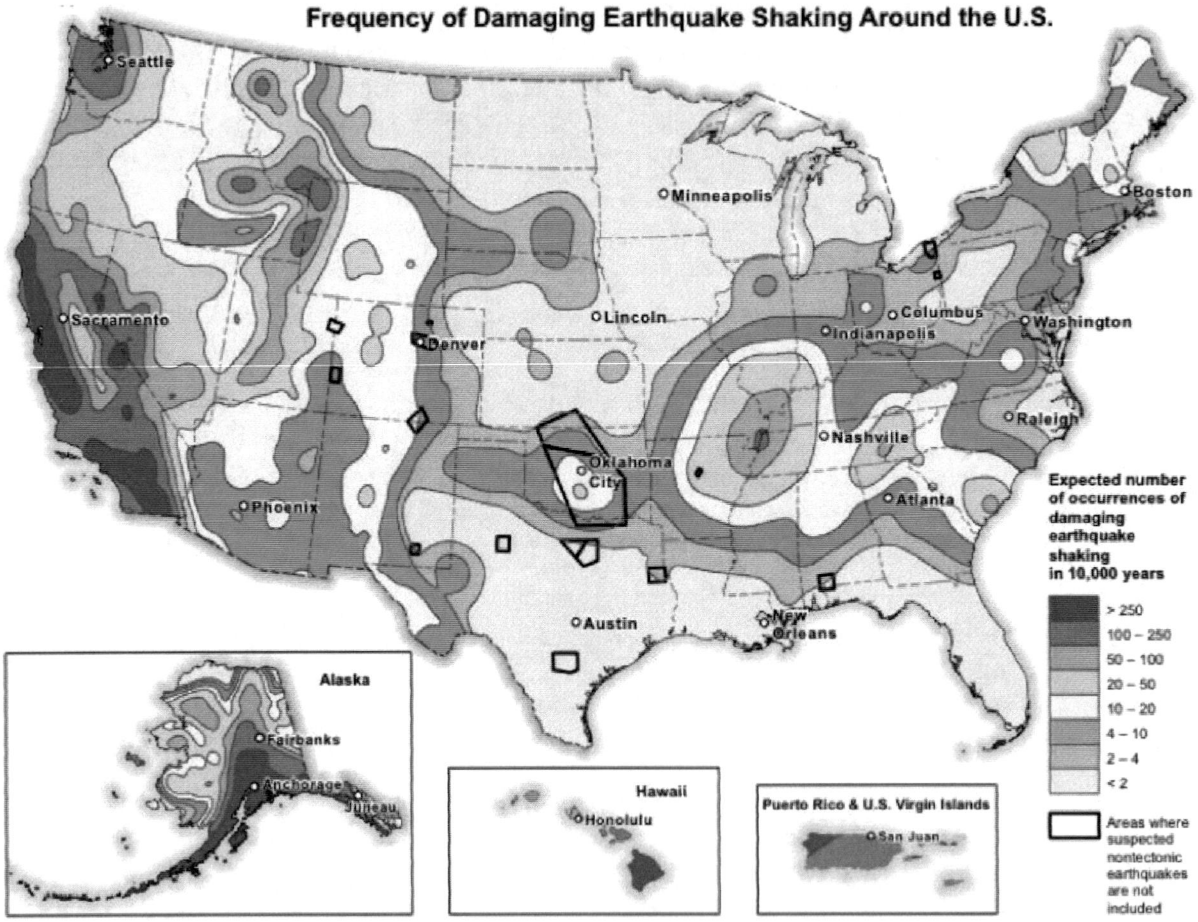

Figure 3. Frequency of Damaging Earthquake Shaking Around the U.S.

2.1.3 Tsunamis

According to the International Tsunami Information Center (ITEC), massive tsunamis that strike vast swaths of the ocean basin where they originate occur about once every fifteen years. Additionally, worldwide there are about two tsunamis per year capable of wreaking severe damage on large areas. Less severe tsunamis capable of inflicting localized damage occur even more often. While tsunamis can and have occurred in almost every large body of water, the Pacific Ocean basin and its associated waters are the most vulnerable [7].

In addition to the frequency of tsunamis, the associated impact zone is of interest to risk analysts. The impact zone is dependent on the size of the tsunami, which is dependent on the energy released by the initiating event—typically an earthquake, but landslides and volcanoes also trigger tsunamis—and the breadth of the initiating event. Tsunamis originate not just at the epicenter of an earthquake, but all along its length. Most data on the size of tsunamis focus on the height of the waves. However, for risk assessment, the breadth of the tsunami and inland incursion are of greater interest.

Inland incursion is a function of the height of the tsunami and the shore topography. Flat shoreline is more vulnerable; for instance, anything less than 30 m (100 ft) above sea level is vulnerable to a 30 m wave. Most tsunamis have a height of 3m (10 ft) or less. As a result, inland flooding as far as 300 m (1000 ft) is not unexpected, although some tsunamis have caused flooding 10 times further inland, with waves 10 times higher [8].

The breadth of tsunamis depends on many factors. Massive tsunamis can impact several hundred to several thousand miles of coastline. As they become bigger, they become less frequent, so that the risk of tsunamis large enough to cause any damage is essentially constant, independent of size. Collectively, it is reasonable to assume 5 tsunamis per year capable of causing damage to 20 km (12 mi) of coastline. This is consistent with data reported by the ITEC [9].

Table 2 shows the distribution of tsunami by body of water. Table 2 also shows the amount of coastline associated with each body of water. (Because of the fractal problem, there is generally little consensus on the length of coastline associated with an irregularly shaped body, whereas area can be calculated with great precision. Therefore, this article estimates coastline using the relationship of $10.5(\text{Area})^{0.5} = \text{Perimeter}$.)

Table 2. Distribution of Tsunamis

Body of Water	Percentage of Tsunamis	Coastline Length
Pacific Ocean	71%	135,000 km (84,000 mi)
Mediterranean Sea	15%	17,000 km (10,000 mi)
Atlantic Ocean and Caribbean Sea	7%	110,000 km (68,000 mi)
Indian Ocean	6%	89,000 km (55,000 mi)
Black Sea	1%	7,000 km (4,000 mi)

Consider a plant on the Pacific coast in California, within 300 m (1000 ft) of the shore and at an elevation of less than 30 m (100 ft) above sea level. The frequency at which a damaging tsunami originating in the Pacific Ocean would strike the 20 km stretch of coastline on which the plant operates is:

$$5 \text{ events/yr} \cdot 0.71 \cdot 20 \text{ km} / 135{,}000 \text{ km} = 0.0005 \text{ events/year} = 5 \cdot 10^{4} \text{ events per year}$$

A similar plant situated on the Atlantic coast in Georgia would be at less risk:

$$5 \text{ events/yr} \cdot 0.07 \cdot 20 \text{ km} / 110{,}000 \text{ km} = 0.00006 \text{ events/year} = 6 \cdot 10^{-5} \text{ events per year}$$

Either plant significantly reduces the hazard by being more than 100 m (330 ft) above sea level, however far back from the shore that must be.

A Note About the Fukushima Nuclear Facility.

The Fukushima Nuclear Facility in Japan sat on coastline in the Pacific Basin. Before the catastrophic tsunami on March 11, 2011, the frequency at which a damaging tsunami would have struck the facility could be estimated at $5 \cdot 10^{-4}$ events per year. In hindsight, pundits and the popular media were quick to say, "they should have known." That destructive tsunamis are possible is not in dispute. The fact that one happened in 2011 is proof of that. The fact that one happened in 2011 is not proof that they happen frequently. After the 2011 tsunami, the frequency at which a damaging tsunami would have struck the facility is unchanged: about $5 \cdot 10^{-4}$ events per year.

2.2 Weather-Related

2.2.1 Hurricanes

The damage by hurricanes is from wind and storm surge. Storm surge is too dependent on the topography of the specific location and other factors for this paper to address, but a reasonable range is 1 to 10 m (3 to 30 ft) of elevation above sea level [10]. The impact zone of hurricane winds is a function of how far inland it travels before winds weaken and the span of the hurricane when it makes landfall. Typically, meteorologists mention 160 km (100 mi) as the distance a hurricane will travel inland before weakening to a tropical storm. The span of hurricane-force winds varies from 50 to 500 km (30 to 300 mi), with an average diameter of hurricane-force winds of 160 km (100 mi) [11].

On average, 1.77 hurricanes per year struck the United States during the years from 1851 to 2010 [12]. Meteorologists categorize hurricanes on the Safir-Simpson scale. While the frequency of hurricanes decreases with increasing strength, the Safir-Simpson scale, unlike the earthquake magnitude scales, is not logarithmic, so there is not an obvious distribution, as Table 3 shows.

Table 3. Hurricane Strength

Safir-Simpson Category	Description	Distribution by Category
1	119 – 153 km/h (74 – 95 mph)	40%
2	154 – 177 km/h (96 – 110 mph)	26%
3	178 – 208 km/h (111 – 129 mph)	26%
4	209 – 251 km/h (130 – 156 mph)	6%
5	> 252 km/h (> 156 mph)	1%

While the Safir-Simpson scale categorizes hurricanes, it is important to keep in mind that the scale is based solely on peak wind speed. The size and destructive energy of a storm and the intensity of the associated storm surge is only loosely correlated to a storm's Safir-Simpson category.

More relevant to a risk analyst considering a hurricane as an initiating event is the frequency at which it will strike a specific location. As was the case with tsunamis, hurricanes can strike any seacoast, but some locations are far more vulnerable than others. In the United States, only the coasts of the Gulf of Mexico and the Atlantic Ocean are vulnerable. Data is available for the eighteen coastal states bordered by these two bodies of water [12], and after normalizing for the length of coastline of each state, it is clear to see that some states are more vulnerable to hurricanes than others, as Table 4 shows, beginning at the southwest coastline in Texas and proceeding east and then north to Maine.

Table 4. Hurricane Frequency

State	**Hurricanes per Century per 160 km (100 mi) of Coastline**
Texas	8
Louisiana	8
Mississippi	16
Alabama	16
Florida	8
Georgia	8
South Carolina	8
North Carolina	8
Virginia	4
Maryland	4
Delaware	4
New Jersey	4
New York	4
Connecticut	4
Rhode Island	4
Massachusetts	4
New Hampshire	4
Maine	2

The estimates of coastline length have the same fractal problem described earlier. In this instance, the coastline length was based on a resolution of 30 minutes of latitude, or about 56.6 km (34.5 mi). Expressing the frequency per 160 km (100 mi) of coastline puts the values in terms of average hurricane diameters so that the values expressed give the frequency at which any location on these coasts can expect a hurricane to strike, per century.

Once the hurricane frequency for a particular location is determined, the facility should be assumed to be vulnerable to storm surge when is stands at an elevation less than 10 m (30 ft) and vulnerable to hurricane-force winds if it is within 160 km (100 mi) of the coast.

A Note About the 2017 Hurricane Season.

Before it was even over, the popular press had dubbed 2017 as The Year of the Hurricane. Three of the five costliest hurricanes on record, adjusted for inflation, occurred in 2017: Harvey (2nd, at $125 billion), Maria (3rd, at $90 billion), and

Irma (5th, at $50 billion). This compares to Katrina in 2005 (1st, at $160 billion) and Sandy in 2012 (4th, at $70 billion) [13]. That the costliest hurricanes are all recent, however, is not a reflection of a new power in hurricanes, but of extent of development on the coastlines, hence the amount of damage a hurricane can do. When a hurricane makes landfall on an undeveloped coastline, it does no damage at all, no matter how powerful it is.

Researchers have noted that there has been a recent uptick in hurricane activity in the Atlantic. It may be the beginning of an upward trend, meaning that past data is no longer relevant, or it may be normal statistical variation. At most, it is too early to tell [14]. After the chemical fires at the Arkema plant in Crosby, Texas, following the unprecedented flooding brought on by Hurricane Harvey in August 2017, some suggested that relying on existing data about hurricanes and flooding was no longer valid [15, 16]. In the aftermath of a rare but devastating disaster, there is a tendency to overestimate how frequent such an event may be. Hence, there is a tendency to overestimate the risk associated with such an event, to declare, "This is the new normal." However, in the absence of new data to replace the existing data, the existing data remains the best data there is. Anything else is pure speculation.

2.2.2 Tornados

Tornadoes are another violent weather-related event that are sometimes a concern. Like earthquakes, tsunamis, and hurricanes, the frequency of tornadoes is highly dependent on location. Most are aware of the region of North America known as "Tornado Alley" which extends northward from Texas through the plain states of Oklahoma, Kansas, Nebraska, South Dakota, North Dakota, and into the Canadian prairie provinces of Manitoba and Saskatchewan. However, nearly every state east of the Rocky Mountains experiences several tornadoes per year, as Figure 4 shows, as do Ontario, Manitoba, Alberta, and Saskatchewan.

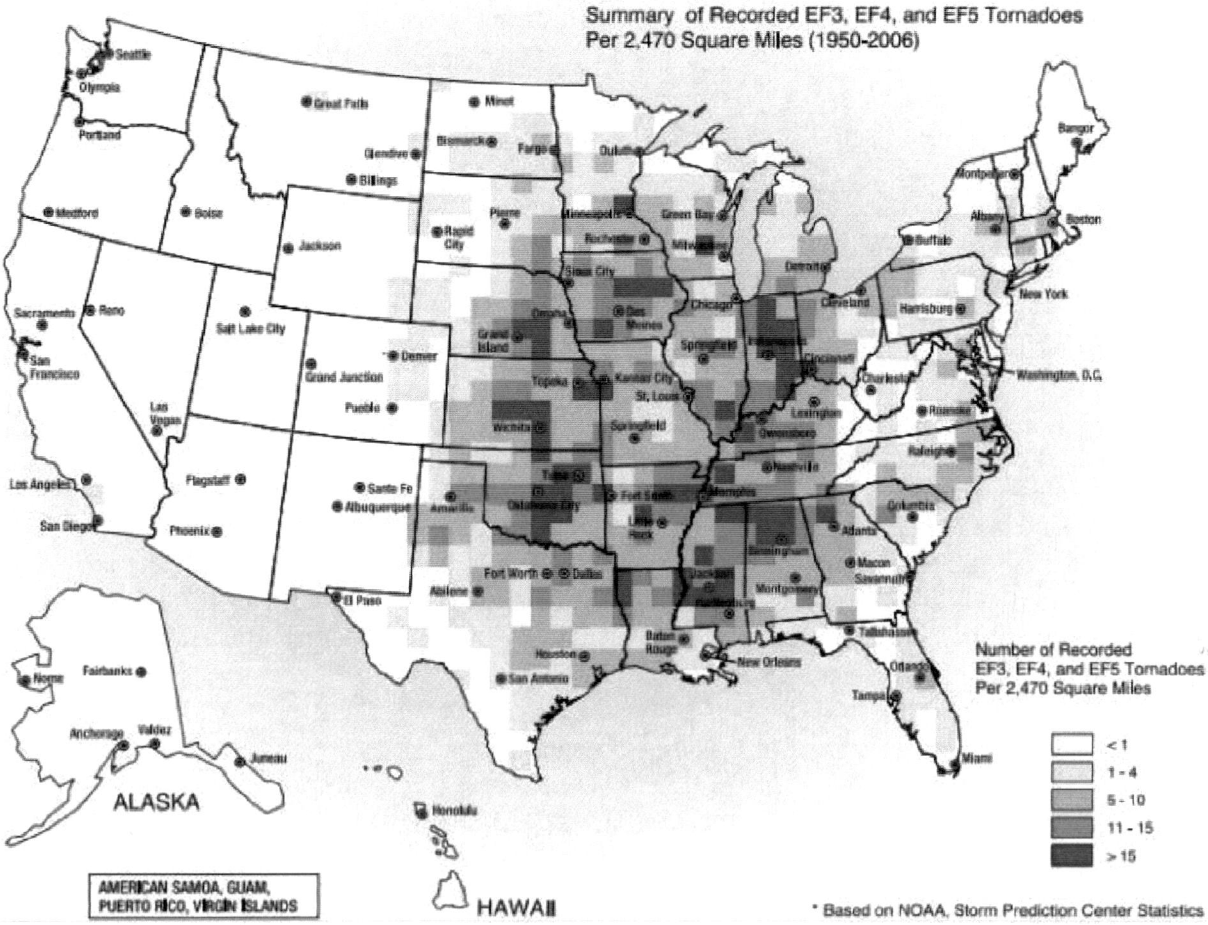

Figure 4. Total Tornado Activity in the United States – 1950 to 2006.

The impact zone of a tornado depends on the width and length of its path. The width of tornado paths ranges from 100 m (300 ft) to 4.2 km (2.6 mi); the typical width is about 1.6 km (1 mi). The longest observed path of a tornado is 352 km (215 mi), but tornado paths are rarely longer than 25 km (15 mi). The average travel speed of tornados is 50 km/h (30 mph). Because they typically last around 10 minutes, the average path of a tornado is 8.3 km (5 mi) [17]. The average impact zone of a tornado, then, is 13 km^2 (5 mi^2).

The blocks in Figure 4 use a color code to show the number of EF3, EF4, and EF5 tornadoes per 2,470 mi^2 per the 57 years covered by the analysis [18]. In other words, tornadoes per $1.4 \cdot 10^5$ mi^2 per year. Expressing the frequency in terms of impact zones is more relevant, however, so the most appropriate units are tornadoes per impact zone per year. The conversion appears in Table 5.

Table 5. Tornado Rates for Figure 4

Legend	NOAA	Frequency at which a tornado is likely to strike a facility
Very Low	< 1 tornadoes per 2470 mi^2 per 57 yr	$< 0.4 \cdot 10^{-4}$ tornadoes per 13 km^2 (5 mi^2) per yr
Low	1-4 tornadoes per 2470 mi^2 per 57 yr	$0.4 - 1.5 \cdot 10^{-4}$ tornadoes per 13 km^2 (5 mi^2) per yr
Medium	5-10 tornadoes per 2470 mi^2 per 57 yr	$1.8 - 3.5 \cdot 10^{-4}$ tornadoes per 13 km^2 (5 mi^2) per yr
High	11-15 tornadoes per 2470 mi^2 per 57 yr	$4 - 5 \cdot 10^{-4}$ tornadoes per 13 km^2 (5 mi^2) per yr
Very High	> 15 tornadoes per 2470 mi^2 per 57 yr	$> 5 \cdot 10^{-4}$ tornadoes per 13 km^2 (5 mi^2) per yr

Figure 4 is very granular, but it only includes the most extreme tornadoes with Enhanced Fujita scale ratings of EF3, EF4, and EF5. Using a state-by-state compilation [19] of confirmed tornadoes of all ratings, EF0 to EF5, Table 6 shows the frequency at with a tornado is likely to strike a facility, again normalizing the tornado frequency to a typical impact area of 13 km^2 (5 mi^2). With tornado frequencies for each state [19] and the area of each state (expressed as tornado impact zones), it is possible to determine tornado rates, as shown in Table 6. The rates shown in Table 6 differ from those shown in Figure 4, in that the Figure 4 rates only include EF3, EF4, and EF5 tornadoes, while the rates in Table 6 include all tornadoes, EF0 to EF5.

Table 6. Tornado Rates

State	**Frequency at which a tornado is likely to strike a facility** [tornadoes per 13 km^2 (5 mi^2) impact zone per year]
Kansas	$5.6 \cdot 10^{-3}$
Oklahoma	$4.7 \cdot 10^{-3}$
Illinois	$4.7 \cdot 10^{-3}$
Mississippi	$4.7 \cdot 10^{-3}$
Alabama	$4.5 \cdot 10^{-3}$
Iowa	$4.4 \cdot 10^{-3}$
Florida	$4.2 \cdot 10^{-3}$
Maryland	$4.0 \cdot 10^{-3}$
South Carolina	$3.6 \cdot 10^{-3}$
Arkansas	$3.6 \cdot 10^{-3}$
Louisiana	$3.6 \cdot 10^{-3}$
Nebraska	$3.5 \cdot 10^{-3}$
Tennessee	$3.5 \cdot 10^{-3}$
Indiana	$3.4 \cdot 10^{-3}$
Missouri	$3.3 \cdot 10^{-3}$
Kentucky	$3.0 \cdot 10^{-3}$
Texas	$2.7 \cdot 10^{-3}$
North Carolina	$2.7 \cdot 10^{-3}$
Georgia	$2.5 \cdot 10^{-3}$
Minnesota	$2.4 \cdot 10^{-3}$
Colorado	$2.4 \cdot 10^{-3}$
North Dakota	$2.2 \cdot 10^{-3}$
Ohio	$2.1 \cdot 10^{-3}$
South Dakota	$2.1 \cdot 10^{-3}$
Virginia	$2.1 \cdot 10^{-3}$

State	**Frequency at which a tornado is likely to strike a facility** [tornadoes per 13 km^2 (5 mi^2) impact zone per year]
Delaware	$2.0 \cdot 10^{-3}$
Wisconsin	$1.8 \cdot 10^{-3}$
Pennsylvania	$1.7 \cdot 10^{-3}$
Connecticut	$1.4 \cdot 10^{-3}$
New Jersey	$1.1 \cdot 10^{-3}$
New York	$0.9 \cdot 10^{-3}$
Michigan	$0.8 \cdot 10^{-3}$
Massachusetts	$0.7 \cdot 10^{-3}$
Rhode Island	$0.6 \cdot 10^{-3}$
Wyoming	$0.6 \cdot 10^{-3}$
West Virginia	$0.5 \cdot 10^{-3}$

The frequency of tornadoes in the states west of the Great Plains and the states in northern New England is at a background level of $< 4 \cdot 10^{-4}$. This is 10 times higher than the "low" level shown in Figure 4, the difference being that Figure 4 only included EF3 to EF5 tornadoes, which account for less than 10% of all tornadoes, while Table 6 includes all tornadoes, EF0 to EF5. As for the 36 states on the list, the tornado leader is Kansas, which is what most would expect.

2.3 Transportation-Related

2.3.1 Pipeline ruptures

Pipeline ruptures result in approximately 25 barrels spilled per billion barrel-miles transported and 5.2 fatalities per trillion barrel-miles transported [20]. Risk analysts, however, are more concerned with the incident rate than spilled volume rate. The Center for Chemical Process Safety (CCPS) has addressed this specifically in its *Guidelines for Initiating Events and Independent Protection Layers in Layer of Protection Analysis* [21]. The committee determined that it would be useful to make two distinctions: between pipelines that are larger and smaller than 150 mm (6 inches) and between full breaches, sometimes called "guillotine" breaches, and partial breaches, or leaks, that amount to 10% of a full breach. The data in Table 7 refers to above-ground pipelines.

Table 7. Pipeline Failure Rates

Piping Size	Piping Failure Type	Frequency
Greater than 150 mm diameter (> 6 inches dia.)	Full breach (Guillotine failure)	$1 \cdot 10^{-5}$ failures per 100 m per year
Greater than 150 mm diameter (> 6 inches dia.)	Partial breach – 10% (Leak)	$1 \cdot 10^{-4}$ failures per 100 m per year
Less than or equal to 150 mm diameter (≤ 6 inches dia.)	Full breach (Guillotine failure)	$1 \cdot 10^{-4}$ failures per 100 m per year
Less than or equal to 150 mm diameter (≤ 6 inches dia.)	Partial breach – 10% (Leak)	$1 \cdot 10^{-3}$ failures per 100 m per year

The CCPS committee determined that there were too many factors to take into account to offer a generic failure rate for underground piping, although in general, underground piping failures are more frequent than above-ground piping failures, in large part because above ground pipelines are considerably easier to inspect.

2.3.2 Train derailments

Most discussions of train derailments report them in rates per train miles or similar metrics. Baldas reported that railroad crude oil spills were at the rate of 431 barrels per billion barrel-miles transported and that fatalities were at the rate of 15 fatalities per trillion barrel-miles transported [20].

Train derailments are quite frequent, but the vast majority occur at speeds of less than 5 mph with little or no damage. This is the sort of derailment that typically occurs within a plant fence line and while it can pose significant operational difficulties, there is generally no reason for this sort of incident to be the concern of a risk assessment team. Train derailments at speeds high enough to result in significant harm, such as the Lac-Mégantic rail disaster of July 6, 2013 that killed 47, are far less frequent. The "List of American railroad accidents" [22] notes 26 major American railroad accidents so far during the 21st century. However, somewhere between derailments that are daily inconveniences for operating facilities and major catastrophes is the number that risk analysts should consider.

The Federal Railroad Administration (FRA) reported that there were almost 19,000 train accidents in the ten-year period from 2008 through 2017. Of those, 71% were derailments [23]. Elsewhere, the FRA reported that there are 140,000 miles of railroad operated in the United States [24]. The rate of derailment, then, is 0.01 derailments per year per mile of railroad, or more relevantly, $6 \cdot 10^{-4}$ derailments per year per 100 meters of railroad.

2.3.3 Truck crashes

On a volume basis, tank truck crashes result in about 660 barrels of crude oil spilled per billion barrel-miles transported. Likewise, tank truck crashes result in about 80 fatalities per trillion barrel-miles transported. Both rates are worse than for transport by train cars and much worse than for transport by pipeline [20].

The incident rate for truck crashes are not reported on the basis of infrastructure—in crashes per mile of roadway—but on the basis of traffic, that is, crashes per 100 million vehicle miles travelled. Counting only those crashes that result in fatalities as sufficiently serious to warrant attention in a facility siting study, the average rate for the period from 2007 to 2015 is 1.24 fatal truck crashes per 100 million miles travelled by large trucks [25].

To determine the incident rate on the basis of crashes per mile of roadway requires that the analyst determine the length of roadway considered and then estimate the volume of truck traffic on that roadway. While there are portions of the National Highway System that carry more than 8,500 trucks per day [26], the vast majority of roadways carry considerably less. As an upper boundary, the incident rate for truck crashes is 0.04 crashes per year per mile of high-volume highway, or $2 \cdot 10^{-3}$ truck crashes per year per 100 meters of highway. For the roadways more likely to be of interest in a risk assessment, the volume is likely to be 10 to 100 times less. However, some sampling is in order before estimating the truck crash rate for a specific roadway.

2.3.4 Airplane crashes

As most would imagine, airplane crashes into process plants are exceedingly rare. A search of the literature only identified three in the United States—one in 1949 in Dallas, Texas [27], one in 1985 in Kansas City, Kansas [28], and one in 2016 in Lubbock, Texas [29]. There are undoubtedly others, but they are so rare that they should always be newsworthy; it is hard to imagine that many go unreported. A recent article discussed the risk of chemical plants near airports [30], but focused more on the consequences of those incidents than on a methodology for estimating the frequency of airplane crashes into chemical plants.

Berg has described a useful methodology for estimating the likelihood of an airplane crashing into a specific facility [31]. Beyond 10 km away from an airport, the likelihood is a function of the area of the facility, the distance from the airport of the facility, and the rate at which aircraft land at and take off from the airport. Within 10 km of an airport, the likelihood remains a function of these factors, but also increases exponentially as the distance decreases.

Berg's method can also account for the weight class of the airplane, the deviation from the landing/take-off axis, additional airports, and whether the aircraft are commercial, military, or other. These distinctions are only necessary, however, when the basic methodology suggests that the frequency is too high and would benefit from a more detailed analysis.

The fundamental relationship is:

$$H = h \cdot f \cdot A_{FACILITY} / (2\pi R \cdot d_{AVG})$$

Where

H	is the frequency of airplane crashes into the facility (crashes/year)
h	is the global crash rate
f	is the frequency of flights from the airport considered (flights/year)
$A_{FACILITY}$	is the area of concern for the facility (L^2)
R	is the distance of the facility from the airport considered (L)
d_{AVG}	is the average flight distance to and from the airport considered (L)

Note: L can be in any units, but it must be in the same units in all three instances.

The global crash rate, h, is $4 \cdot 10^{-7}$ crashes/flight. This is derived from a recent report from the International Air Transport Association (IATA) [32] which indicated a five-year average of 86 aviation accidents per year, of which ~20% were severe enough to result in the loss of the aircraft, or "hull loss" accidents. Hull loss accidents are the kind of loss that would affect a chemical plant [32]. For a facility occupying 250 acres (1.0 km^2) at a distance of 50 km (31 mi) from Saint Louis Lambert International Airport (STL) in Missouri, where there are 260 flight departures a day (95,000 flights/year) at an average flight distance of 1,200 km (750 mi), the annual frequency of airplane crashes into the facility would be $1 \cdot 10^{-7}$ crashes/year.

Facilities within 10 km of an airport experience an exponential climb in the crash rate the closer they are to the airport. Berg's work addresses this as well. The relationship, based on flight operations—take-offs and landings—rather than flights, is:

$$H = h \cdot e^{(-0.5/\text{km} \cdot R)} \cdot f \cdot A_{\text{FACILITY}} / (2\pi \text{ km} \cdot R)$$

Where

H	is the frequency of airplane crashes into the facility (crashes/year)
h	is the global landing/take-off crash factor ($1.3 \cdot 10^{-7}$ crashes/flight)
R	is the distance of the facility from the airport considered (km)
f	is the frequency of flights from the airport considered (flights/year)
A_{FACILITY}	is the area of concern for the facility (km^2)
2π km · R	is the area of the 1 km wide annular segment at distance R from the airport considered (km^2)

Considering an identical process facility, but only 5 km (3.1 miles) from STL, the annual frequency of airplane crashes into the facility would be $3 \cdot 10^{-5}$ crashes/year.

Strictly speaking, the relationship above only applies to facilities that are within 45° of the flight path for landing and take-off. A conservative assessment would treat all facilities as though they are within 45° of a flight path, particularly since most large airports have more than one runway.

3 Conclusions and Recommendations

Risk has two components, consequence and likelihood. It follows, then, that an analysis that only addresses consequences does not address risk. In the absence of likelihood, a consequence is not particularly meaningful. Decisions should be driven by risk assessments, not consequence assessments. Otherwise, every plant in the world should have an anti-meteor task force. Unlike "black swans," which are inconceivable until they happen, "meteors" are rare, but conceivable. However, most people would not know how to quantify how rare these "meteors" are.

When we considered nine types of conceptually rare events in terms of the likelihood of them impacting a process facility, it was clear that they were not only conceivable, but calculable. Except in the case of actual meteors, the frequency is a function of a facility's location and in some cases, other factors. Rather than use the numbers in Table 8, which compiles the examples used throughout this article (rounding to the nearest order of magnitude to make comparison easier), analysts are advised to do the simple math necessary to determine the likelihood for their specific facility.

Table 8. Example Event Frequencies

Facility Event Description	**Frequency (nearest order of magnitude)**
> 20 m meteor strike	$1 \cdot 10^{-8}$ per year
Damaging earthquake in upper Midwest	$1 \cdot 10^{-4}$ per year
Damaging earthquake on California coast	$1 \cdot 10^{-2}$ per year
Damaging tsunami on Pacific coast	$1 \cdot 10^{-3}$ per year
Damaging tsunami on Atlantic coast	$1 \cdot 10^{-4}$ per year
Hurricane on Gulf of Mexico coast	$1 \cdot 10^{-1}$ per year
Hurricane on New England coast	$1 \cdot 10^{-2}$ per year
Tornado in Tornado Alley	$1 \cdot 10^{-2}$ per year
Tornado in Rocky Mountains	$1 \cdot 10^{-4}$ per year
Rupture of large pipeline (> 150 mm, > 6 inches)	$1 \cdot 10^{-5}$ failures per 100 m per year
Significant train derailment	$1 \cdot 10^{-3}$ failures per 100 m per year
Significant truck crash on high volume highway	$1 \cdot 10^{-3}$ failures per 100 m per year
Significant truck crash on medium volume highway	$1 \cdot 10^{-4}$ failures per 100 m per year
Significant truck crash on low volume highway	$1 \cdot 10^{-5}$ failures per 100 m per year
Plane crash on plant 50 km from medium-sized airport	$1 \cdot 10^{-7}$ per year
Plane crash on plant 5 km from medium-sized airport	$1 \cdot 10^{-4}$ per year

When U.S. Supreme Court Justice Stevens wrote the majority opinion in the 1980 OSHA benzene case, he said, "Some risks are plainly acceptable and others are plainly unacceptable," and then went on to add that the odds of fatality of one in a billion could not be considered significant but that for odds of a fatality of one in a thousand "a reasonable person might well consider the risk significant" [33]. Given a range of one in a thousand years as plainly unacceptable and one in a billion years as plainly acceptable, none of these "meteors" meet the criteria of "plainly acceptable," not even meteors. On the other hand, there are several that meet the criteria of "a reasonable person's significant risk."

It is in that gap between once every thousand years and once every billion years that we will find the definition of "credible". Analysts should consider the frequencies of the events discussed here, and define "credible" for their organization. With that definition in hand, they will be able eliminate scenarios that are "not credible."

4 References

[1] Taleb, Nassim Nicholas, *The Black Swan: The Impact of the Highly Improbable (First Edition)*. Random House, New York. 2007.

[2] "Impact Event," *Wikipedia – The Free Encyclopedia*, The Wikimedia Foundation, San Francisco, CA, last edited 01-Feb-2018. Accessed 14-Feb-2018 at https://en.wikipedia.org/wiki/Impact_event#Frequency_and_risk .

[3] Marcus, R., H.J.Melosh, and G. Collins, "Earth Impact Effects Program: A Web-based computer program for calculating the regional environmental consequences of a meteoroid impact on Earth," *Meteoritics and Planetary Science*, 40:6, June 2005. pp. 817-840. Accessed 14-Feb-2018 at http://impact.ese.ic.ac.uk/ImpactEffects/

[4] Shearer, P.M. and P.B.Stark, "Global Risk of Big Earthquakes Has Not Recently Increased," Proceedings of the National Academy of Sciences of the United States of America, 109:3, November 12, 2011. pp. 717-721. Accessed 14-Feb-2018 at http://www.pnas.org/content/109/3/717.full .

[5] NEIC, "Earthquakes," *Earthquake Hazards Program*, National Earthquake Information Center, United States Geological Survey, U.S. Department of Commerce, Denver, CO. Accessed 14-Feb-2018 at https://earthquake.usgs.gov/earthquakes/ .

[6] NEIC, "Introduction to the National Seismic Hazard Maps," *Earthquake Hazards Program*, National Earthquake Information Center, United States Geological Survey, U.S. Department of Commerce, Denver, CO. Accessed 14-Feb-2018 at https://earthquake.usgs.gov/earthquakes/ .

[7] ITIC, "Where and how frequently are tsunamis generated?" *International Tsunami Information Center Web Site*, International Tsunami Information Center, Honolulu, HI. Accessed 14-Feb-2018 at http://itic.ioc-unesco.org/index.php?option=com_content&view=article&id=1163:where-and-how-frequently-are-tsunamis-generated&catid=1340&Itemid=2055 .

[8] National Geographic Staff, "Tsunamis: Facts About Killer Waves, National Geographic, 28-Dec-2004. Accessed 14-Feb-2018 at https://news.nationalgeographic.com/news/2004/12/tsunamis-facts-about-killer-waves/ .

[9] ITIC, "List of Tsunamis," *International Tsunami Information Center Web Site*, International Tsunami Information Center, Honolulu, HI. Accessed 14-Feb-2018 at http://itic.ioc-unesco.org/index.php?option=com_content&view=category&layout=blog&id=1160&Itemid=1077&lang=en .

[10] Masters, J., "Characteristics of Storm Surges, *Weather Underground*, The Weather Company, LLC, Atlanta, GA. Accessed 14-Feb-2018 at https://www.wunderground.com/hurricane/surge_characteristics.asp .

[11] Hasling, J.F., "What is the Size of a Hurricane?" *WRC Texas Hurricane Handbook*, Weather Research Center, Houston, TX. Accessed 14-Feb-2018 at http://www.wxresearch.org/family/pg5.html .

[12] Blake, E.S., C.W. Landsea, and E.J. Gibney, "The Deadliest, Costliest, and Most Intense United States Tropical Cyclones from 1851 to 2010," *NOAA Technical Memorandum NWS NHC-6*, National Hurricane Center, National Oceanic and Atmospheric Administration, Miami, FL. August 2011. 52 pp. Accessed 14-Feb-2018 at https://www.nhc.noaa.gov/dcmi.shtml .

[13] NHC Public Affairs, "Costliest U.S. Tropical Cyclones Tables Updated," Addendum to *NOAA Technical Memorandum NWS NHC-6*, National Hurricane Center, National

Oceanic and Atmospheric Administration, Miami, FL. January 26, 2018. 52 pp. Accessed 14-Feb-2018 at https://www.nhc.noaa.gov/news/UpdatedCostliest.pdf .

[14] Ramaswamy, V., “Global Warming and Hurricanes: An Overview of Current Research Results,” *GFDL Web Page*, Geophysical Fluid Dynamics Laboratory, Princeton University, Princeton, NJ. 24-Jan-2018, Accessed 14-Feb-2018 at https://www.gfdl.noaa.gov/global-warming-and-hurricanes/ .

[15] Dempsey, M. and J. Carpenter, “Arkema documents: Planning, mechanical failures led to Harvey chemical fires,” Houston Chronicle, 15-Nov-2017. Accessed 14-Feb-2017 at https://www.houstonchronicle.com/news/houston-texas/houston/article/Arkema-documents-show-planning-mechanical-12358188.php .

[16] Dunklin, R. “Agency: Texas chemical plant unprepared for Hurricane Harvey,” Chicago Tribune, 17-Nov-2017. Accessed 14-Feb-2017 at http://www.chicagotribune.com/news/nationworld/sns-bc-us--harvey-chemical-plant-20171115-story.html .

[17] “About Tornadoes,” *Weather Underground*, The Weather Company, LLC, Atlanta, GA. Accessed 01-Feb-2017 at https://www.wunderground.com/resources/education/tornadoFAQ.asp?MR=1 .

[18] FEMA, Taking Shelter From the Storm: Building a Safe Room for Your Home or Small Business, FEMA 320, 3rd Ed., Federal Emergency Management Agency, Department of Homeland Security, Washington, DC. August 2008. p. 3. Accessed 14-Feb-2018 at http://www.weather.gov/media/bis/FEMA_SafeRoom.pdf .

[19] Livingston, I., “Annual and monthly tornado averages for each state (maps), *U.S.Tornadoes*, 06-Apr-2016. Accessed 14-Feb-2018 at http://www.ustornadoes.com/2016/04/06/annual-and-monthly-tornado-averages-across-the-united-states/ .

[20] Baldas, A., “The Crude Truth: Assessing the Associated Risks of Rail, Tanker, and Pipeline Crude Oil Transport,” *International Mechanical Engineering Congress and Exposition,* Safety Engineering and Risk Analysis Division, ASME. Montreal, Canada. November 2014.

[21] Chastain, W., et al., *Guidelines for Initiating Events and Independent Protection Layers in Layer of Protection Analysis*, Center for Chemical Process Safety, American Institute of Chemical Engineers, New York. 2015. pp. 105-112.

[22] “List of American railroad accidents,” *Wikipedia – The Free Encyclopedia*, The Wikimedia Foundation, San Francisco, CA, last edited 04-Feb-2018. Accessed 14-Feb-2018 at https://en.wikipedia.org/wiki/List_of_American_railroad_accidents .

[23] FRA, “Ten Year Accident/Incident Overview,” *FRA Office of Safety Analysis Web Site*, Federal Railroad Administration, U.S. Department of Transportation, Washington, DC. November 2017. Accessed 14-Feb-2018 at http://safetydata.fra.dot.gov/OfficeofSafety/publicsite/Query/TenYearAccidentIncidentOverview.aspx .

[24] FRA, "Freight Rail Overview," *FRA Web Site*, Federal Railroad Administration, U.S. Department of Transportation, Washington, DC. Accessed 14-Feb-2018 at https://www.fra.dot.gov/Page/P0362 .

[25] FMSCA, "Trends Table 4. Large Truck Fatal Crash Statistics, 1975-2015," *Large Truck and Bus Crash Facts*, Federal Motor Carrier Safety Administration, U.S. Department of Transportation, Washington, DC. February 2017. Accessed 14-Feb-2018 at https://www.fmcsa.dot.gov/safety/data-and-statistics/large-truck-and-bus-crash-facts-2015 .

[26] FHA, "Peak-Period Congestion on High-Volume Truck Portions of the National Highway System," Freight Analysis Framework, V4.3. Federal Highway Administration, U.S. Department of Transportation, Washington, DC. 2017. Accessed 14-Feb-2018 at https://ops.fhwa.dot.gov/freight/freight_analysis/nat_freight_stats/nhsconghvtrk2045.htm

[27] "American Airlines Flight 157," *Wikipedia – The Free Encyclopedia*, The Wikimedia Foundation, San Francisco, CA, last edited 19-Jun-2017. Accessed 14-Feb-2018 at https://en.wikipedia.org/wiki/American_Airlines_Flight_157 .

[28] McGuire, M.J. "January 9," *This Day in Water History*. Accessed 14-Feb-2018 at https://thisdayinwaterhistory.wordpress.com/2015/01/09/january-9-1985-plane-crashes-into-kansas-water-treatment-plant-2012-first-haitian-cholera-victim/

[29] AP, "Small plane that crashed into Texas cement plant was stolen from Midland," Lubbock Avalanche-Journal, Lubbockonline.com, 8-Jun-2016. Accessed 14-Feb-2018 at http://www.lubbockonline.com/filed-online/2016-06-08/small-plane-crashed-texas-cement-plant-was-stolen-midland .

[30] Konersmann, R. "Chemical Plants in the Vicinity of Airports—Assessing the Risk," *Chemical Engineering and Technology*, 32:2. 2009. pp.232-240.

[31] Berg, H.-P., "Risk Assessment of Aircraft Crash Onto a Nuclear Power Plant," *Reliability: Theory and Applications*, 01(20) (Vol. 2). March 2011. pp. 38-51.

[32] Gani, A., "How common are plane crashes?" *The Guardian*. 25-Jul-2014. Accessed 14-Feb-2018 at https://www.theguardian.com/world/2014/jul/24/how-common-are-plane-crashes .

[33] Stevens, J.P. writing for the majority, "Industrial Union Department AFL-CIO v. American Petroleum Institute et al." *Case No. 78-911*, U.S. Supreme Court, Washington, D.C. July 2, 1980. 448 U.S. 655.

Concept Risk Comparison of Transportation Means

Cynthia Spitzenberger
DNV GL
1400 Ravello Drive, Katy, TX 77449
cynthia.spitzenberger@dnvgl.com

Lindsay Deal
DNV GL

Prepared for Presentation at
American Institute of Chemical Engineers
2018 Spring Meeting and 14th Global Congress on Process Safety
Orlando, Florida
April 22 – 25, 2018

Concept Risk Comparison of Transportation Means

Cynthia Spitzenberger
DNV GL
1400 Ravello Drive, Katy, TX 77449
cynthia.spitzenberger@dnvgl.com

Lindsay Deal
DNV GL

Keywords: risk, transportation, QRA, highly volatile liquid

Abstract

Geographical distances between production and consumption of highly volatile liquids are increasing transport volumes across the continental United States. Supply chain planning requires comparison across transportation means. The complex process of selecting a transport means and route involves a range of parameters targeting both cost and feasibility. Risk to the public along the chosen route factors into both cost and feasibility comparisons.

This paper provides a framework to compare the risk to the public from a highly volatile liquid loss of containment event from pipeline, rail, truck and barge transportation. The risk-based framework normalizes the frequency on volume and distance transported and addresses consequences in terms of potential volume lost and population density to provide a direct, lateral comparison.

The presented risk results vary across transport modes and population densities, and are presented on an annual volume shipped basis. Given an annual volume amount and a specified population density, comparison of the different transport modes is possible on a coarse basis. However, route length likely varies between the different transport modes and would factor the expected risk value. Population density would also likely vary along the length of a transport route and would change the expected risk result.

Generally, for shipment of volumes less than 7 million gallons annually, the risk impact to public from the different transport modes in increasing order is barge, pipeline, rail and truck. For shipment volumes over 7 million gallons annually, pipeline transport presents the least risk exposure of the four transport modes.

1 Background

The transport of highly volatile liquids (HVL) within the US is a timely topic. The purpose of this evaluation is to provide a simplified and generic framework for the comparison of possible transport routes of HVL from a public risk perspective. To create a generic framework, a case study was performed to evaluate the HVL transport options. The postulated transport route selected is from the Commonwealth of Pennsylvania to a theoretical Gulf Coast refinery. Such a route allows comparison for the four transport options.

- Pipeline
- Rail
- Barge
- Truck

The case study considered specific transit locations for the paths of the different transport modes (existing pipeline and rail networks; rivers for barge transit; highways for hazardous transport). However, the details of the case study are not elaborated within the paper as they form only the basis to generalize the calculations to develop the framework. A normalized risk result is estimated for the comparison evaluation. This avoids calculation of societal risk metric which would be specific to location and frequency of operation. The key result aim is to draw conclusions from the generic framework regarding risk comparisons between the four modes of transit by volume transported. The generic result also provides a simplified comparison tool that could be used for coarse initial evaluations independent of the route.

The loading / unloading operations that would occur at either end of the route for the different transport modes are not considered. Cost evaluations for the different transport means are not included or considered.

2 Framework Basis

Parameters and assumptions that are common across transport modes include:

- Material
- Population categories
- Meteorological data
- Consequences of a release
- Limitations to the framework

DNV GL's Phast consequence tool was used for the consequence modeling.

2.1 Material

This study uses pure propane as a representative material for liquefied petroleum gas (LPG). However, highly volatile liquids can have a range of composition. For example, LPG is a mixture of materials highly dependent on its source.

2.2 Population Categories

Population densities drive both the public exposure to a consequence along the route and the probability of a delayed ignition.

Population densities are available from U.S. Census data by zip code allowing any route to be subdivided and quickly categorized with the application of geographic information system (GIS) software. This framework applies the Nuclear Regulatory Commission (NRC) definitions of Low, Medium and High population density categories (1) and develops continuous population density ranges from the 2010 U.S. Census data (2) to fit those definitions as shown in Table 1.

Table 1 Population Density Range

Population Density Category	**NRC Representative Persons per square mile** (1)	**Population Density (persons per square mile) Range Applied**
Low	15	0 – 47
Medium	1,862	48 – 6,142
High	10,000	6,143 + (max 156,667 (2))

2.3 Meteorological Data

The dispersion of a cloud of hazardous material is governed primarily by wind speed, wind direction and the atmospheric stability. Factors, which increase the dilution of a hazardous cloud with respect to distance traveled, are increasing wind speeds and decreasing stability of the atmosphere. However, high winds may transport hazardous materials far downwind before they become sufficiently diluted to no longer pose a hazard. An unstable atmosphere, typically experienced on a sunny day, causes increased vertical mixing, which further dilutes the hazardous clouds as they disperse downwind. The effect of wind direction is obvious in that only receptors downwind of the release are affected.

The meteorological data used in the Phast model consist of wind speed, humidity, solar radiation flux and ambient temperature.

Meteorological data vary along a given transport route in addition to the typical seasonal and weather pattern variations. Detailed analysis along a given route requires consideration of these variations. Initially, an average could be applied along the entire route with sensitivities used to bound how the risk results change at the extreme end of parameter ranges allowing a more detailed focus on the high-risk segments.

For simplicity, this framework applies a single combination of wind speed, humidity, solar radiation and ambient temperature since the goal is to compare similar routes across transport types. The weather condition applied is D stability with 5 m/s (11mph) wind speed, average atmospheric temperature of

50 degF. The maximum downwind distance of consequence effects is applied in all directions to account for variations in wind direction.

2.4 Consequences

Consequences are initially split between instantaneous (less than 20 seconds release duration) and continuous releases. Immediately ignited instantaneous releases result in a fireball. Immediately ignited continuous releases result in a jet or pool fire depending on the discharge rate.

All delayed ignition events are modeled as flash fires or vapor cloud explosions (VCE), where pool fires will accompany the flash fires/VCE for liquid spills. Most delayed ignition events also burn-back to form jet or pool fires that follow the initial flash fire or VCE, although the impacts to personnel are dominated by the initial flash fire/VCE effects.

Trucks, barges and railcars are assumed to stop upon an instantaneous release or upon ignition. However, small continuous releases may occur during transit. BLEVE (Boiling Liquid Expanding Vapor Explosion) hazards are considered as escalation events possible for the rail and truck transits.

Table 2 presents a summary of the different consequence hazards with the applied endpoint metrics and fatality fractions. Note that populations along the route are not assessed to be indoor or outdoor. Instead all population densities consider the same potential hazard exposure. The fatality fraction is meant to be a conservative average to reflect that some populations would be protected if indoors.

Table 2 Summary of Consequences

Release Type	Ignition Type	Consequence Type	Endpoint Metric	Fatality Fraction
Instantaneous	Immediate	Fireball	Fireball radius	0.7
Continuous		Jet Fire	Distance to $35 kW/m^2$	1
		Pool Fire	Distance to $35 kW/m^2$	1
Either Instantaneous or Continuous	Delayed	Flash Fire	Lower flammability limit (LFL)	0.7
		VCE	None	--
Escalating	Piloted	BLEVE	Distance to 0.3 bar	1

2.4.1 Fireballs

Fireballs are short-lived flames, generally resulting from ignition and combustion of turbulent vapor / two-phase fuels in air. They are usually caused by instant or short duration releases related to catastrophic failure of pressurized vessels or pipelines.

Persons indoors are generally protected from the short-duration radiation effects from fireballs. This framework applies a 0.7 fatality probability within the fireball radius to account for any people outdoors.

2.4.2 Jet Fire

The widely-used Cone model is applied as the basis for the jet fire modeling within Phast, which describes the shape of a jet flame as a frustum of a cone. The parameters describing the frustum, accounting for choked flow, are derived from comparisons with experimental data from laboratory and field tests. The key input parameters in defining jet fires are release rate, velocity, material, and release elevation.

This framework assumes all jets are oriented horizontally. A horizontal jet fire typically results in a larger hazard zone than a vertical or angled release and is generally more hazardous for personnel and equipment. The jet flame lengths and the subsequent radiation hazard ranges are primarily driven by the release rate and the material. Immediate fatality is assumed for all personnel within the 35 kW/m^2 radiation contour of a jet fire or a pool fire.

2.4.3 Pool Fire

The pool fire model in Phast calculates the shape and intensity of the flame, and a range of radiation results. A pool fire flame is modeled as a cylinder sheared in the direction of the wind, with diameter, height, and tilt angle (measured from the vertical). The flame shape gives input to the radiation calculations. The pool diameter is calculated by:

- Continuous liquid leak - The stable burning size is calculated, where the mass burning rate balances the mass release rate of liquid; the pool diameter is, then, the lesser of the stable burning size or the bund diameter if a bund is defined.
- Unbunded leak - If a bund area is not defined, the pool fire model takes into account any physical barriers to the spread of the liquid pool. As such, the pool is allowed to spread at a uniform depth until it attains a minimum thickness at a steady state. Factors such as sloping, drainage, and curbing in the immediate area are therefore not considered in determining the pool diameter.

The surface area of a pool is a critical parameter for fire calculations. Models are available for spills onto flat ground for both continuous spills (3) and instantaneous spills (4). Consequences from ignition of an "infinite" spreading pool are overly conservative and are not applied in the study.

2.4.4 Flash Fires

A flash fire is effectively the advancing flame front of an ignited vapor cloud. Although it presents significant personnel hazards (any outdoor personnel caught within the flash fire envelope are considered immediate fatalities), flash fires do not cause significant structural damage. There is little radiation outside of the lower flammable limit (LFL) contour, and damage done by the flash fire is typically restricted to ignition of easily ignitable materials such as cabling and plastic. Furthermore, flash fires do not generally create overpressures and as such their damage is limited to thermal impacts only. Persons indoors are generally protected from the short-duration radiation effects from flash fires. This framework applies a 0.7 fatality probability within the LFL contour to account for any people outdoors.

2.4.5 Vapor Cloud Explosions

The gas dispersion model within Phast requires as inputs: material, phase, release rate, duration, and velocity. Where the cloud is ignited without being in contact with any area of congestion, a *flash fire* is assumed to occur. The predicted overpressure caused by a VCE is associated with the volume (mass) of the flammable cloud confined within the obstructed region(s), which needs to be differentiated from the entire volume of the vapor cloud or the total released inventory.

While most roadways, railways, rivers and pipeline right of ways are not congested at levels comparable to industrial facilities, congested regions can locally exist along any transit way (tree or hedge rows, crowded railyards, tunnels). The dimensions, make-up and proximity of congested regions drive the predicted overpressure from a release. For simplicity, fatalities from VCE are excluded from this framework.

2.4.6 Escalation/BLEVE

BLEVE hazards are considered as escalation events. BLEVE scenarios require heat or flame impingement on a storage container to create the increase in internal pressure and weakening of the container structure. BLEVEs are considered for the rail and truck transport modes. If a railcar has an ignited hazard, it will likely impact a neighboring railcar which would eventually fail as a BLEVE scenario. If there is an ignited leak from the truck, then the scenario may escalate to a BLEVE scenario with the remaining inventory in the truck tanker.

2.5 Framework Limitations

The intent of the framework is to compare transport by mode. The framework does not include:

- Loading and offloading risk which are often risk drivers for barge, truck and rail transport modes. Loading and offloading hazards would be at the endpoints of the route and the risk impact would be localized to these areas and highly dependent on the local operations.
- Flow within a river channel and other effects of terrain. Hazard zones may be expanded by these elements; material carried downstream or channeled by ground terrain.
- Shrapnel. This is another potential hazard impact; but the potential zone is uncertain. Fatality estimates from the other hazards should be sufficient to form the basis of the framework calculation.

3 Transport Modes

3.1 Pipelines

The pipeline operation is assumed to be continuous and the loss of containment frequencies are on an annual basis. The framework risk estimate for the pipeline release is used as the base value and to normalize the other results; thus the pipeline release is independent of the volume transferred.

3.1.1 Frequency Source

Previous DNV GL evaluation (5) of U.S. Department of Transportation Pipeline and Hazardous Materials Safety Administration (PHMSA) pipeline incident records indicate that a 10-year average of liquid pipeline data (entire system) results in a leak rate of 2E-03 per mile per year. 90% of the incidents are a leak of the pipeline represented by a hole of 10% of the diameter. The remaining 10% is a rupture of the pipeline diameter. The pipeline size distribution indicates that the average pipeline carrying highly volatile liquids is roughly 14-inches in diameter. The frequencies presented in Table 3 are irrespective of population density.

Table 3 Pipeline Scenarios and Frequencies

Spill Size Category	Representative Release Scenario	Release Type	Frequency (per mile per year) by Population Density Category		
			Low	Medium	High
Leak	1.4-in release	Continuous	2E-3	2E-3	2E-3
Rupture	14-in pipeline rupture	Continuous	2E-4	2E-4	2E-4

The PHMSA estimated release frequencies are generic averages for liquid pipelines in the US. Although not applied in the current study, DNV GL has methods to assess pipeline route loss of containment frequencies highly refined by failure mechanism and local conditions (soil type, coverage, pipe thickness, etc.). This level of detail is not applied in the case study and framework as the goal is for concept comparison of transport modes.

3.1.2 Release Conditions and Consequence Sizes

Table 4 presents the estimated consequence hazard zones for the pipeline release scenarios. All hazard zones within the discussion are presented with respect to the downwind distance from the release location.

Table 4 Pipeline Consequence Zones

Spill Size Category	Downwind Distance of Consequence Zone (m)			
	Fireball radius	35kW/m^2	LFL	0.3 bar
Leak	--	80	90	--
Rupture	--	430	810	--

3.1.3 Ignition Probabilities

The TNO Purple Book is used to define both immediate and delayed ignition probabilities (6) for pipeline releases. Immediate ignition is often a result of the energy that caused the loss of containment and is a function of the reactivity of the material. TNO direct ignitions for liquids, 0.065 for both continuous and instantaneous releases, are applied for immediate ignition of pipeline releases.

Delayed ignition probabilities are a function of proximity and strength of ignition sources and their exposure to the flammable region of the cloud. For pipeline routes, the framework applies the TNO delayed ignition probabilities of 1.7E-4 per person per second to account for population. Total ignition probabilities are capped at 1.

3.2 Rail

Rail transport is assumed to be performed by unit trains, which means dedicated LPG rail tanker transport. The rail transport is assumed to consist of 120 railcars per transit. The rail transport volume within the 120 railcars is applied as the base volume to normalize the barge and truck tanker transport.

3.2.1 Frequency Source

A recent 10-year railroad accident history (2005-2015) published by the Federal Railroad Administration (FRA) Office of Safety Analysis is analyzed to estimate the railcar release frequencies (7). The FRA database updates the accident records monthly and specifies each railroad related accident by region, state, type of accident, type of track, track class, cause of accident, casualty subset, hazard material involved or not, and asset damage level. FRA data indicate that the frequency of having a railcar accident along the transit route is about 2.1 accidents per million transit mile.

Since additional precautions are taken when a rail car is carrying a hazardous material, like LPG, the fraction of accidents resulting in a release is estimated from a subset of the FRA data where hazardous materials are present. The probability of release given an accident is also a function of the number and type of cars involved in the accident. FRA data do account for the number of cars involved in the accident. When hazardous material is being carried, approximately 6.1% of transit accidents with 120 or fewer HAZMAT cars result in a hazardous material release. While accidents are less frequent during transit than in the yard, they are more likely to occur at higher speeds and, thus, more likely to result in a release given an accident. An overall release rate of 1E-07 per mile is used in the analysis.

The NRC has estimates of the breakdown of probabilities by spill size category and population density category (1) in Table 5. FRA data are also used to derive:

- The probability distribution for the range of potential release sizes for transit accidents and the size of the releases shown in Table 5 and Table 6, respectively.
- The number of cars derailed from the 3,159 transit accidents in the FRA database where there were at least 120 cars carrying hazardous material. Data show a maximum of 32 cars out of 120 cars releasing in such accidents (8).

Table 5 Rail Release Probability Distribution by Size and Population Density

NRC Spill Size Category	Probability by Spill Size (7)	Probability by Population Density Category (1)		
		Low	Medium	High
Minor	0.80	0.1	0.1	0.8
Moderate	0.12	0.3	0.4	0.3
Severe	0.07	0.6	0.3	0.2
Extra Severe and Extreme	0.01	0.9	0.05	0.05

Table 6 Rail Scenarios and Frequencies

Spill Size Category	Representative Release Scenario	Release Type	Frequency (per mile) by Population Density Category		
			Low	Medium	High
Minor	3-in release from fitting	Continuous	1E-08	1E-08	8E-08
Moderate	1 car rupture	Instantaneous	5E-09	6E-09	2E-08
Severe	6 car rupture	Instantaneous	5E-09	3E-09	2E-09
Extra Severe and Extreme	20 car and rupture	Instantaneous/ Escalating	1E-09	6E-11	6E-11

3.2.2 Release Conditions and Consequence Sizes

Table 7 presents the estimated consequence hazard zones for the rail release scenarios. The consequences are extracted for the ground level impact. Note the Minor spill scenario's LFL impact distance is small as it is modeled as an elevated release and only a small impact distance is estimated at grade.

Table 7 Rail Consequence Zones

Spill Size Category	Downwind Distance of Consequence Zone (m)			
	Fireball radius	35kW/m^2	LFL	0.3 bar
Minor	--	50	1	--
Moderate	--	60	340	--
Severe	--	140	880	--
Extra Severe and Extreme	90	230	1630	65

3.2.3 Ignition Probabilities

The TNO Purple Book is used to define both immediate and delayed ignition probabilities (6) for barges, rail and trucks. Immediate ignition is often a result of the energy that caused the loss of containment and is a function of the reactivity of the material. LPG is classified as a K0 material as the flash point temperature is approximately -104°C. TNO road tanker direct ignitions (0.1 for continuous releases and 0.4 for instantaneous releases) are applied for immediate ignition of rail releases.

Delayed ignition probabilities are a function of proximity and strength of ignition sources and their exposure to the flammable region of the cloud. For rail routes, the framework applies the TNO delayed ignition probabilities of 6.7E-03 per second accounting for the proximity of the train and 1.7E-4 per person per second to account for population. Total ignition probabilities are capped at 1.

3.3 Barge

Barge transport would consist of a single riverboat pushing 2-3 tank barges.

3.3.1 Frequencies and Scenarios

A U.S. Department of Defense (DoD) study estimated in 1980 a barge accident rate of 9.66E-06 per mile with approximately 4% of accidents resulting in a loss of cargo (9). While this data is older, waterborne commerce rates along the rivers of the evaluated transit route have been relatively flat (10); so an overall release rate of 4E-07 per mile based on the 1980 accident data is used in the analysis.

The NRC has estimates of the breakdown of probabilities by spill size category and population density category (1) in Table 8.

Table 8 Barge Release Probability Distribution by Size and Population Density

NRC Spill Size Category	Probability by Spill Size	Probability by Population Density Category		
		Low	Medium	High
Minor	0.44	0	0.5	0.5
Moderate	0.48	0	0.9	0.1
Severe, Extra Severe and Extreme	0.08	0.1	0.9	0

Although the NRC suggests spill size probabilities, no indication is provided of the potential associated release size. The U.S. Coast Guard (USCG) keeps casualty data (11) along the river route which is applied in the case study. In the period between 2001 and 2011, there were 141 recorded casualties of tank barges, 27 (19%) of which resulted in discharged cargo. The amount of recorded cargo discharge ranged from 1 gallon to 18,700 gallons averaging 715 gallons. Based on DNV GL past experience with the review of historical incident databases, it is likely that there is an under-reporting of the no release events. So the frequency estimates from the DoD study are applied in the study instead of the USCG data.

Table 9 presents the definition of the representative scenarios and the associated frequencies.

Table 9 Barge Scenarios and Frequencies

Spill Size Category	Representative Inventory Released Scenario (gal)	Release Type	Frequency (per mile) by Population Density Category		
			Low	Medium	High
Minor	715	Continuous	0	9E-08	9E-08
Moderate	18,700	Continuous	0	2E-07	2E-08
Severe	672,000	Continuous	3E-09	3E-08	0

3.3.2 Release Conditions and Consequence Sizes

Table 10 presents the estimated consequence hazard zones for the barge release scenarios. The resulting hazard zones are idealized as there is no inclusion in the modeling made for river movement. Note that the moderate and severe spills have the same hazard zones; this results from the same discharge rate but different durations of the release event, and the reported hazard zones are the steady state impacts.

Table 10 Barge Consequence Zones

Spill Size Category	Downwind Distance of Consequence Zone (m)			
	Fireball radius	35kW/m^2	LFL	0.3 bar
Minor	--	20	15	--
Moderate	--	110	170	--
Severe	--	110	170	--

3.3.3 Ignition Probabilities

The TNO Purple Book is used to define both immediate and delayed ignition probabilities (6) for barges, rail and trucks. TNO road tanker direct ignitions (0.1 for continuous releases and 0.4 for instantaneous releases), are applied for immediate ignition of barge releases.

Delayed ignition probabilities are a function of proximity and strength of ignition sources and their exposure to the flammable region of the cloud. For barge routes, the framework applies the TNO delayed ignition probabilities of 5.0E-03 per second accounting for the proximity of the push boat and 1.7E-4 per person per second to account for population. Total ignition probabilities are capped at 1.

3.4 Truck

For truck tanker transport, there are required weight limits. Therefore, the inventory is dependent on the material; pure propane within a standard tanker truck would contain about 10,238 gallons at approximately 89% volumetric capacity. Roughly 395 trucks would equate to a 120-car train load.

3.4.1 Frequency Source

Data on vehicle-km was collected in the Vehicle Inventory and Use Survey, which is now discontinued. The latest exposure data is for 2007 (12). The survey data indicates an exposure of 800 million truck miles. The incident database (13) includes 14 releases of Class 2.1 gases from cargo tank motor vehicles in transit in 2007. This gives a leak frequency of 2E-08 per tanker mile.

The NRC has estimates of the breakdown of probabilities by spill size category and population density category (1) in Table 11. Table 12 presents the definition of the representative scenarios and the associated frequencies.

Table 11 Truck Release Probability Distribution by Size and Population Density

NRC Spill Size Category	Probability by Spill Size (7)	Probability by Population Density Category (1)		
		Low	Medium	High
Minor and Moderate	0.995	0.1	0.1	0.8
Severe and Extra Severe	0.004	0.3	0.4	0.3
Extreme	0.001	0.5	0.3	0.2

Table 12 Truck Scenarios and Frequencies

Spill Size Category	Representative Release Scenario	Release Type	Frequency (per mile) by Population Density Category		
			Low	Medium	High
Minor and Moderate	3-in release from fitting	Continuous	2E-09	2E-09	1E-08
Severe and Extra Severe	Tank rupture	Instantaneous	2E-11	3E-11	2E-11
Extreme	Tank rupture	Escalating	5E-12	3E-12	2E-12

3.4.2 Release Conditions and Consequence Sizes

Table 13 presents the estimated consequence hazard zones for the truck release scenarios.

Table 13 Truck Consequence Zones

Spill Size Category	Downwind Distance of Consequence Zone (m)			
	Fireball radius	35kW/m^2	LFL	0.3 bar
Minor and Moderate	--	110	360	--
Severe and Extra Severe	80	90	330	--
Extreme	80	--		55

3.4.3 Ignition Probabilities

The TNO Purple Book is used to define both immediate and delayed ignition probabilities (6) for barges, rail and trucks. TNO road tanker direct ignitions (0.1 for continuous releases and 0.4 for instantaneous releases), are applied for immediate ignition of truck releases.

Delayed ignition probabilities are a function of proximity and strength of ignition sources and their exposure to the flammable region of the cloud. For truck routes, the framework applies the TNO delayed ignition probabilities of 6.7E-03 per second accounting for the proximity of the truck and 1.7E-4 per person per second to account for population. Total ignition probabilities are capped at 1.

4 Transport Mode Comparison

The different transport routes are assessed for their respective fatal risk impact along their respective transit locations and distances in the case study. To generate the framework result, the per shipment volume risk is calculated.

Table 14 presents the framework result. The total maximum release frequency for the respective transport mode is presented. The pipeline frequency is an annual frequency per mile, while the rail, barge and truck values are per shipment per mile. The volume transported is indexed across the transport modes, where the pipeline total volume is assumed constant, and the 120 car rail train is used as the basis for the other modes. Thus, to achieve the same volume transported as the rail option, the barge has a multiplier of 2.4 and the truck transit has a multiplier of 395.

The societal risk impact for the respective transport modes is calculated and then normalized to the annual pipeline risk result on a per shipment basis. The rail result is 5 times greater than the pipeline. However, the barge and truck normalized per shipment risk are less than the risk from pipeline and rail. The barge risk is about 11 times lower than the pipeline. Note though that these results are by shipment and not total volume.

Figure 1 presents a graph of the risk results normalized to the Pipeline – Low Population Density risk result versus the annual volume shipped.

Table 14 Transit Mode Comparisons

Transport Mode	Release Frequency (per mile)	Multiplier for Volume Transported	Per Shipment Risk (normalized to annual pipeline risk)
Pipeline	2E-03 annually	1 (filled pipeline for a year)	1
Rail	1E-07	1 (120 car train)	5
Barge	4E-07	2.4 (2-3 barges per trip)	0.09
Truck	2E-08	395	0.4

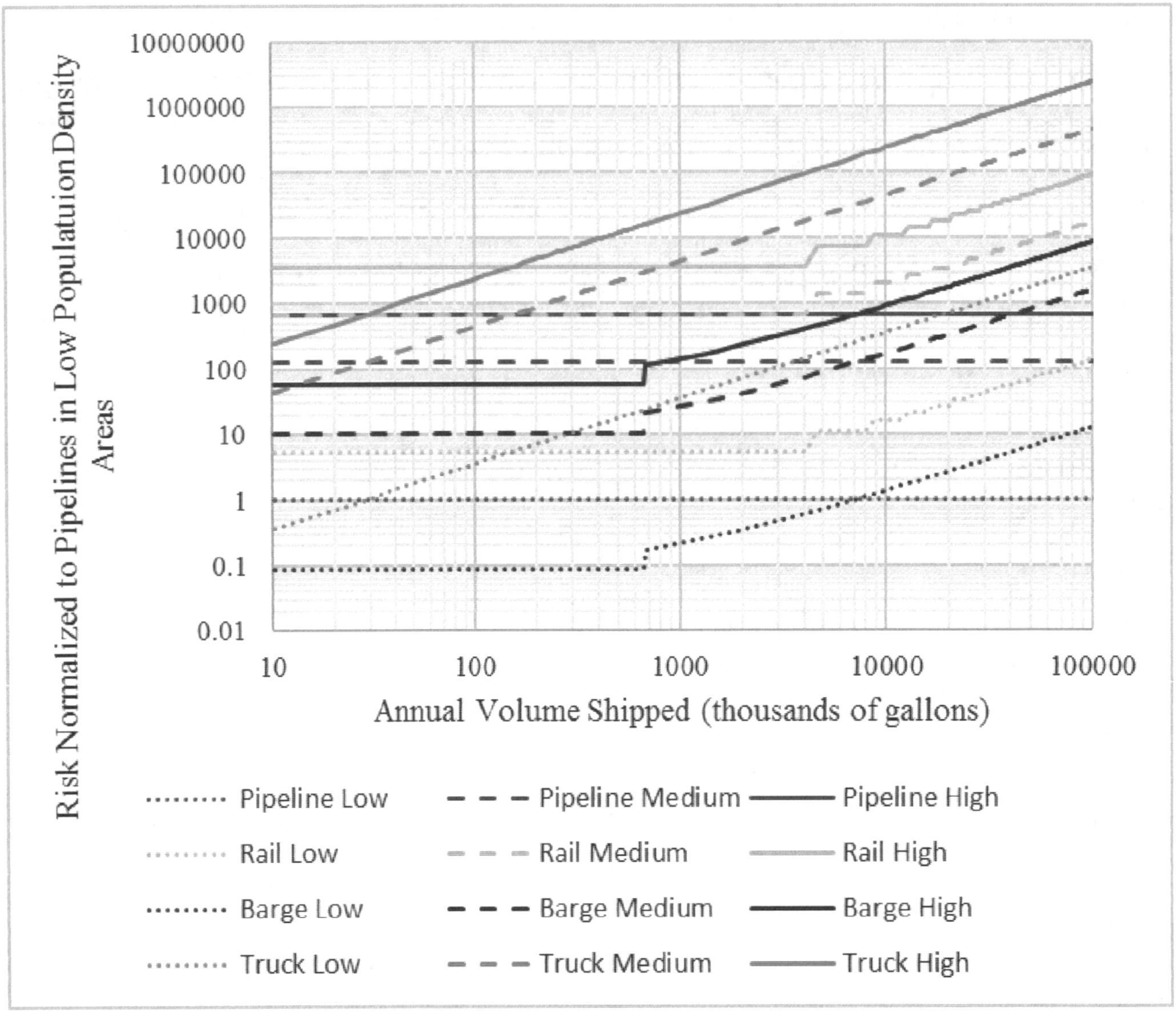

Figure 1 Transit Mode Risk Comparisons

Figure 1 presents the results for all four transit modes and three population densities. The following comparison discussions are made between the different transport modes assuming that the route length is the same across modes, and the population density is constant along the route length. However, both length and population density will vary by route even with the same end points. Route length will modify the risk linearly. Population density changes along the route has a more pronounced effect on the result.

4.1 Pipeline

As the results are normalized to Pipeline – Low Density, its curve is constant at value 1. As the pipeline can operate continuously, and the frequency is already on an annual basis, the pipeline results are constant across the x-axis of the annual volume shipped amounts.

As population density changes from Low to Medium to High, the pipeline result curves naturally increase in risk value. The Low to Medium result is a multiplier of 124; the Medium to High result is a multiplier of 5.

4.2 Rail

The rail result is constant up to the unit train 120 car volume, 4000 thousands of gallons. The slope of the risk result increases from the unit train volume as additional shipments are required to deliver the volume. The rail risk is always greater than the respective pipeline risk.

4.3 Barge

The barge result is constant up to the 700 thousands of gallons volume, and then the risk slope increases as additional transports are required to deliver the volume. The Barge – Low Density risk result is about 10 times less than the pipeline.

If the required volume to be shipped is in the order of 6700 thousands of gallons, which is 11 barge trips annually, then barge transport is the lowest risk option across the four modes and three population densities. However, if more than 11 barge equivalent volumes are required, pipelines present the lowest risk option.

4.4 Truck

The truck risk lines are constantly increasing with volume shipped given that additional truck shipments are required to achieve the volume. If only 3 or fewer truck shipments are required in a year, then the risk result is less than the pipeline option for the same volume equivalent.

5 Conclusions

The presented risk results vary across transport modes and population densities, and are presented on an annual volume shipped basis. Given an annual volume amount and a specified population density, comparison of the different transport modes is possible on a coarse basis. However, route length likely varies between the different transport modes and would factor the expected risk value. Population density would also likely vary along the length of a transport route and would change the expected risk result.

Generally, for shipment of volumes less than 7 million gallons annually, the risk impact to public from the different transport modes in increasing order is barge, pipeline, rail and truck. For shipment volumes over 7 million gallons annually, pipeline transport presents the least risk exposure of the four transport modes.

6 References

1. **U.S. Nuclear Regulatory Commission.** *Final Environmental Statement on the Transportation of Radioactive Material by Air and Other Mode.* 1977.

2. **U.S. Census Bureau.** [Online] 2010. [Cited: March 28, 2014.] http://www.census.gov/geo/www/gazetteer/files/Gaz_zcta_national.txt.

3. **Croce, Mudan K. S. and P. A.** *"Fire Hazard Calculations for Large Open Hydrocarbon Fires". SFPE handbook of Fire Protection Engineering,.* Boston, MA : Society of Fire Protection Engineers, 1988.

4. **Raj, P.P.K. and A.S. Kalelkar. Eds. D. H. Slater, E. R. Corran and R. M. Pitblado.** *Assessment Models in Support of Hazard Project. Major Industrial Hazards Project.* Sydney, Australia : Warren Center, University of Sydney, 1974.

5. *Using Past Incidents as Leading Indicators for Pipeline Integrity Management.* **Chen, Hao, et al.** Houston, Texas : American Institute of Chemical Engineers, 12th Global Congress on Process Safety, 2016.

6. **TNO Reference Manual .** *Bevi Risk Assessment Version 3.2, New "Purple Book". 2009.*

7. **U.S. Department of Transportation.** Federal Railroad Administration Office of Safety Analysis. [Online] 2015. http://safetydata.fra.dot.gov/OfficeofSafety/Default.aspx.

8. —. Federal Railroad Administration Office of Safety Analysis. [Online] http://safetydata.fra.dot.gov/OfficeofSafety/Default.aspx..

9. **Louis Berger & Associates, Inc. et. al.** *Final Environmental Impact Statement for the Safe Collection, Transportation, and Final Disposal of U.S. Department of Defense Stocks of DDT.* 1981. DACA45-80-0032.

10. **Department of the Army Corps of Engineers.** *Waterborne Commerce of the United States, Part 5, National Summaries.* 2016. CEIWR-WCUS-16-5.

12. **US Census (2009).** 2007 Economic Census: Transportation Commodity Flow Survey. [Online] http://www.rita.dot.gov/bts/sites/rita.dot.gov.bts/files/publications/commodity_flow_survey/2007/hazardous_materials/html/table_06b.html.

13. **Office of Hazardous Materials Safety (OHMS).** Incident Reports Database Search. [Online] 2014. https://hazmatonline.phmsa.dot.gov/IncidentReportsSearch/search.aspx.

AFPM/API Advancing Process Safety – Practical Tools for Process Safety Performance Improvement

Lara Swett
AFPM
1800 M St., NW Suite 900 North
Washington, DC 20036
lswett@afpm.org

Prepared for Presentation at
American Institute of Chemical Engineers
2018 Spring Meeting and 14th Global Congress on Process Safety
Orlando, Florida
April 22 – 25, 2018

AFPM/API Advancing Process Safety – Practical Tools for Process Safety Performance Improvement

Lara Swett
AFPM
1800 M St., NW Suite 900 North
Washington, DC 20036
lswett@afpm.org

Keywords: industry programs, process safety performance, process safety metrics, learning from incidents, lessons learned,

Abstract

American Fuel and Petrochemical Manufacturers (AFPM) and American Petroleum Institute (API) have been managing the Advancing Process Safety (APS) Programs since 2010. Through these programs, the associations have been able to collect a large amount of process safety data through API RP 754 collection, the event sharing database, and process safety site assessments. This paper will present the analysis of that data, the improvement opportunities identified, and the tools developed to assist sites in improving their process safety performance. One example of this is the identification of an industry opportunity to improve process safety events caused by open ended lines. Approximately 25% of process safety events have been caused by the failure to isolate energy during equipment commissioning, draining, and unloading/loading activities. AFPM in conjunction with several member companies have developed a suite of practice sharing tools called "Walk the Line" to assist our members in developing a program to address these types of events.

1. Background and Development of Advancing Process Safety Programs

The petroleum refining and petrochemical manufacturing industry is committed to protecting the health and safety of its workers, contractors and neighbors as well as the environment in which it operates. For decades, industry has worked together to develop safe work practices and standards aimed at reducing risks of major industry hazards and preventing process safety incidents. Despite these efforts, process safety incidents still occur.

In 2010, a group of refining and petrochemical industry executives belonging to the then National Petrochemical and Refiners Association (NPRA) now AFPM, tasked process safety professionals within their companies to develop a working group to explore ways to collectively improve process safety performance across the entire refining and petrochemical industry. The goal was to develop programs that would create a significant improvement in the collective industry process safety performance. Over a two-year period, the group developed a suite of programs. Over time, these programs have evolved and changed based on industry needs.

Today, the six programs are; Industry Learning and Outreach, Hazard Identification & Practices Sharing, Process Safety Regional Networks, Human Reliability, Mechanical Integrity, and Process Safety Site Assessment. Together these programs are called the Advancing Process Safety Programs (APS).

The objective of APS is to reduce process safety risk in industry by collaborating and leveraging industry knowledge.

This paper will discuss the process to collect and analyze process safety data being collected through APS, the observations from the data, and tools available to industry to address those observations.

2. Advancing Process Safety Programs

APS has a formal governance that sits between the two trade associations, AFPM and API. Each working group has a formal charter, operating guidance document, and accountability to the Process Safety Advisory group through the Process Safety Workgroup. This formal structure and oversight by a group of senior executives is one of the contributing factors to the long term sustainability and success of APS. Figure 1 shows the relationship between API and AFPM for APS, and the governance structure of the program.

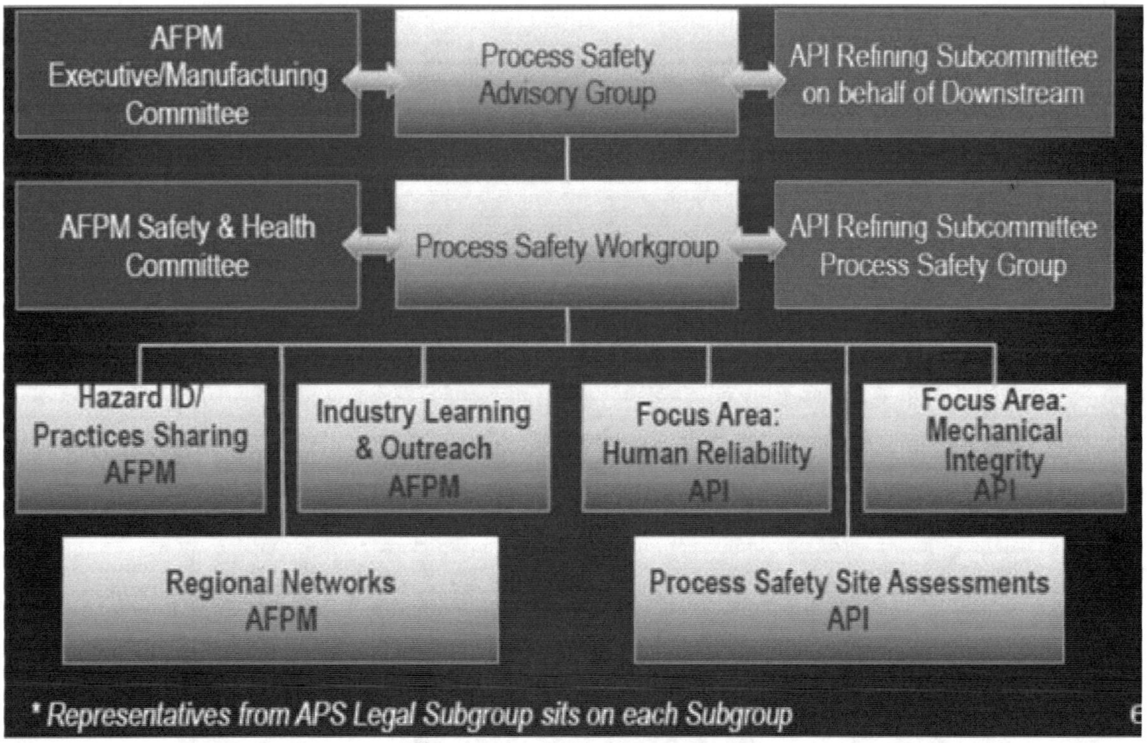

Figure 1: APS Governance

2.1 Safety Portal – Easily Accessible Tools

The Safety Portal[1] is an online database/website that houses all of the APS materials and is available to all AFPM members. It is searchable by keywords to allow for easy access to a variety of documents that address similar topics. The Safety Portal has been another key factor in the success of these programs. The site is currently undergoing a redesign which will make it more user friendly and will incorporate an improved search function.

The following document types are located in the Portal:

- Benchmarking application from *ANSI/API RP 754 Process Safety Indicators for the Refining and Petrochemical Industries*
- Event Sharing Database (Over 700 events)
- Hazard ID and Practice Sharing Documents
- Walk the Line Program page
- Industry Learning Reports
- Safety Bulletins
- Regional Network resources
- Presentations and Papers from AFPM Conferences

On average, there are about 2000 unique individuals that access the portal on a quarterly basis. This is one metric used to measure the program's success. Figure 2 shows the number of log-ins by unique visitors since its inception in 2012. Each individual is only counted once regardless of how many times they logged on. On quarter 4, 2017, we published several documents on a monthly basis and rolled out an outreach campaign that we believe led to a spike in log-ins.

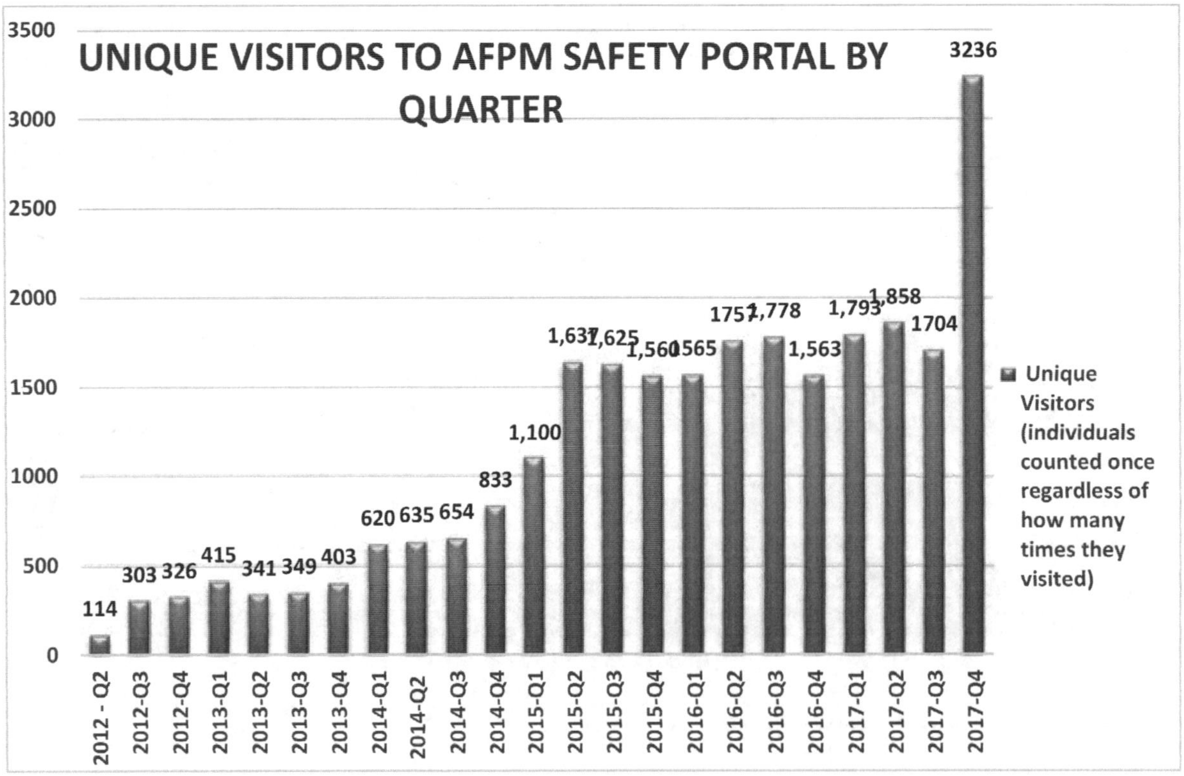

Figure 2: Unique Visitors to the Safety portal each Quarter

3. Advancing Process Safety Data Collection and Analysis

Through APS, AFPM and API member companies are asked to provide on an annual basis their Tier 1 and 2 Process Safety Events (PSEs) per *ANSI/API RP 754 Process Safety Indicators for the Refining and Petrochemical Industries* [2], to the event sharing database which includes cause and learning information and the option to conduct an API Process Safety Site Assessment [2]. Through the Industry Learnings and Outreach Subgroup, these data streams are analyzed and are the basis for the AFPM Learnings Report [3], outreach activities, and identification of opportunities for improvement.

3.1 Data Analysis

Included in calendar year 2016 data analysis are 688 event sharing submittals, approximately 3500 Tier 1 and 2 events (between 2012-2016 for refining and petrochemical), and results from 56 site assessments.

This analysis provides insight into opportunities for improvement that will be discussed later, however for discussion purposes each input, API RP 754, Event Sharing Database, and Site Assessment results are discreet data sets and therefore some observations might vary from data set to data set.

3.1.1 General Observations from 2016 Data Analysis

In general, key trends did not change from previous year's data analysis. However, with the addition of a more granular collection process that includes causal factors and a break out of types of normal operation (recent upgrade that provides additional trending opportunities) in the API RP 754 data, there is additional insight on point of release, cause and mode of operation observations.

AFPM has been collecting, API RP 754 Tier 1 and 2 data for the past 5 years. Figure 3 showing the 5-year process safety event incident rate trends for refining and petrochemical facilities. Over that time, there has been a downward trend in industry PSEs. Through additional analysis of refinery size, the small sites, < 190,000 bbls/day have achieved the most significant decrease in PSE rates. Antidotal information suggests parallels between the APS programs and their performance improvement. We do not have the same antidotal evidence on the petrochemical side.

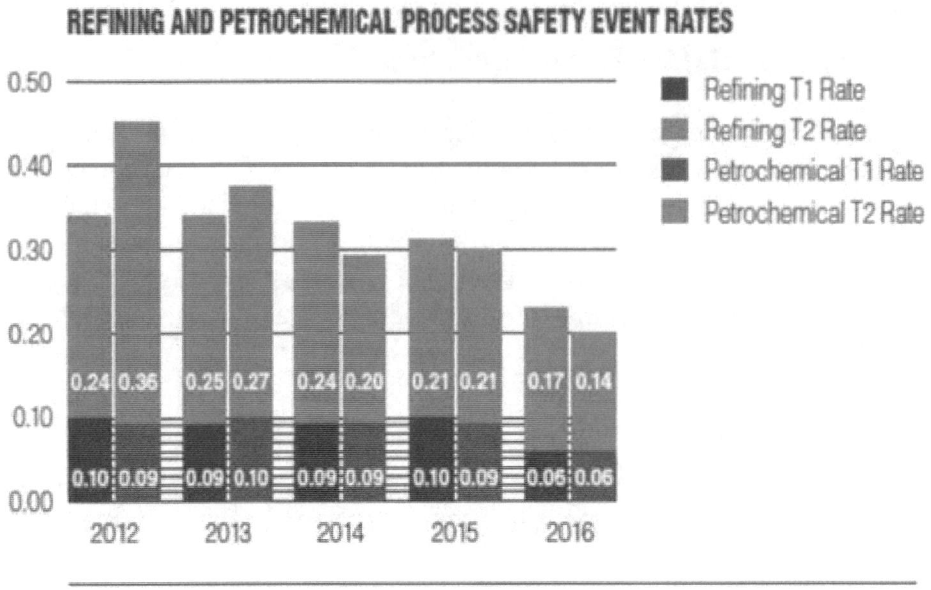

Figure 3: API RP 754 PSE trends for 2012-2016

Figure 4 illustrates where these Tier 1 and 2 releases are occurring in the Refining and Petrochemical industries. Combined, Piping Systems and Atmospheric Tanks categories comprise over half of the PSEs.

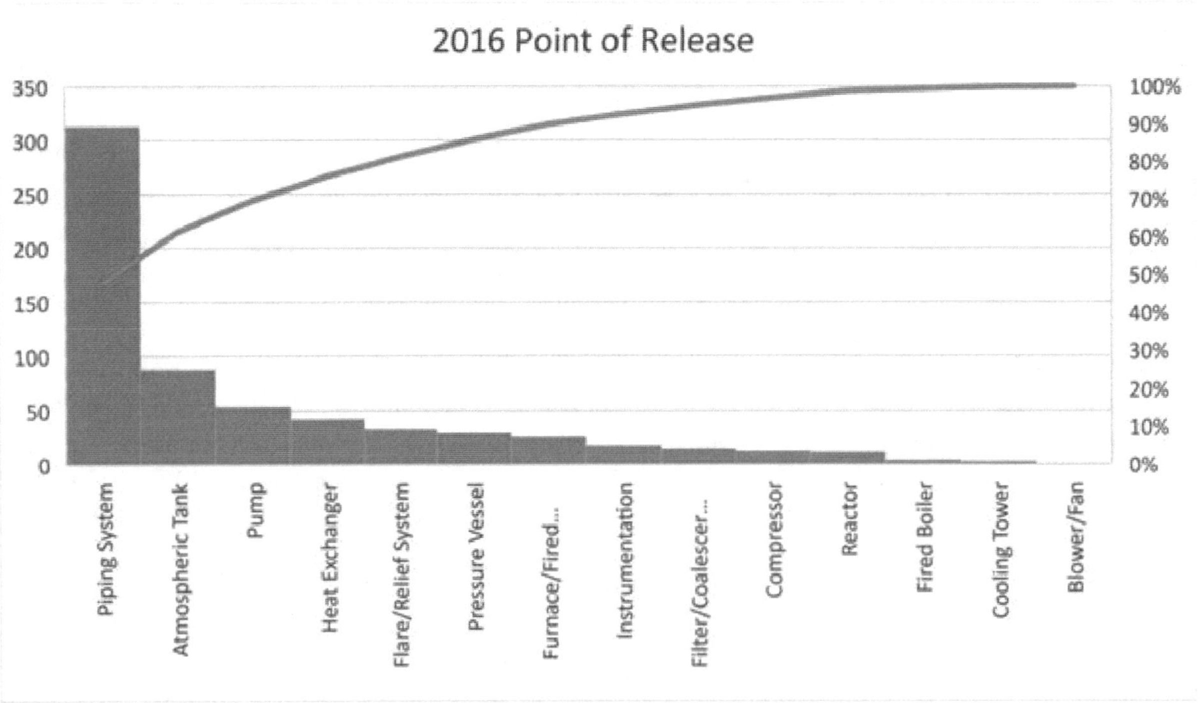

Figure 4: Equipment Point of Release, 2016 API RP 754 Data, Tier 1 and 2, Refining and Petrochemical

Upon further breakdown of that data:

Piping Systems - 43% of 2016 reported API RP 754 event data involved piping systems. Of those events, 57% were Large Bore Piping (>2 inch) and 42% were Small Bore Piping (<2 inch).

- Piping Leak Type:
    - 48% Piping Leaks
    - 24% Valve Leaks
    - 12% Flange/Gasket Leaks
    - 4% Tubing Leaks
    - 4% Fitting Leaks
    - 3% Open ended line
- Valve Leak Causes:
    - 54% Valve left open (Walk the Line Event)
    - 20% Valve Body Leak
    - 15% Valve did not hold/ leaked through

Atmospheric Tanks – 13% of events involved Atmospheric tanks, this percentage has remained consistent at 12-13% over the past 4 years. Forty four percent of PSEs related to atmospheric tanks were overfills and 87% of those events occurred during Normal Operating, Filling/Draining activities.

Of all the 2016 PSEs submitted, 59% occurred during Normal Operation and 50% of these events had at least one cause of Equipment Reliability and 62% had a cause of Human Error. Note, the submitter can choose more than one cause per event.

Human Error was the cause of 23% of all Tier 1 and 2 events submitted through API RP 754 data collection. Sixty two percent of those events occurred during Normal Operations. The following activities had at least one cause of Human Error:

- 83% Equipment Prep/Taking Equipment Out of Service
- 57% Loading/Unloading
- 56% Equipment Commissioning/Putting in Service After Maintenance
- 36% Filling/Draining
- 36% Steady State
- 25% Changing Line-ups

By looking at the Event Sharing database, we can observe some additional information that can provide additional information on causes:

- Fixed Equipment Mechanical Integrity – internal and external corrosion, erosion, cracking, and inspection less than adequate
- Equipment Reliability – premature failure and maintenance/repair less than adequate
- Human Factors – valves left open, open ended lines, loading, unloading, and tank filling

- Design – winterization, specs not adequate, followed by gaps in the specifications and construction

Further information can be pulled out from the Process Safety Site Assessment results. Figure 6 represents the results from the API Process Safety Site Assessments which show that the lowest scoring protocols are from Mechanical Integrity, specifically process hoses, integrity operating windows, and critical check valves. There is also a wide distribution of scores around conduct of operations topics like operator rounds and shift handover.

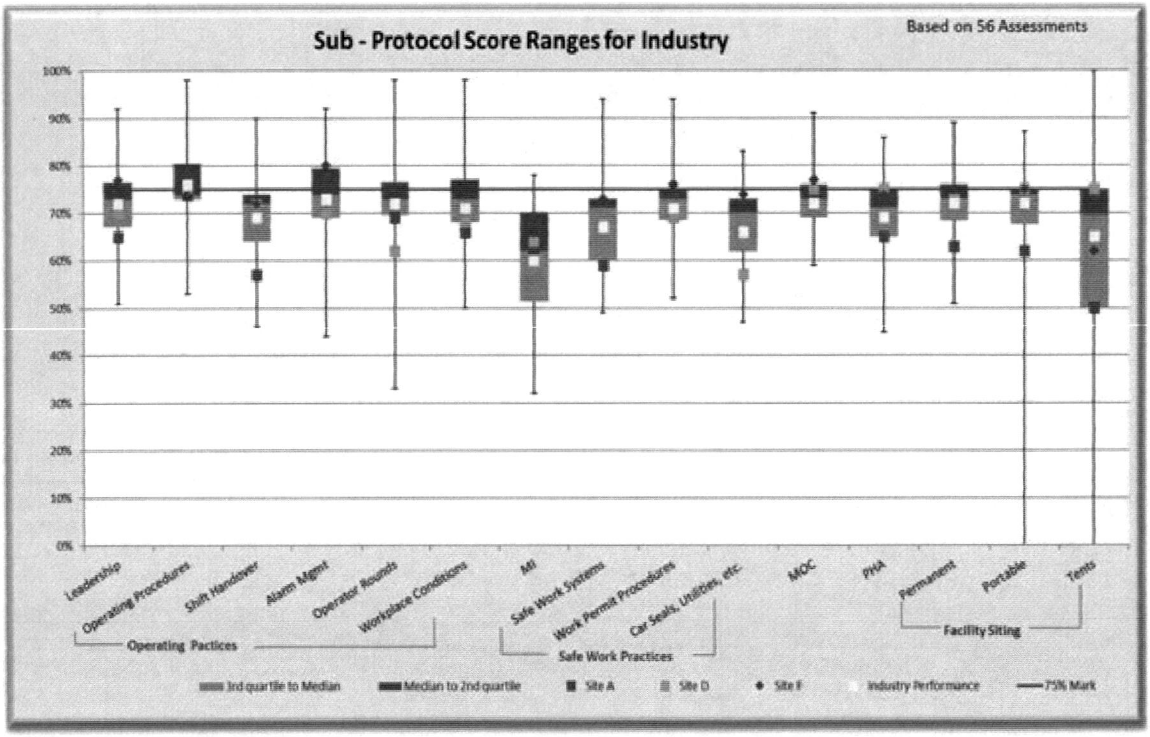

Figure 6: API Process Safety Site Assessment Score results based on 56 assessments

3.1.2 Data Conclusions

Over the past several years of data analysis, there have been two reoccurring themes among the reported causes; human reliability and mechanical integrity account for a majority of events that occur within industry. If you break down both categories a little further, human reliability as it relates to procedures is an improvement opportunity as well open-ended lines. On the mechanical integrity side, there are opportunities around operating integrity windows and critical check valves.

4. Opportunities for Industry Improvement

As part of APS, the trade associations have had the opportunity to work with the Legal Subgroup and develop several tools to enhance sharing and learning across the industry. These tools include;

- Safety Bulletins – A summary of process safety events with common causes
- Company Alerts – Companies share their alerts for broad distribution on a specific incident
- Practice Sharing documents – sharing the summary of a site's practice
- Hazard ID documents – a list of process safety hazards associated with a specific type of equipment or maintenance/operating activity
- Learnings Report – Annual summary of cumulative data analysis

Over the years of data review, several topics have been identified that have warranted additional sharing and outreach to industry. As mentioned above, the data points to a needed focus on topics related to Mechanical Integrity and Human Reliability. Below is a listing of tools developed to date, the section after will focus on the AFPM Walk the Line Program that is aimed to eliminate operator line-ups.

Table 1. APS Tools

Topic	Tools
Winterization and Dead legs	Hazard Identification: Winterization Hazard Identification: Dead legs
Management of Temporary Repairs	Safety Bulletin: Hazards of Installing Temporary Repairs for Piping Systems - Clamps Hazard Identification: Temporary Repair of Piping and Piping Components Practice Share: Temporary Repair Checklist Practice Share: Energy Isolation Procedure
Corrosion Under Insulation	Safety Bulletin: Hazards of Corrosion Under Insulation
Floating Roof Tanks	Safety Bulletin: Hazards of Floating Roof Tanks
Atmospheric Tanks	Hazard Identification: Atmospheric Tank Preparation for Out of Service Maintenance Hazard Identification: Atmospheric Storage Tank Operation
Piping Vibration	Safety Bulletin: Hazards of Piping Vibrations
Integrity Operating Windows	Practice Share: Integrity Operating Window Alarm Management

	White Paper on Integrity Operating Windows
Injection Point	Hazard Identification: Injection Point and Process Mixing Points
Hose Hazards	Hazard Identification: Hose Hazards
Critical Check Valve	Hazard Identification (in development)
Permit-to Work (Human Reliability)	Practice Share: Permitted vs Non-Permitted Work Practice Share: Standards for Work Permit Execution: Energy Isolation (LOTO); Joint Job Walks, Safeguard Verification & Stop Work Triggers Practice Share: Permit-Related Auditing Process Practice Share: Permit Issuer/Receiver Roles (Competence, Training & Execution)
Fired Heaters (Human Reliability)	Hazard Identification: Fired Heaters Practice Share: Fired Heater Risk Management Practice Share: Start-up Checklist Practice Share: Transient Operations HAZOPs Safety Bulletins: Cold Start-up, Hot Restart, Identifying and responding to bogged/flooding heater
Open-ended Lines (Human Reliability)	*See Walk the Line below*
Risk Assessment	Practice Share: Higher Risk Work Procedure Practice Share: HAZOP Vulnerabilities List
Other	Practice Share: Control and Defeat Review Practice Share: Radio Protocol

In addition, AFPM members have access to the Event Sharing database that contains over 700 process safety events to perform their own trending and analysis or to review when conducting a HAZOP or PHA.

4.1 Walk the Line an Industry Practice Sharing Program

Over the years, the concept of human reliability has been a reoccurring theme during the data analysis, many events pointed to some level of human factors that caused an event to occur. Specifically, a majority of the events were attributed to open valves during equipment commissioning, draining, and loading/unloading activities; many are “high risk” potential where human action is the last barrier to prevent. We have started referring to these events as Walk the

Line events and label them as such in the Event Sharing database. Based on the 2016 API RP 754 data analysis, 29% of events can be directly attributed to a Walk the Line causes.

Since 2013, a group working on this topic has developed an industry outreach and practice share program called, Walk the Line that is leveraged from a Celanese program that addresses the human reliability associated with conduct of operations and specifically, operator discipline. It reinforces the non-negotiable fundamental responsibility of an operator to know with 100% certainty where a material will flow between any two points in a process. Walk the Line addresses the culture of operator line-up and provides tools such as a hazard identification and practice sharing documents, workshops and checklists that help the operator become successful in meeting this responsibility. A paper describing Walk the Line was presented by Jerry Forest, Celanese, at the 2014 Process Plant Safety Symposium (PPSS). The paper and presentation won the annual PPSS Best Paper Award.

Walk the Line's philosophy on how to address human behavior

1. Change the Culture by raising awareness: Constantly set and reinforce the expectation to "Walk the Line" (know where a material will flow in a process). Tools provided include newsletters describing Walk the Line and a video that summarizes the concept.
2. Provide operational continuity and discipline tools to help the operators understand the current operating state of the plant. These include templates for shift instructions, shift notes, shift relief, operator reading sheets, and design practices.
3. Provide operational readiness tools to help operators be successful in line-up situations that have the highest probability of error including startup and returning equipment to service from maintenance. These tools include commissioning practices, Safe Operating Procedure design, and Pre Start-up Safety Review templates and practices.

Since the AFPM Walk the Line rollout in 2014, it is estimated that eleven companies have shared at least one practice and approximately thirty-five companies have implemented at least one aspect of Walk the Line.

AFPM has held a Walk the Line Workshop for the past two years with 200 people attending the 2017 Workshop. The 2018 Workshop is currently being schedule for, July 25-26 at the Waterway Marriott in the Woodlands, Texas. Contact the author for registration information.

All of these tools are available on the AFPM Safety Portal by searching Walk the Line or WTL.

Table 2. Walk the Line tools

Topic	**Tools**
Toolbox Topics:	Bleeders

	Coker Switch Instrument Lineup Barge Loading Wrong Valve Operated LOTO Error Maintenance Operations Handover Similar Valve Operation More Bleeders Line Open for Maintenance
Training:	Walk the Line Training Training for Front Line Supervisors Training for Secure Transfers Abbreviated Training
Hazard Identification:	Shift Turnover Maintenance Operations Turnover Operator Line ups
Practice Sharing:	First Line Break Point Labeling Open Valve Labeling and Management Refinery tank Farm and Terminal Product Transfer Shift Handover Evaluation Rounds Informal Unit Walk Through Operating Instructions Operator Shift Notes Shift Change Communications Shift Change Meeting Critical Bleeder Line Labeling Spring Loaded Valves Plug Program Checklist SOP Maintenance/Operations Turnover P&ID Walk down

	MOC Tagging Pre-Startup Safety Review Soap Testing IR Cameras Tracking Walk the Line Incidents
Presentations:	Walking Down the Line Barrier Analysis Technology to Support Standard Work Reliability-Centered Operations Rounds Operating Discipline and Walk the Line Reboot Flavor of the Month

One of the most notable successes of this program is how industry has adopted the terminology, Walk the Line, to identify similar types of events. Sites have posted signs, created hard hat decals, and even if they have not adopted any tools, they recognize the terminology and what it stands for. Based on anecdotal feedback, many sites that have adopted a form of Walk the Line have seen positive results.

5.0 Industry Outreach

Throughout the past eight years of working on APS, we have learned that continued outreach is a large factor in continued sustainability and industry engagement in the programs. For that reason, we have a designated Subgroup the responsibility for continuous outreach and industry education, the Learnings & Outreach Subgroup. To be added to the distribution list to receive information on upcoming webinars, documents, and events, please email Lara Swett at AFPM, lswett@afpm.org. You do not have to be an AFPM member to participate in webinars workshops.

5.1 Webinars

What started as a quarterly webinar for API 754 Question and Answer assistance has evolved into a webinar series that covers all elements of APS tools. For 2018, five webinars are planned. Each webinar will spend some time reviewing new published documents and tools, review API RP 754 questions, and also address one specific focus area such as Mechanical Integrity (April 10th), Human Reliability as it relates to Fired Heaters (June 19th) and Permit to Work (September 12th), and Regional Networks (November 13th). Contact the author for registration information.

5.2 Industry Presentations

The Learnings and Outreach Subgroup continues to look for opportunities to present the learnings from APS to a variety of audiences; senior leaders, process safety, maintenance and operations. The long-term goal of APS outreach is to share the learnings and available tools to disciplines broader than process safety as safety touches all disciplines.

5.3 Testimonials

One of the ways we measure effectiveness of the programs is by collecting testimonials from participants. The ultimate metric of successes is a reduction of process safety events across the industry which we are starting to see. However, from the testimonials below, it is clear that APS is having a positive impact on the industry at the site level.

"Site made changes on Energy Isolation procedures based on a discussion at a previous Regional Network Meeting."

"During shutdown and startup, this site pulls events from the safety portal event sharing database and shares with operators. They have also discovered learnings from some of these documents and as a result have made modifications to one of their units."

"Based on a presentation from a Regional Network meeting on high-risk work procedures, a site gathered new ideas to take back." (Note: document available in practices sharing section)

"Network participant shared the "Tank Preparation for Out of Service Maintenance" Hazard Identification document with a team of specialists at the site. The group integrated parts of the document with their site's procedures and work practices. Specifically, they leveraged information about roof support and tank cleaning from the document which resulted in an improved program."

"Took information shared at AFPM Regional Network on "Dangers of Corroded Tanks" back to management and will now inspect their tanks for microbial corrosion in the future."

"One company used the results from a site assessment, as well as the benchmarking results from the overall petroleum industry site assessments to drive a mechanical integrity subgroup at their company. This group presented data to the Board of Directors which resulted in the Board approving the hiring of a corporate MI Director to drive corporate MI initiatives."

"Company drafted a PHA checklist on Clamps/Temporary Repairs based on the Temporary Repairs Hazard Identification document."

"Took Safety Bulletin on "Overpressure of an Exposure-Proof Enclosure," and input that document into their company's High Learning Value Event process. The document was sent globally to all sites."

6.0 Conclusion

The AFPM and API have been working together on an industry led initiative, APS, to continuously improve process safety performance through enhanced information, communication and collaboration. What started as a commitment to share information broadly in order to improve all sites, has now become a movement within industry. Through the help of the Legal Subgroup, many barriers to sharing have been remove. This has helped companies feel comfortable sharing through APS, knowing they are helping another site or company. We are seeing improved process safety performance within the industry and there is antidotal evidence that APS has helped several sites and companies improve their performance. As the program matures, we will start reaching out to additional disciplines such as training, operations, and maintenance to further share learnings and tools. We will continue to measure effectiveness on an ongoing basis and are excited to see what this program can do for industry in another five years.

1 References

[1] AFPM Safety Portal, Available at https://safetyportal.afpm.org/ Accessed on March 20, 2018.

[2] American Petroleum Institute. (2016). *Process Safety Indicators for the Refining and Petrochemical Industries, (ANSI/API RP 754)*

[3] API Process Safety Site Assessment Site, Available at http://www.api.org/products-and-services/site-safety, Accessed March 20, 2018

[4] Forest, J. (2014, April) . Walk the Line. Paper presented at the meeting of the Center for Chemical Process Safety, New Orleans, LA

513044 A Novel Prediction Method of the Chemical Leakage Range with Artificial Intelligence: Machine Learning for Chemical Species

Abstract:

This study focuses on the prediction of the chemical leakage range by using artificial intelligence (AI). A multi-variable regression model is developed by deep neural network (DNN) method. The DNN model is formulated by Python with Tensorflow library. The model contains 12 input variables, 70 hidden layers, and 3 output variables. Each layer has Rectified Linear Unit (ReLu) function as an activation function and layers are fully connected. All weights are initialized by Xavier initializer. The chemical leakage range data set is generated by the commercial software, ALOHA. The simulation results show the hazardous zones (AEGL-1, AEGL-2, AEGL-3) under various conditions. To train the AI, the data set is configured with variables: chemical species, wind speed (1~7 m/s), temperature (10~35 â„ƒ), ground roughness (open country, open water), stability classes (A~F), and tank storages (200~4,000 kg). Total date sets of each chemical are 30,240, which are divided into 25,920 training data and 4,320 test data. Training data set is the given information to artificial intelligence and the test data set is unknown information. As the results, the developed model performs high accuracy over 95 % on average for all tested chemicals. The result shows that artificial intelligence figure out the regularity of the data on its own and make predictions with high accuracy. The developed model can be extended to predict chemical leakage ranges for unknown chemicals as well as chemical mixtures by using chemical property information as the input data. The same method could also be used for quick estimation of fire and explosion accidents. This kind of research is expected to contribute to real-time response of corresponding organization in the chemical process safety field.

This research was supported by Disaster-Safety Platform Technology Development Program of the National Research Foundation of Korea(NRF) funded by the Ministry of Science and ICT. (NRF-2017M3D7A1085361)

512180 Estimating Physical Properties of the Productsof an Atmospheric Distillation Columnby Support Vector Regression

Abstract:

Atmospheric distillation column is one of the most important units in an oil refinery where crude oil that is little use as it is, is fractioned into its more valuable constituents. Different fractions of the crude distillation unit are then further processed in downstream conversion and purification units to produce final products such as gasoline, diesel and jet fuel. The physical properties and the quantity of the final products vary depending on the physicochemical properties of the crude oil being processed and the operation parameters of the atmospheric distillation column. The process operators should keep the physical properties of hydrocarbon products in specified limits and operate the crude distillation unit according to instructions from the production planning department for maximizing the profit from operations. The physical properties of different crude distillation unit fractions must be measured by taking a sample from the stream periodically and analyzing these samples in a laboratory with appropriate equipment or by online analyzers that are very expensive to install, operate and maintain. Almost all of the state-of- the art online equipment has a time lag to complete the analysis in real time due to complexity of the analyses. Therefore, the laboratory or on-line measurements become available to decision makers and operator with a time lag. The intermittent nature of the measurements leads to sub-optimal control of the unit or in some cases leads to off-spec products that have considerably less economic value than the products that are within specified limits. As a result, estimation of the physical properties from online plant data and implementation of a soft sensor has a potential benefit in improving the profit. Among different approaches, the support vector regression shows great promise in machine learning due to its ability in generalizing well to unseen test data.

We aim to estimate the physical properties of the hydrocarbon products of an atmospheric distillation column using the large volume of data obtained from real-time and online sensors installed in the system. Linear, Polynomial and Gaussian Radial Basis Function (Gaussian RBF) kernels have been tested and the SVR parameters have been optimized by embedding k-fold cross validation using a variety of algorithms including genetic algorithm (GA), grid search (GS) and non-linear programming (NLP). The performance of SVR has been compared against artificial neural network (ANN) method and robust quality estimator (RQE) already functioning in the ADC. The testing results suggest that SVR with any of the integrated kernels performs well in estimating the property and generalizes well to blind test data. Compared to RQE, the mean testing error of estimation has been improved by 31% with SVR from 6.4ËšC to 4.4ËšC and the standard deviation of estimation error is improved by 23% with SVR from 7.3 ËšC to 5.6 ËšC.

513440 Genetic Process Visualization Using Parametric t-Sne

Abstract:

Distributed control systems (DCS) are essential for modern chemical plants. Plant operators use the DCS to monitor the operation in real time and keep the process within safe and productive domains of operation. Commonly, the monitoring interface in the DCS is designed for single variable analysis, which is expressed in two ways: tags with current values on the process diagram or a line chart on a time axis. Based on the thresholds of each variable, alarms notify personnel if any single variable drifts out of its safe limitations. These strategies are clear to operators and are also simple to implement, but they can only provide limited information of the process due to the handicap of single variable statistics. In such a system, the DCS is blind to any correlation information among multiple variables.

In our work, we proposed a generic process visualization method using dimensionality reduction techniques based on deep learning, which is the state-of-the-art technique that uses deep neural networks (DNN) to extract features hidden within large data sets [1]. It has already achieved notable success in many areas, including image analysis [2], natural language processing [3] and even human-level intelligence for control tasks [4]. For the objective of process monitoring and data visualization, we adopted a deep learning model, namely parametric t-SNE [5]. As a dimensionality reduction method derived from t-SNE [6], it uses a DNN to optimize the projection of the data into the latent space by minimizing the Kullback-Leibler divergence from original space. In this method, different measurements in each data record are considered as a whole, allowing correlation features to be learned parametrically by the DNN. The objective of utilizing the parametric t-SNE in process visualization is to project the high-dimensional measurement onto a 2D map, where different normal operating regions and faults can be separated. Thus, a notable change can be observed in the latent space which can provide qualitative information regarding the process operation to the plant operator.

In practice, we observed that the nonuniformed variance in each variable can create bias, causing the DNN to greedily learn the features from only the most contributing variables in the training set. In other words, variables that show less variance in the training set are muted in the model, which is misleading for visualizing unknown conditions of the process in the future. Therefore, beyond following the two-stage training procedure of the parametric t-SNE, we used combinatorial variation creation method to introduce variance equally to each variable. Process data from a pyrolysis reactor was used as a case study for the proposed method. Data from one operating cycle was used to training the DNN and the trained model was tested on another cycle. The visualization model is especially effective in collaboration with data-driven fault detection methods (e.g. PCA-T2), where it can provide more in-depth information about process operation on the 2D map. This can relieve the operators from staring at the numerical values and line charts of each variable in interest and move their work to a more intuitive interface.

Reference

[1] LeCun, Yann, Yoshua Bengio, and Geoffrey Hinton. "Deep learning." *Nature* 521.7553 (2015): 436-444.

[2] Krizhevsky, Alex, Ilya Sutskever, and Geoffrey E. Hinton. "Imagenet classification with deep convolutional neural networks." *Advances in neural information processing systems*. 2012.

[3] Graves, Alex, Abdel-rahman Mohamed, and Geoffrey Hinton. "Speech recognition with deep recurrent neural networks." *Acoustics, speech and signal processing (icassp), 2013 ieee international conference on*. IEEE, 2013.

[4] Mnih, Volodymyr, et al. "Human-level control through deep reinforcement learning." *Nature* 518.7540 (2015): 529-533.

[5] van der Maaten, Laurens. "Learning a parametric embedding by preserving local structure." *RBM* 500.500 (2009): 26.

[6] Maaten, Laurens van der, and Geoffrey Hinton. "Visualizing data using t-SNE." *Journal of Machine Learning Research* 9.Nov (2008): 2579-2605.

514287 Low-Touch Machine Learning Is Fulfilling the Promise of APM

Abstract:

New methods and cutting-edge technologies are driving asset performance management (APM) well beyond historical capabilities, rapidly increasing its bottom-line value. Technologies such as cloud computing, data science and machine learning are now being integrated with automated methodologies directly into APM solutions.

This wave of integration firmly places advanced analytical techniques into the hands of operators and engineers with previously unimagined scale and ease of use. The incremental progress in APM over the last 20 years pales in comparison to what's now possible through digital transformation.

Low-touch machine learning is the key catalyst to scale APM's potential well beyond existing first principles-based solutions and "armies" of consultant engineers and data scientists. A widespread integration of machine learning in APM will mark a transition from estimated engineering and statistical models towards measuring actual patterns of asset behavior.

Manufacturing facilities staff can now readily extract value from decades of existing design and operations data to better manage and optimize asset performance. This "low-touch" machine learning method continuously embraces changes in asset behavior, empowering real-time APM value creation. Vetted and tested across diverse industries, scalable across multiple assets and powered by cloud and parallel computing, low-touch machine learning ushers in a new era of performance and optimization for all industries.

514056 Big Data Analytics Platforms: The Key to Enabling the Analytic Enterprise

Abstract:

Analytics and the Industrial Internet of Things are the new hot buzzwords these days, but analyzing data is nothing new, and manufacturers have been generating data off the line for decades. So why all the hype? The last few years have seen the evolution and advancement of several technologies that are converging to allow for unprecedented analysis and insights into business problems. But leveraging these takes more than just smart coders. Analytic insights need to be deployed into production and action has to be taken to truly realize the benefits. Find out why analytic platforms are the key to enabling the Analytic Enterprise.

513880 Industry 4.0 and Driving Better Organizational Decisions

Abstract:

Industry 4.0 is more than just about the Industrial Internet of Things, it is also about leveraging the Industrial Internet of People (IIOP). This talk will focus on using the tools now available to bring sensors, data, and analytics together to drive better organizational understanding and continuous improvement. Who are the key stake holders, and what do they need to improve to improve their contribution to the way a chemical plant is operated? How do you structure these projects to ensure successful deployments? This talk will focus on practical use cases and examples of Industry 4.0 in action.

Optimization of extraction condition for phytic acid from peanut meal by response surface methodology

*Hong Ren *, Ting Li , Huijie Wan, Jinping Yue*

Beijing Technology & Business University, School of Food and Chemical Engineering, Beijing Higher Institution Engineering Research Center of Food Additives and Ingredients, Beijing Key Laboratory of Flavor Chemistry, Beijing, China

Abstract Phytic acid (PA), a molecule with high commercial value, is one of the important component in peanut meal. However, PA has not yet been isolated from peanut meal to exploit and utilize. This paper reported the extraction conditions of PA from peanut meal after removed protein. The independent variables were hydrochloric acid (HCl) concentration, solid to liquid ratio, extraction time and extraction temperature. Response surface methodology (RSM) was used to optimize the extraction conditions based on the extraction yield of PA. The results show that the second-order polynomial models derived from responses well with the experimental ($R^2 = 0.9783$). The optimal extraction condition was obtained with solid to liquid ratio of 1:16 (g:mL), HCl concentration of 0.02 mol/L, extraction time of 105 min, and extraction temperature of 30°C. At this condition, PA with higher purity were obtained. the extraction ratio was 6.12%, and the content of PA was 182.7 mg/g dry PA extract. The experimental values under optimal condition were in good consistent with the predicted values. The PA extracted from peanut meal was verified qualitatively by IR spectra. The extraction technology of PA from peanut meal has a strong potential for realized high-value utilization of peanut meal.

Keywords: Phytic acid, Peanut meal, Response surface methodology, independent variables, optimization

1. Introduction

Myo-inositol hexakisphosphate, also called phytic acid (PA), is considered as an anti-nutrient component because of its strong ability to combine with multi-charged metal ions, specially Zn^{2+}, Ca^{2+} and Fe^{3+} [1,2]. Recently, a variety of pharmacological activity of PA have been reported. S. Norazalina found the potential value of PA extracted from rice bran in reducing colon cancer risk in rats [3]. The finding showed treatment with 0.2% (w/v) PA extract gave the greatest reduction in the formation of aberrant crypt foci. Yukako Okazaki's study showed that PA may improve the composition of cecal organic acids, microflora, and mucins, and it may decrease the levels of serum proinflammatory cytokines in rats fed a high-fat, mineral-sufficient diet[4]. PA is a potential absorption enhancer of flavonoid components on tight junction integrity in Caco-2 cell monolayers[5]. PA is a naturally occurring constituent which exhibits protective action in Parkinson's disease and it has been shown to lower blood glucose levels[6,7]. Moreover, PA has been widely used in metallic material field as a new generation of green corrosion inhibitor for copper [8-11]. cupronickel B30 [12], brass [13] and Mg-Li alloys[14]. PA is widely used in food and light industry.

PA mainly distributes in seeds of plants, such as cereal grains, legumes, nuts, oilseeds, and so on, acting as a main source of phosphorus. Many cereal grains and oilseeds contain about 1~3% PA [15]. Peanuts is one of the most widely grown oil seeds in the world with 29 million tons produced every year [16]. In China peanuts are the most important oil seeds, ranking first among the world's peanuts producing countries[17]. Nowadays, a majority of peanuts are crushed to produce oil and this process generates a large amount of peanut meal as the by-product, which is about 9 million tons every year in china [18-20]. A large number of studies showed that peanut meal is a good source of active ingredients, which contains about 47~55% protein, 20~30% carbohydrates, 8~10% crude fiber, 2~3% fat, and 1.0~1.2% PA [21]. However, most of the available peanut meals are used as feed ingredient for animals, and only a small portion was used to recover proteins, causing great waste [22]. Therefore, obtaining the active ingredients and high-value utilization of peanut meal has become increasingly important task for effectively reducing peanut resources waste.

Compared with other plant materials, peanut meal is rich in PA, but there are no good process to isolate PA from peanut meal reported so far. The effective extraction procedure and method to separate PA from peanut meal are very necessary. In this study, we firstly removed proteins from peanut meal, and use response surface methodology (RSM) to optimize the extraction conditions. The objective of present study was to find the optimum conditions for the extraction of PA, including hydrochloric acid concentration, extraction temperature, extraction time and ratio of liquid to material, to maximize the extraction rate of PA from the peanut meal after removing proteins.

2. Materials and methods

2.1 Materials

Peanut meal was provided by Qinglonghu Wanchunyuan Vegetable Center in Beijing. The peanut meal was dried at 50°C in an oven after removing water-soluble proteins. The processing steps of removing proteins from peanut meal is shown in **Fig.1**. The dried peanut meal after removing proteins was ground using a mill.

Phytic acid sodium salt hydrate was purchased from Sigma-Aldrich Co. (St. Louis, MO, USA). Ferric trichloride was purchased from Tianjin Samtec Chemical Reagent Co., LTD. 5-Sulfosalicylic acid dehydrate was purchased from West Gansu Chemical Plant in Shantou city. HCl was purchased from China National Pharmaceutical Group Chemical Reagent Co., LTD.

2.2 Determination of PA content

The determination of PA content was evaluated by the Ferric trichloride combined 5-Sulfosalicylic acid method according to the procedure described previously[22]. 0.15g Ferric trichloride and 1.5 g 5-Sulfosalicylic acid was dissolved in 500 mL waters to be the Ferric trichloride and 5-Sulfosalicylic acid reaction solution. Inhale 0.0, 1.0, 2.0, 3.0, 4.0, 5.0 mL of 0.1 mol/L Phytic acid sodium salt hydrate solution in 6 colorimetric cylinders respectively, and trickle water into the colorimetric cylinders to 5 mL. A 5 mL Phytic acid sodium salt hydrate solution was combined with 4 mL Ferric trichloride and 5-Sulfosalicylic acid reaction solution. The absorbance of the solution was measured using a UV-vis spectrophotometer (754 UV-vis Spectrophotometer, Shanghai Jinghua Co, LTD) at a wavelength of 500 nm. Standard curve was draw with the absorbance as the ordinate and PA content as the ordinate. The regression equation was obtained as the following equation:

$$Y = -4.126x + 0.878 \quad (R^2 = 0.9991) \qquad (1)$$

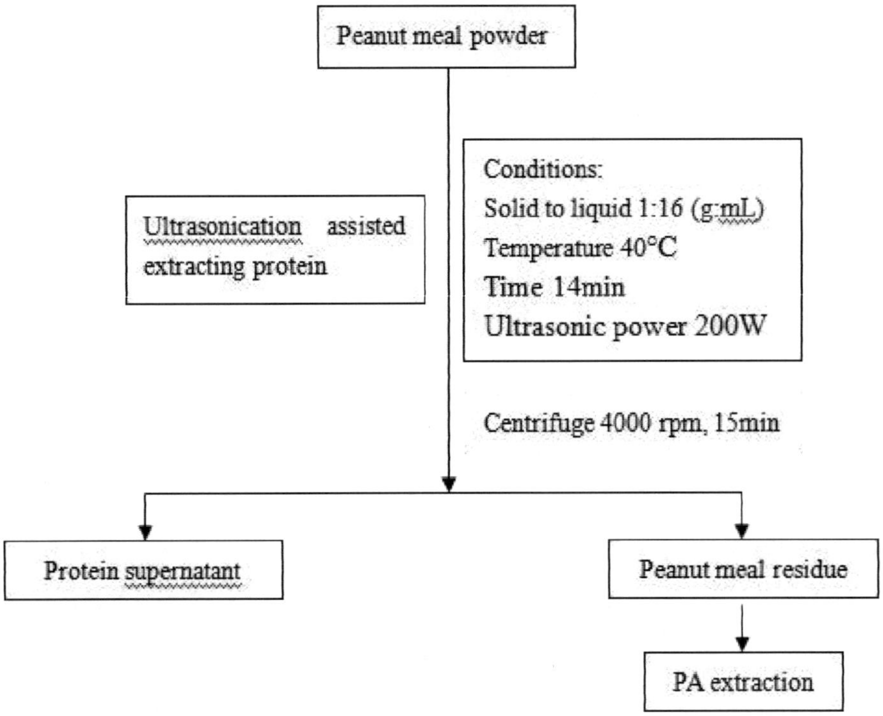

Fig. 1 The protein removal steps from peanut meal

2.3 Preparation of solid-liquid extracts

The influence of the solid-to-liquid ratio on the extraction of PA was investigated by using the following six ratios (1:10, 1:12, 1:14, 1:16, 1:18, 1:20; g:mL; sample powder: solvent). The hydrochloric acid concentration was fixed at the concentration of 0.01mol/L. The mixtures were extracted at 32°C for 83 min. The extraction solution was centrifuged at 4000 rpm for 10 min to obtain the supernatant. The content of PA (mg PA/g dry peanut meal) was then determined. The ratio that gave the highest value of yield was chosen for RSM.

2.4 Experimental designs

A d-optimal RSM experiment was designed using Design Expert (8.0.6, Stat-Ease Inc., USA). The independent variables were hydrochloric acid concentration (mol/L, X_1: 0-0.04), extraction time (min, X_2: 20-140), extraction temperature (°C, X_3: 25-60). The operating conditions were selected after single factor experiment and shown at **Table1**. The response variable was fitted by a second-order polynomial as follows:

$$Y = \beta_0 + \sum_{i=1}^{3} \beta_i X_i + \sum_{i=1}^{3} \beta_{ii} X_i^{\,2} + \sum_{i=0}^{2} \sum_{j=i+2}^{3} \beta_{ij} X_i X_j \qquad (2)$$

Y : the predicted response; β_0: the intercept coefficient; β_i: the linear coefficient; β_{ii}: the squared coefficient; β_{ij}: the interaction coefficient; X_i, X_j: the coded independent variables; X_iX_j: the interaction terms; X_i^2: the quadratic terms.

The optimal conditions for the extraction of PA from peanut meal were then carried out using the equations of RSM.

Table 1 Independent variables and their levels used for Box-Behnken rotatable design

Independent variable	**Level**		
	-1	**0**	**1**
HCl concentration (X_1)	0.01	0.02	0.03
Time (X_2)	80	100	120
Temperature (X_3)	25	30	40

2.5 The extraction rate and purity of PA

The extraction rate of PA is the ratio of the weight of PA crude extract to the weight of peanut meal, and the purity of PA is the ratio of the pure PA to the PA crude extract. The calculation formulas of the extraction rate and purity are as follows:

The extraction rate (%) = (The weight of PA extract / The weight of peanut meal) × 100% (3)

The PA purity (%) = (The weight of PA / The weight of PA extract) × 100% (4)

2.6 Fourier Infrared spectrum analysis the PA of peanut meal

FT-IR measurement was recorded by a AVATAR 370 (Thermo Nicolet). 1mg dried PA sample was mixed with 200 mg KBr. The mixture was milled 5 min in the agate mortar and tablet. The spectra of each sample were used to analysis in the range of 400-4000 cm^{-1}.

3. Results and discussion

3.1 Determination of single factor experiment

3.1.1 Effect of solid/liquid ratio

In order to optimize the extraction of PA from peanut meal, solid to liquid had to be selected to study [23]. The ratio of solid to liquid varied from 1:5 to 1:20 (g:mL) while other extraction variables were fixed. In **Fig.2a**, as the decrease of the solid to liquid ratio, the content of PA increased. The highest yield of PA was found at 1:16 (g:mL). This indicated that more volume of solvent was required to sufficiently solubilize the target PA when the higher driving force of molecules was supposed to be higher in a high solid to liquid ratio. However, there was no significant differences in extraction yield from 1:16 to 1:20 (g:mL) ($p > 0.05$). Therefore, 1:16 (g:mL) was chosen as a compromise and fixed in the RSM experiment.

3.1.2 Effect of HCl concentration

In acid solutions, the complexing power of PA is declined, and the solubility of PA is increased. The effect of HCl concentration on the content of PA is shown in **Fig.2b**. The highest extraction yield of PA was observed at HCl concentration of 0.02 mol/L. From 0 to 0.02 mol/L, the content of PA increased. While from 0.02 to 0.04 mol/L, the content of PA decreased. As increased of HCl concentration, the PA was dissolved in the HCl solution because of the decline of complexing capacity of PA. When the HCl concentration was beyond the optimized concentration, other acid soluble components, such as acid soluble saccharide, amino acids, could be dissolved, may recombine with PA and form complexes, resulting in the decrease of the content of PA.

3.1.3 Effect of extraction time

The effect of extraction time on the content of PA was studied varying time from 20 to 140 min. The content of PA was substantially increased with the time rose from 20 to 100 min (**Fig.2c**). It had a decrease at longer time. It could be that other acid soluble components were dissolved and recombined with PA. The optimized extraction time was 100 min.

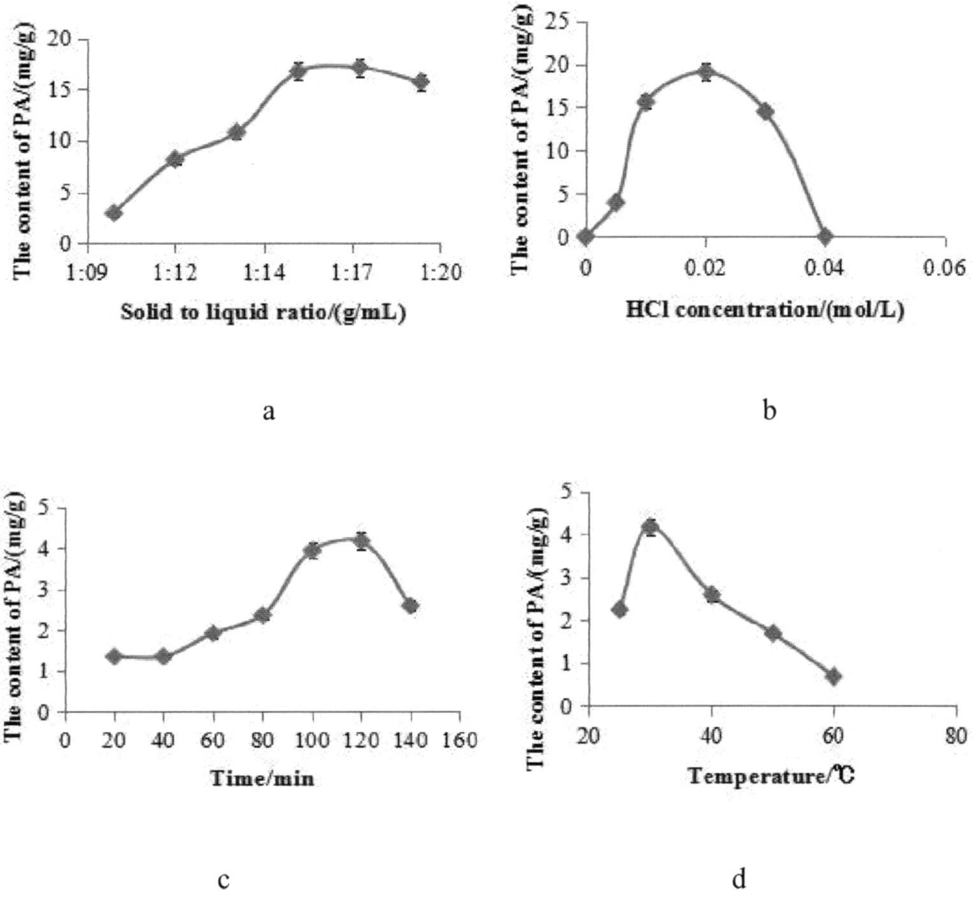

Fig.2 Single factor experiment

Effect of solid to liquid ratio (a), HCl concentration (b), extraction time (c) and Temperature (d) of PA from peanut meal. Different letters in the same sample indicate significant differences ($p < 0.05$) between treatments.

3.1.4 Effect of extraction temperature

As shown in **Fig.2d**, the yield of PA was increased with increased of extraction temperature from 25 °C to 30 °C. The maximum value of extraction temperature was at 30°C, having an obviously decrease with a further increase of extraction temperature. As the temperature increased, other ingredients were dissolved and combined with PA, causing the decreased of extraction yield.

According to the findings of above studies, the HCl concentration of 0.02 mol/L, extraction time of 100 min, extraction temperature of 30°C were selected as a central point for RSM. The ratio of solid to liquid was fixed at 1:16 (g:mL) in the following experiment.

Table 2 Coded levels, conditions runs and measured responses used in experimental design for response surface methodology

Run	X_1	X_2	X_3	Yield (mg PA/g peanut meal)
1	0.02	100.00	32.50	13.48
2	0.02	80.00	40.00	9.30
3	0.02	120.00	40.00	10.27
4	0.03	100.00	25.00	1.85
5	0.02	100.00	32.50	13.36
6	0.01	100.00	40.00	4.26
7	0.02	100.00	32.50	11.92
8	0.02	80.00	25.00	9.09
9	0.03	120.00	32.50	0.90
10	0.02	100.00	32.50	12.59
11	0.01	100.00	25.00	9.87
12	0.01	120.00	32.50	8.52
13	0.03	100.00	40.00	6.45
14	0.03	80.00	32.50	1.80
15	0.02	120.00	25.00	10.56
16	0.01	80.00	32.50	5.87
17	0.02	100.00	32.50	13.22

Table 3 Analysis of variance (ANOVA) for the fitted quadratic polynomial model for optimization of extraction parameters

Source	Sum of Square	Degree of freedom	Mean square	F value	p-value p>F	Significance [a]
Model	279.20	9	31.02	35.12	<0.0001	**
X_1	38.36	1	38.36	43.43	0.0003	**
X_2	2.20	1	2.20	2.49	0.1586	
X_3	0.15	1	0.15	0.17	0.6930	
X_1X_2	3.15	1	3.15	3.56	0.1010	
X_1X_3	26.05	1	26.05	29.50	0.0010	**
X_2X_3	0.062	1	0.062	0.071	0.7982	
X_1^2	173.67	1	173.67	196.60	<0.0001	**
X_2^2	20.78	1	20.78	23.52	0.0019	**
X_3^2	3.30	1	3.30	3.74	0.0945	
Residual	6.18	7	0.88			
Lack of Fit	4.47	3	1.49	3.47	0.1301	Not Significant
Pure Error	1.72	4	0.43			
Cor Total	285.38	16				

R^2=0.9783, Adj R^2=0.9505, Pred R^2=0.7401, adeq precisior=16.526.

[a] *significant ($p < 0.05$), **extremely significant ($p < 0.01$)

3.2 Response surface analysis

3.2.1 Analysis of the adequacy of the fitted model

RSM was applied to determine the effect of the HCl concentration (X_1), extraction time (X_2), and temperature (X_3) on the content of PA (Y). The responses of each independent variable are listed in **Table 2**. The mathematical model for describing the content of PA was given by the following second-order polynomial equation:

$$Y=12.91-2.19X_1+0.52X_2-0.14X_3-0.89X_1X_2+2.55X_1X_3-0.12X_2X_3-6.42X_1^2-2.22X_2^2-0.89X_3^2$$

The response surface quadratic model was drawn up, and the analysis of variance for the fitted quadratic polynomial model for optimization of extraction parameters are presented in Table 3. The p-value for the model was less than 0.0001, which indicated the model was extremely significant and could be used to monitor the optimization. X_1, X_1X_3, and quadratic terms (X_1^2 and X_2^2) were extremely significant on the PA extraction value.

As **Table 3** shown that the "Lack of Fit" is not significant relative to the pure error. The coefficient of multiple determinations (R^2) for the quadratic regression model was 0.9783. The R^2 was higher than 0.7, indicated that the model was suitable for use in the experiment [24]. The value of adjusted determination coefficients (R^2_{adj}) was 0.9505, closed to the R^2. Therefore, this model can be used to navigate the design space [25].

3.2.2 Response surface and contour plots

The response surface plots and contour plots shown in Fig.3. It demonstrated the effect and interaction of independent variables on the yield of PA. Fig.3 a-1 and a-2 shows the effect of HCl concentration and extraction time on the content of PA. when the extraction time was fixed, the content of PA gradually increased with the increase of HCl concentration, up to the HCl concentration of 0.02 mol/L , a large increment of PA content was observed. This due to the increase of HCl concentration improved the solubility of the acid soluble components. When the HCl concentration was fixed, an obvious quadratic effect of extraction time was observed. Fig. 3 b-1, b-2, and Table 3 demonstrates the interaction of HCl concentration and extraction temperature is significant ($p < 0.05$). The content of PA as affect by HCl concentration and extraction temperature, exhibiting a clear increase in content with the raise of HCl concentration at a fixed extraction temperature. While an obvious quadratic effect of extraction temperature was observed. In **Fig.3** c-1 and c-2, a similar linear increase in content with the increase of extraction time ranged from 80 min to 104 min at a fixed temperature. While a decrease in temperature from 104 min to 120 min led to marked decrease in content of PA. This indicated that the extraction yield of PA depended on the medium level of time. These occurrences might be due to the recombination of PA and acid soluble proteins.

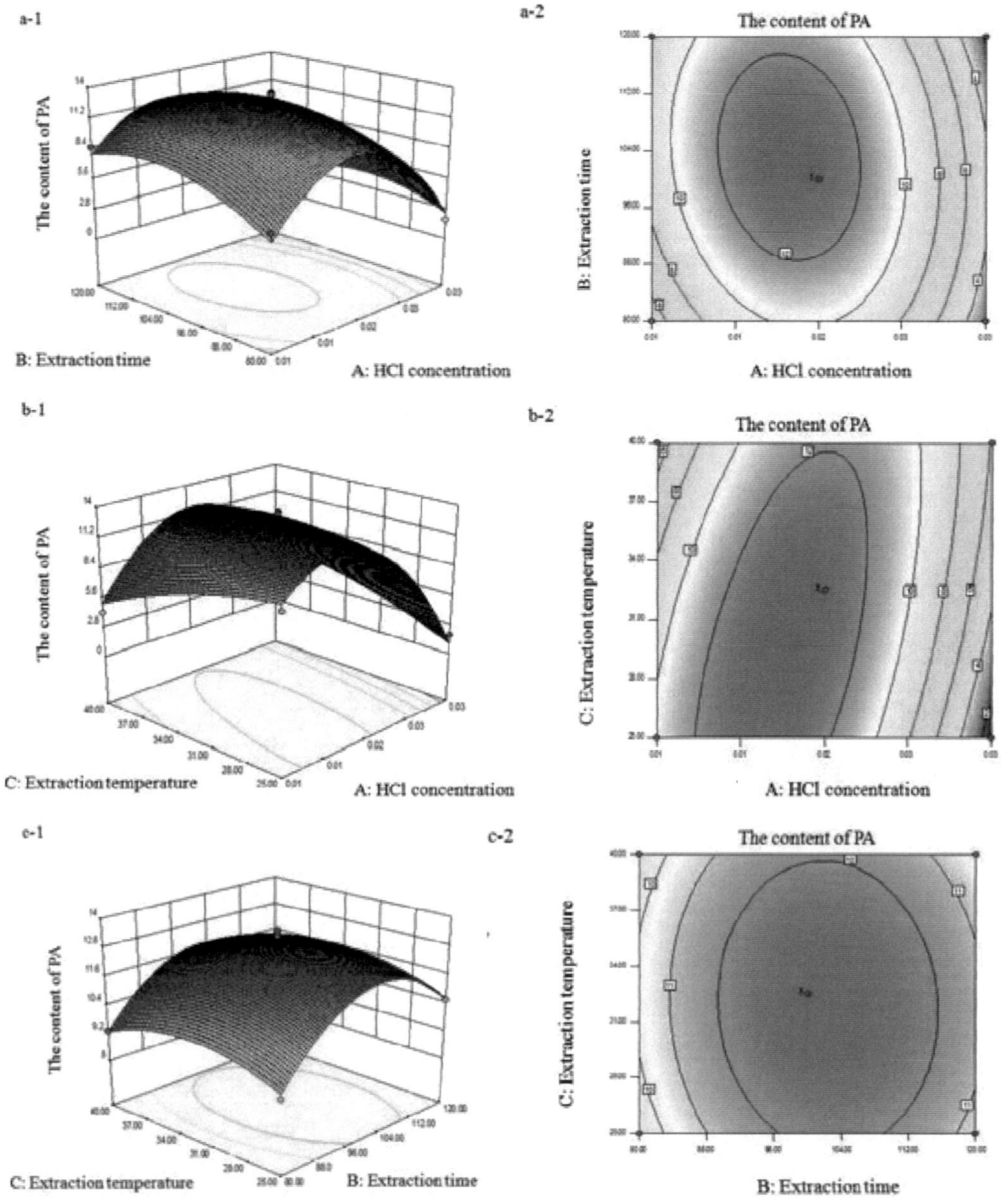

Fig. 3 Response surface plot and contour plot showing the effects of independent variables on the content of PA. a. HCl concentration and extraction time, b. HCl concentration and extraction temperature, c. extraction time and extraction temperature

3.2.3 Optimization of extraction craft parameters and the model validation

The optimal extraction conditions for PA were determined in order to simultaneously maximize the extraction. It was found that the optimal conditions were an HCl concentration of 0.02 mol/L, extraction time of 105 min, extraction temperature of 30 °C. The predicted and experimental values for the extraction of PA from peanut meal were 13.31 mg/g and 11.18 mg/g, respectively. In the optimal conditions, the experimental value was close to the predicted value, and the extraction ratio and content of PA was 6.12 ± 0.51% and 182.70 ± 2.35 mg/g, respectively. Removing the protein firstly and using the optimum extraction conditions can improve the extraction yield and purity. These results confirmed the predictability of the model to give the maximum yield of PA in the experimental process used.

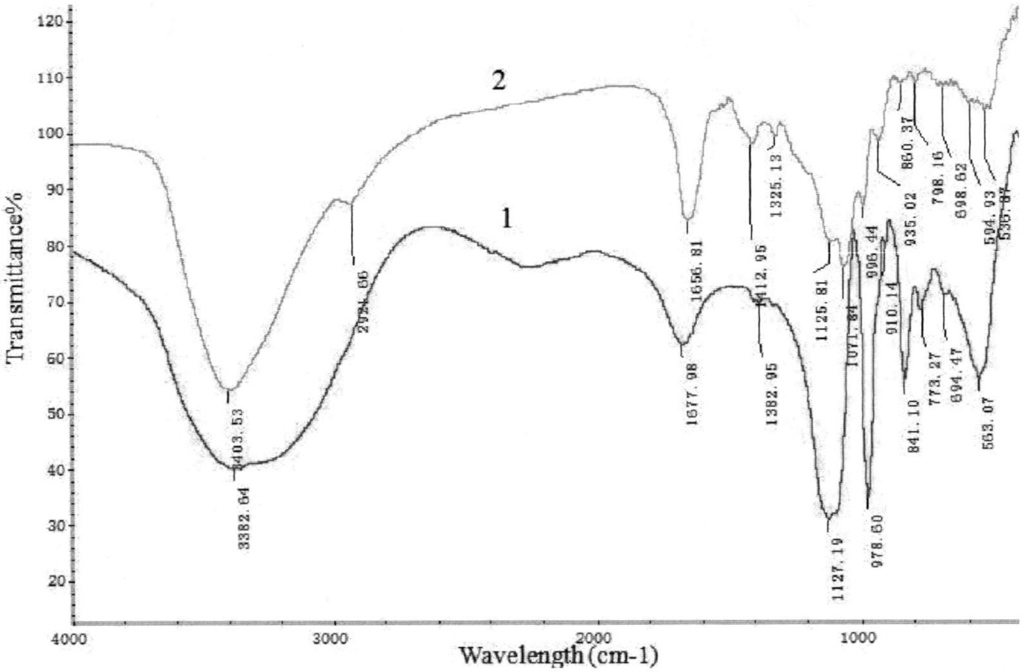

Fig.4 FT-IR spectra of PA extract.

1: Reference material phytic acid sodium salt hydrate from Sigma; 2: PA extract from peanut meal

3.3 Fourier Infrared spectrum analysis the PA of peanut meal

Each FT-IR spectra site of a compound provides qualitative information. PA contains phosphoester bond (P=O, P-O, C-O, the region of 1330~900 cm^{-1}; -OH, the region of 3650~3200 cm^{-1}). **Fig.4** was shown that reference material phytic acid sodium salt hydrate and PA extract from peanut meal have same peaks. Phytic acid sodium salt hydrate has the peaks at the wavelength of 3382 cm^{-1} (-OH), 1127 cm^{-1}, 978 cm^{-1} (P=O, P-O, C-O). PA extract has peaks at the wavelength of 3403 cm^{-1} (-OH), 2921 cm^{-1} (-CH), 1125 cm^{-1}, 1071 cm^{-1}, 996 cm^{-1} (P=O, P-O, C-O). Therefore, the PA extract from peanut meal was verified qualitatively by comparing with phytic acid sodium salt hydrate in the IR spectra.

4 Conclusion

In this study, we first removed proteins from peanut meal, and RSM was used to optimize the experimental variables including HCl concentration (mol/L), extraction time (min), and extraction temperature (°C). Solid to liquid ratio had significant effect on the extraction yield at the ratio of 1:16 (g:mL). This was chosen as the optimum parameter. The optimal conditions to obtain the highest extraction of PA were determined as follows: 0.02 mol/L of HCl concentration, 105 min of extraction time, 30°C of extraction temperature. At this condition, the extraction ratio was 6.12 ± 0.51%, which is far higher than the extraction ratio of traditional method, and the content of PA was 182.70 ± 2.35 mg/g dry PA extract. The PA extract derived from peanut meal could be verified qualitatively by IR spectra.

PA is the most important anti-nutrient ingredient in peanut meal. Peanut nutrient ingredients removing PA is one of the most attractive and promising sources of plant proteins or plant polysaccharides. On the other hand, peanut meal is the important renewable source of PA, PA extracted from peanut meal will be better applied in medicine and light industry. This study could not only be useful in science research and industry production of PA from peanut meal, but also conducive to high value utilization of peanut meal and reduce waste of peanut resources.

Acknowledgements

This work was supported by the National Key Research and Development Program of China (2016YFD0400802), National high technology research and development program of China (863 Program) (2011AA060701).

References

[1] Wang H., Zhou Y., Ma J., Zhou Y Y., Jiang H (2013). The effects of phytic acid on the Maillard reaction and the formation of acrylamide. Food Chemistry, 141 (1) , 18-22.

[2] Lopez H.W., Leenhardt F., Coudray C., Remesy C. (2002). Minerals and phytic acid interactions: is it a real problem for human nutrition? Int. J. Food Sci. Technol., 37, 727-739.

[3] Norazalina S., Norhaizan M.E., Hairuszah I., Norashareena M. S. (2010). Anticarcinogenic effcacy of phytic acid extracted from rice bran on Azoxymethane-induced colon carcinogenesis in rats. J. Experimental and Toxicologic Pathology, 62 (3), 259-268.

[4] Yukako O., Tetsuyuki K. (2014). Dietary phytic acid modulates characteristics of the colonic luminal environment and reduces serum levels of proinflammatory cytokines in rats fed a high-fat diet. Nutrition Research, 34 (12) , 1085-1091.

[5] Fu Q. X., Wang H. Z., Xia M. X. (2015). The effect of phytic acid on tight junctions in the human intestinal Caco-2 cell line and its mechanism. J. European Journal of Pharmaceutical Sciences, 80, 1-8.

[6] Lv Y. Q., Zhang Z., Hou L., Zhang L., Zhang J. Y., Wang Y. H., Liu C., XuP . P., Liu L., Gai X. Y., Lu T. X. (2015). Phytic acid attenuates inflammatory responses and the levels of NF-kB and p-ERK in MPTP-induced Parkinson's disease model of mice. J. Neuroscience Letters, 597, 132-136.

[7] Jin N. K., SungN. H., Kim H. K. (2014). Phytic acid and Myo-inositol support adipocyte differentiation and improve insulin sensitivity in 3T3-L1 cells. Nutrition Research, 34, 723-731.

[8] Notoya T., Otieno-Alego V., Schweinsberg D.P. (1995). The corrosion and polarization behaviour of copper in domestic water in the presence of Ca Mg and Na-salts ofphytic acid, Corros. Sci. 37 (1), 55-65.

[9] Yang H.F., Yang Y., Yang Y. H.(2005). Formation of inositol hexaphosphate monolayers at the copper surface from a Na-salt of phytic acid solution studied by in situ surface enhanced Raman scatteringspectroscopy Raman mapping and

polarization measurement. Anal. Chim.Acta, 548 (1-2), 159-165.

[10] Wang Y. .H., He J. B. (2012). Corrosion inhibition of copper by sodium phytate in NaOH solution: cyclic volt absorptometry for in situ monitoring of soluble corrosion products. Electrochim. Acta. 66, 45-51.

[11] Li C. C., Guo X.Y., Shen S. (2014). Adsorption and corrosion inhibition of phytic acid calcium on the copper surface in 3 wt% NaCl solution, Corros. Sci., 83, 147-154.

[12] Hao C., Yin R. H., Wan Z. Y., Xu Q J., Zhou G. D. (2008). Electrochemical and photoelectrochemical study of the self-assembled monolayer phytic acid on cupronickel B30, Corros. Sci., 50 (12), 3527-3533.

[13] Li Y., He J. B., Zhang M. (2013). Corrosion inhibition effect of sodium phytate on brass in NaOH media. Potential-resolved formation of soluble corrosion products. Corros. Sci., 74, 116-122.

[14] Gao L. L., Zhang C. H., Zhang M. L. (2006). Phytic acid conversion coating on Mg-Li alloy. Journal of Alloys and Compounds, 485 (1-2) (2009), 789-793.

[15] H. R. Park, H. J.Ahn, S. H.Kim, C. H.Lee, M. W. Byun, G. W. Lee. Determination of the phytic acid levels in infant foods using different analytical methods. Food Control, 17 (9), 727-732.

[16] Nader J., Fawaz N., Afif C., Louka N. (2016). A novel process for preparing low-fat peanuts: Optimization of the oil extraction yield with limited structural and organoleptic damage. Food Chemistry, 197, 1215-1225.

[17] Su G. W., Ren J. Y., Yang B. (2011). Comparison of hydrolysis characteristics on defatted peanut meal proteins between a protease extract from Aspergillus oryzae and commercial proteases. Food Chemistry, 126 (3) , 1306-1311.

[18] Wang Q., Shi A., Liu H., Liu L., Zhang Y., Li N., Gong K., Yu M., Zheng L. (2016). Chapter 5-Peanut By-Products Utilization Technology. Peanut: Processing Technology and Product Development, Copyright © 2016 China Science Publishing & Media Ltd. Published by Elsevier Inc., 211-325.

[19] Reddy N., Chen L.H., Yang Y. Q. (2013). Thermoplastic films from peanut proteins extracted from peanut meal. Ind. Crop. Prod. 43, 159–164.

[20] Wu H. W., Wang Q., Ma T. Z. (2009). Comparative studies on the functional properties of various protein concentrate preparations of peanut protein. Food Research International, 42 (3), 343-348.

[21] Zheng L., Ren J. Y., Su G. W., Yang B., Zhao M. M. (2013). Comparison of *in vitro* digestion characteristics and antioxidant activity of hot- and cold-pressed peanut meals. Food Chemistry, 141 (4), 4246-4252.

[22] Kiassos E., Mylonaki S., Makris D. P., Kefalas P. (2009). Implementation of response surface methodology to optimise extraction of onion (Allium cepa) solid waste phenolics. Innovative Food Science and Emerging Technologies, 10 (2), 246-252.

[23] Sahin S., Samli R. (2013). Optimization of olive leaf extract obtained by ultrasound-assisted extraction with response surface methodology. Ultrason Sonochem, 20 (1), 595-602.

[24] Wong K. H., Li G. Q., Li K. M., Valentina R. N., Chan K. (2017). Optimisation of Pueraria isoflavonoids by response surface methodology using ultrasonic-assisted extraction. Food Chemistry, 231 (15), 231-237.

[25] Bancha Y., Nuttapum S. (2015). Optimization of process parameters for phenolics extraction of Cratoxylum formosum ssp. formosum leaves by response surface methodology. Food Science Technology, 52 (1), 129-140.

512826 Influences of Geometric Factors on a Solar Thermochemical Reactor for Two-Step CO_2 Splitting

Abstract:

Solar thermochemical processes utilize the inexhaustible solar energy as thermal driving force to provide heat and motivate the reactions, which offer pathways to store solar energy to chemical fuels. Some elevated temperature reactions, like synthesis of syngas and methane reforming, are anticipating to accomplish in solar reactors, which have the ability to provide extremely high temperatures. Presently, the dominating factors in the design of solar reactors are heat and mass transfer with reaction mechanisms. A novel partition cavity-receiver reactor concept is proposed in this paper. For the purpose to provide a longer pathway of interaction between catalyst and reactants, a partition is introduced in this cavity-receiver reactor. A numerical computational fluid dynamics (CFD) analysis is performed to study on the influences of geometric factors, like length-diameter ratio, partition length and thickness, the transparency of partition, to fluid flow and heat transfer under constant solar power input for both two-dimensional and three-dimensional models. A two-step solar thermochemical redox reaction using ceria as catalyst to split CO_2 is modeled in the partition cavity-receiver reactor to investigate the relationship between geometric factors and reaction rates. Based on the comparisons and analysis of results, optimized geometric factors and corresponding operating conditions are discussed.

Keeping the balance in technical and human competencies to create a high-performance process safety culture

Coen van Driel
TÜV SÜD Process Safety
Mattenstrasse 24
CH-4002 Basel - Switzerland
Coen.vandriel@tuev-sued.ch

Prepared for Presentation at
American Institute of Chemical Engineers
2018 Spring Meeting and 14th Global Congress on Process Safety
Orlando, Florida
April 22 – 25, 2018

Keeping the balance in technical and human competencies to create a high-performance process safety culture

Coen van Driel
TÜV SÜD Process Safety
Mattenstrasse 24
CH-4002 Basel - Switzerland
coen.vandriel@tuev-sued.ch

Keywords: Process Safety Competency, Leadership, balance, high performance organization, culture, risk management, human factor,

Abstract

The core of high-performance in process safety is managing risks to prevent incidents from happening. In a high-performance process safety culture, there a balance between process safety management system with technical soundness and human factor which is enabled by leadership. An essential element in the journey to a high-performance process safety culture is competency management. In other words, having and continuously working on the right and balanced mix of competencies at all levels of the organization. Realizing that right and balanced mix of competencies is an experience driven transformation journey resulting in a high-performance process safety culture.

Content

1 Introduction

As the industry, we can realize technically advanced plants and great innovation. But in the approach towards optimizing the performance in process safety as well as asset integrity, the industry tends to keep a silo mentality, approaching from an occupational safety perspective, an asset integrity/reliability perspective, or a process safety (process safety and asset integrity do deal not only with safety but cover health, safety, environment and security issues with the primary goal to avoid loss of containment) perspective. With the advanced plants, the chemical production becomes more complex, the need to drive continuous improvement in process safety — not just for industry, but for all stakeholders — becomes increasingly urgent. Continuous improvement cannot only be realized with better rules and regulation resulting in procedures (management systems). The success of continuous improvement lies in the hands of the behavior and competences of the people in the organization. Despite rigorous rules and regulation, severe incidents continuous to happen. Is the focus on compliance out of balance to improve?

An example that perfectly shows the unbalance between system focus and human focus is the challenge of climbing a summit like the Mont Blanc which rises 4,808 m (15,774 ft) above sea level.

Nowadays the summit is ascended by an average of 20,000 mountaineer-tourists each year. It could be considered a technically easy, yet arduous, ascent for someone who is well-trained and acclimatized to the altitude. From l'Aiguille du Midi (where the cable car stops), Mont Blanc seems quite close, being 1,000 m (3,300 ft) higher. But while the peak seems deceptively close, La Voie des 3 Monts route (known to be more technical and challenging than other more commonly used routes) requires much ascent over two other 4,000 m mountains, Mont Blanc du Tacul and Mont Maudit, before the final section of the climb is reached and the last 1000 m push to the summit is undertaken.

Each year climbing deaths occur on Mont Blanc, and on the busiest weekends, normally around August, the local rescue service performs an average of 12 missions, mostly directed to aid people in trouble on one of the normal routes of the mountain. Some routes require knowledge of high-altitude mountaineering, a guide (or at least an experienced mountaineer), and all require proper equipment. All routes are long and arduous, involving delicate passages and the hazard of rock-fall or avalanche. Climbers may also suffer altitude sickness, occasionally life threatening, particularly if they do not acclimatize to it. (1)

The mayor of one of the towns near the Month Blanc is responsible for what is happening on the mountain. One of the things he is requesting is, for the safety of the climber's, more stringent rules and regulations. For example, the guides that lead groups of climbers to the summit must follow the rules and regulation, as well as the people that do the attempt on their own. As in the industry, if something happens with one of the climbers the guide must defend his responsibility for taking the action in court. Despite rigorous systems in place, incidents and casualties should be accounted for. On the other hand, that same mayor sees that the culture of climbing the Mont Blanc is changed and more and more driven by money greed of the guides. Also, the perception of the climbers has changed from seeing the Mont Blanc as a challenging mountain to climb to an easy mountain to walk up, which is influenced by declining competencies of climbers and guides. This trend cannot be changed by focusing just on the system but also requires attention to change the mindset of the climbers (human factor). There is a need in changing the trend around the Mont Blanc as well in the chemical industry to get on a journey with people with the right competencies to high-performance in these settings.

2 Process Safety high-performance organization

2.1 Definition of high-performance

In the business world and in the chemical sector, there is continuous strive for improving performance. Like great athletes, even high-performing organizations must continuously strive to improve and "work on their game." Without the passion for improvement, they are unlikely to remain high performers for long. After all, there's no shortage of ambitious business leaders who are taking notes on their fellow high performers and working hard to ensure that their own companies eventually reach the top echelons of organizational excellence (2).

In the view of Katzenbach and Smith (3) the essence of a high-performance organization is the power of teamwork. *" Wisdom lies in recognizing a team's unique potential to deliver results and in understanding its many benefits--development of individual members, team accomplishments, and stronger companywide performance".* As Basketball legend Michael Jordan slam dunks the same point: "Talent wins games, but teamwork and intelligence win championships."

Although teamwork is an essential element of high-performance, the picture is not complete to make high-performance successful. Dr. André de Waal MBA (HPO Center) (4) has formulated the five success factors for high-performance organization:

- Quality of Management/Leadership
- Open and action driven culture
- Long term focus for the organization
- Continuous improvement as the base for the organization
- Quality of the employee's

Many views of high-performance organization, as the ones above, are focusing on the human aspect of high-performance. This is not yet the complete picture. People in society and organizations also need boundary conditions, for example knowing what their responsibility is, what the structure is, and the rules and regulations. In a high-performance process safety culture the balance between the system and the human factor is embedded in the organization and continuously facilitated by leadership (see figure 1). (5) On the one hand, there is focus needed on the technical soundness of the installation and a management system to manage the barriers. On the other hand, having a culture in which there is focus on continuously improving with good set of values and beliefs. This supported by craftsmanship (good balanced set of competencies) at all levels of the organization. This all will not work if the client and stakeholders do not appreciate the effort of the organization.

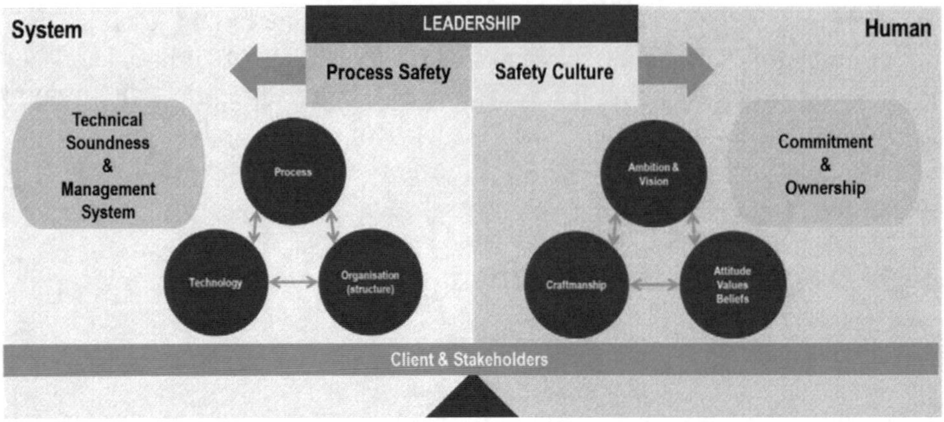

Figure 1 System and human related elements. Both are mutually dependent.

In addition to make high performance work in the chemical industry risk thinking should be at the core of the system and culture (human behavior) (figure 2). In the integrative system, there is a strong base of a risk-based safety culture supporting the different processes of design integrity, technical integrity, operational integrity and emergency response. Those processes handle the issues and threats for safety, process safety, and asset integrity using the risk approach. It enables one focus in the organization on continuously managing the issues and threats on the likelihood and consequences, not only the ones that are easily in the view, but also the unlikely ones with high consequence. The performance management system, discussed in more detail in section 2.2, based on the Plan – Do – Check – Act (PDCA) circle and using measurement & metrics (KPIs) enables a high performance for all three aspects. Viewing them as part of the same system allows for greater risk mitigation and added operating value.

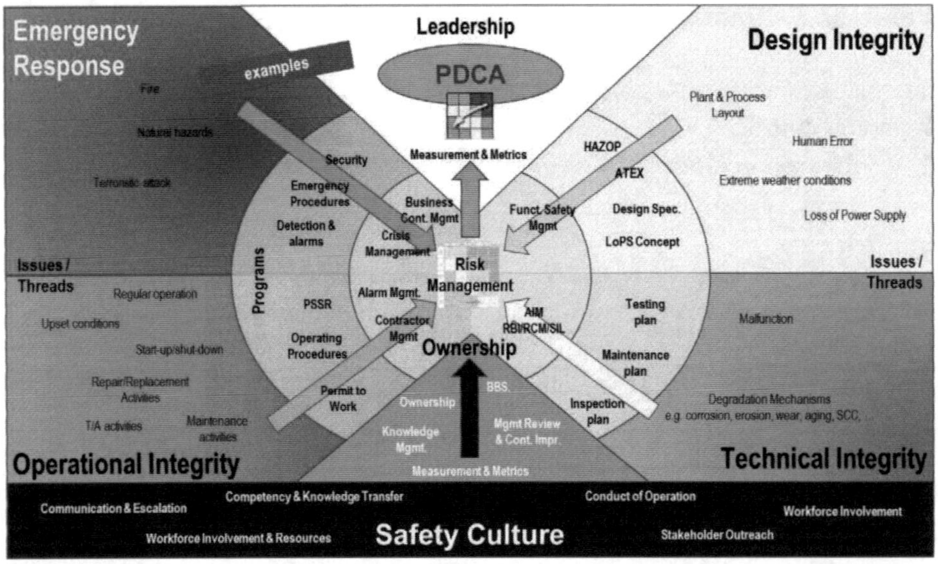

Figure 2 Risks management at the core of the performance of the organization.

All elements, system and human facilitated by leadership, to create an environment of high-performance are there to realize a value creation for the internal and external stakeholders. The value creation is not only to realize benefits for the shareholders (the financial results) but more importantly the non-financial benefits like sustainability/continuity development, innovation, satisfied external stakeholders (nearby residents, local authorities, etc.), brand name, process safety, and competencies/talent of the organization.

2.2 Measurement of high-performance

The essence of why the chemical industry has survived is that the products are produced with good quality, with optimal margin and with minimum risk of an incident happening. People should be able to understand that and convert that into an integrative risk approach using all aspects of process safety, occupational safety and asset integrity (reliability). Striving for excellence in operational performance (high-performance) using a risk management approach is crucial for further success.

Business leaders crave knowledge about the key elements of organizational performance. After all, in the corporate world, these are the elements that make or break companies every day. How can excellence in operational performance be achieved using a risk management approach? The methodology, seen in Figure 3, integrates the existing knowledge of the hazard analysis into the work processes and links it in to performance management. The advantage is that everybody in the organization gets to know their most relevant risk and have that integrated into their daily business. Secondly, the performance management is then built upon the risk meaning the indicators are related to the risk in the plant.

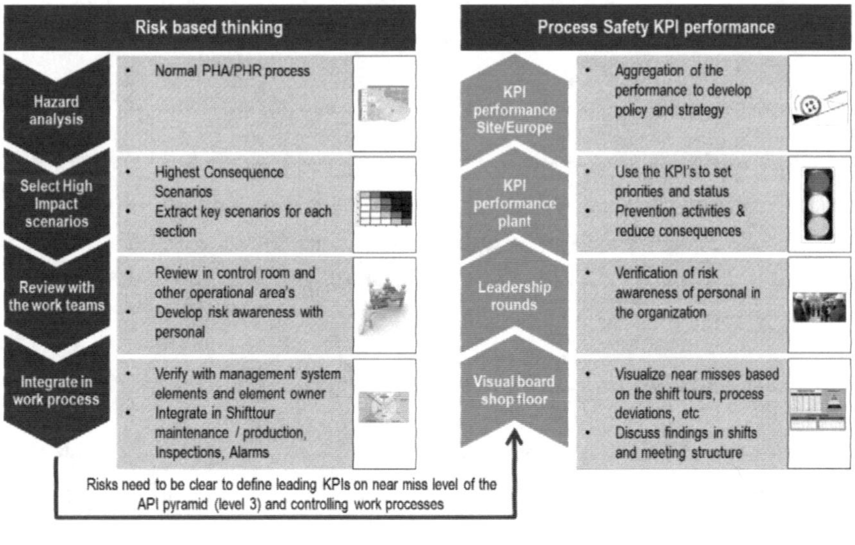

Figure 3: risk based thinking as foundation for Safety performance

The core of performance management is closing the loop of the continuous improvement cycle Plan-Do-Check-Act. As mentioned in the previous paragraph, one of the fundamentals of high-performance organization is continuous improvement in combination with leadership taking the ownership to put energy in the cycle to excel. This energy of leadership balances between using the performance management system and the human aspects as having a vision, the facilitating and process safety competencies and values and beliefs. The elements of the performance management system are defined process, strategy and planning, short interval control, leading and lagging Key Performance Indicators (KPI's), reporting, and formulating actions to improve performance. Those elements can become successful by the commitment of all people in the organization from all levels.

As is known in the industry, risk comes from multiple points and all should be proactively managed as major incidents rarely result from a single cause or failure. Rather, multiple failures coincide and collectively result in an exceptional event with severe, and potentially fatal, consequences. The model, which is based on the Swiss cheese model, in figure 4, visualizes that when different barriers fail in an organization a major incident can happen. Unfortunately, there are many incidents in the industry where this has happened with catastrophic consequences.

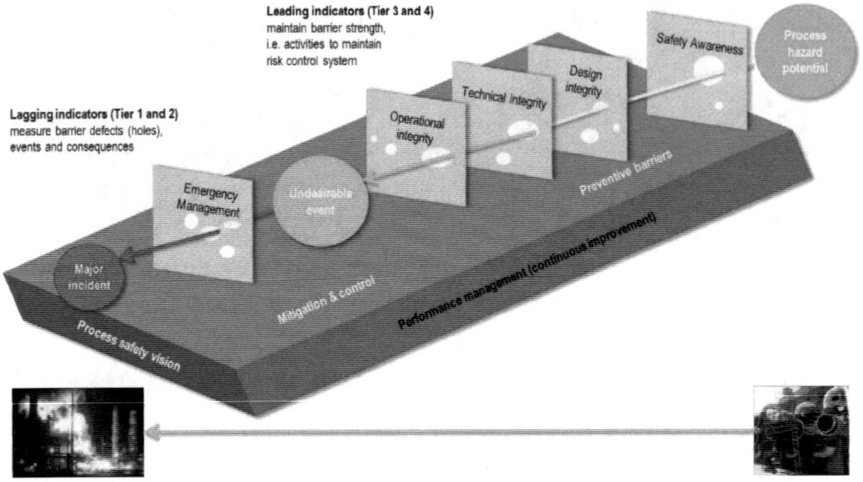

Figure 4: process safety control system

As a human being, we prefer no risks, while we continuously take risks in our daily lives: by crossing the street, driving to work, going on ski holiday and as mentioned in the introduction, climbing the Mont Blanc. Whatever we invest in safety, a zero risk is unachievable goal. Yet it is not only coincidence, when things are going well. With balanced leadership, coincidence can be organized.

The success of an organization depends on the financial performance and even so important, the success in the field of occupational safety and process safety. Even though the frequency of the incidents in the industry is low, the impact of such an incident is very high. After the severe incidents, the governments insist on more and stricter control mechanisms. On the other hand, we all know that the risk of zero is only achievable in theory.

The biggest risk of risk management is blindly trusting the rules and regulations. Stricter control and tighter procedures often lead to a counterproductive effect. From research is seen that with extra control and supervision the risk on incidents increases. Debit is that nobody in the end is taking responsibility because everybody thinks somebody else is responsible for the overall picture. Beside this, in the execution behavior develops 'as long I am following the rules I am doing well'.

The challenge for organizations is to find the balance between using the risk management systems, safety technical solutions and the human factor. The human factor having the proper competencies, driven by values and beliefs and facilitated by leadership at all levels in the organization.

3 Competency management in process safety

3.1 Process Safety Competency definition

In process safety, human factors have contributed to the causes of several incidents in the past, in a variety of safety critical industries. For example: On March 23rd, 2005, the BP Texas City refinery experienced a catastrophic process accident. It was one of the most serious U.S. workplace disasters of the past two decades, resulting in 15 deaths and more than 170 injuries. Both the BP and the U.S. Chemical Safety and Hazard Investigation Board reports (6) identified numerous technical and human failings at the refinery and within corporate BP.

As Stephen Pryor mention in his quote in the Vision 2020 the human factor is about the knowledge and using the knowledge to understand the risk and act to it.

"Process safety is not something for the leader...and the safety experts to know; everybody in the workforce has to become more and more knowledgeable about understanding what the risks are, helping us to identify risks, and making sure that they understand and execute all of our procedures properly and consistently to avoid those risks." — Stephen Pryor, ExxonMobil Chemical (quote from Vision 2020).

Not only those incidents create a sense of urgency for developing process safety competency in the organizations and industry. It is, also, getting attention for many other reasons like the brain drain through aging of the workforce, the loss of knowledge in acquisitions, and too much focus on compliance that people forget to think. In addition, regulators and industry associations like the CCPS are more and more putting process safety competence on the agenda. Therefor companies begin to realize that investing in process safety competence is an asset for the company and correlates with the process safety performance.

The following definition of competency give the boundaries of what is meant by a competency:

The definition of the European Qualifications Framework (EQF): 'Competence' means the proven ability to use knowledge, skills and personal, social and/or methodological abilities, in work or study situations and in professional and personal development. In the context of the EQF, competence is described in terms of responsibility and autonomy (EC, 2008).

In other words, competency has two sides the measurable technical/ management system competencies and the non-measurable competencies as social, human and personal competencies. In this context, the balance between system/ technical related competencies and human related competencies makes a distinctive difference in the success. This is also expressed in the Vision 2020 of the CCPS:

Intentional Competency Development to ensure that all employees who impact process safety are fully capable of meeting the technical and cultural requirements for their jobs. The bottom line: no matter how good the culture or management system is, or how well the company adheres to standards, it takes highly competent employees to implement those systems or standards. And that requires intentional competency development.

Another perspective on competencies and the framework that has been developed by the UN is looking at three elements: core values, core competencies, and managerial competencies. In this framework, the technical competencies or craftsmanship are missing. But the different angle is the core values which people are evaluated on. This fits nicely with the model of balanced leadership.

In the definition of Prof P. Dehnbostel (7) there are three components that form the base for occupational development of the mix of competencies of people.

- Technical Competence constitutes a person's ability and readiness to process tasks and problems in an autonomous, professionally appropriate and methodical manner and to evaluate the result.
- Social Competence describes a person's ability and readiness to work together with others in a target oriented manner, understand the interests and social situations of others, deal with and communicate with others in a rational und responsible way.
- Personal Competence describes a person's ability and readiness to develop further and to shape his or her own life in an autonomous and responsible manner within the respective social, cultural or occupational context.

Measuring competencies can be done across a 4-tier scale, based on a combination of the tiers used by both Woodside and Santos (8). This allows a more granular determination of the competency required across a varied workforce. These tiers are defined as follows:

Awareness

Has knowledge of theory and displays conceptual understanding. Actively participates in discussions regarding the skill. Performs routine tasks with significant supervision. Learns how to do things.

Basic application

Performs fundamental and routine tasks. Requires occasional supervision. Increases functional expertise and ability. Works with others.

Skilled application or proficiency

Independent contributor. Integrates work with other disciplines. Frequently mentors or coaches' others. Assesses and compares alternative options.

Mastery or expert

Advanced experience in the skill. Applies creative solutions to complex problems. Defines and drives critical business opportunities and needs. Represents the organization internally and externally on critical issues. Sets standards within the organization. Recognized as subject matter expert.

The process safety competencies framework should consist of a mix of Craftsmanship competencies, ranging from technical/management system competencies to personal and social competencies, and core values and beliefs. This defined for all levels of the organization from the workforce on the shop floor to the board of directors. In the IchemE framework (8) the roles are defined and the technical/ management system competencies. This is a starting point for developing the complete process safety competencies framework including the personal and social competencies and core values/ beliefs. The TÜV SÜD Process Safety competency model starts from the same technical/ management systems competencies but in the transformation process based on competency development the personal/ social competencies and core values/ beliefs will be jointly developed with the client. The current DNA of the company is considered and the vision the company.

3.2 Contribution of Process safety competency to high-performance

As craftsmanship and attitude/behaviors are essential aspects to make a high-performance process safety culture successful, competency development is making a difference in the success of a high-performance culture. Naturally in the process industry there is more attention to develop the technical skills of the people in the organization. In this context, the balance of competencies – technical/ management system competencies, social/ personal competencies and core values/ beliefs makes the difference.

There are some actuals threats impacting the process safety performance. For example, as in recent crises years, there has not been enough attention to competency development and/or a lot of senior people are leaving the companies. Another threat is that people do not know the assets anymore because the assets are too reliable.

The opportunity is there to invest in the competency development in an experience driven transformation journey facilitated by leadership. Such an experience driven transformation journey for competency development has two components:

A) Individual development: current competencies – experience driven development plan – execute journey – evaluate and update picture
B) Organizational development: current culture, knowledge, and structure (roles and succession planning) – organizational journey – execute journey – evaluate and adjust

Integral embedded in the first component (A) is the learning process supported by the different pedagogic tools like coaching, experience learning settings, training on the job, classroom training, e-learning, and learning by doing. The learning process depends on: prior knowledge, motivation, memory, understanding, intellectual engagement. Leadership can support the learning process by continuously challenging to prevent normalization from happening. The Organizational development (B) is supported by the transformation approach and leadership.

4 Role of Leadership

4.1 Leadership balance for high-performance

The role of leadership is continuously developing, as the organization is developing part of its environment. The way value is created by leadership is changing. From a top down approach to a facilitating approach. The main role of leadership is to create an environment in which others can become successful, resulting in a high-performance organization. Also, the appreciation of scarcity in the world makes that we should value more what the world can offer us. Make optimal use of the talents and competencies of the individual and organization.

The conceptual model of balanced leadership as visualized in figure 4 can create an environment in which people will become successful. Role of leadership at all levels in this process is to create an environment in which the experience driven transformation journey to high performance process safety culture is possible by keeping the balance between the left side of the model "system/technical soundness" and the right side of the model "human commitment and ownership".

The management system is the organization in which boundaries conditions, in which people are daily working at all levels, are formally described. It contains the description of the core process and secondary processes with roles and responsibilities. The organization has the opportunity here to jointly formulate what is expected from the organization and the individuals in the organization to become successful. The formal communication chain of planning and reporting based on the formulated KPI's is the performance management system that enables an organization to visualize the pathway to success and the realization of that success. The risk based performance management system, as described in section 2.2, having the KPI's based on the risk scenarios to continuously improving the performance. The CCPS looks at management systems in the vision 2020 as "Vibrant Management Systems" which are engrained throughout the organization. Vibrant systems readily adapt to the organization's varying operations and risks. The core of the system side of the organization is having and maintaining a technical sound asset base. The pots and the pans for a chemical organization or in other words the technically advanced plants that produce the products with optimal quality for the customers in a safe way.

In a certain way, you can say that the left side of the model is about the structure and the right side about culture. The right side of the model is going a step further. It is about the mix of competencies (craftsmanship), the total behavior of people, value and beliefs, the results or value creation and ambitions of the stakeholders in relation to the environment. The human side of the model shifts the focus to the "why" (vision) of the organization. To whom that develop and propagate the vision and the ones who comply with the vision. The elements of the model that bring live to an organization. The employees and the leaders at all levels use the system elements to make the organization successful without losing the "why", the what and how of the organization. In the values and beliefs of the organization a sound risk appetite is there to operate the organization is a safe way for all stakeholders.

According to the vision 2020 of the CCPS leadership has a major impact on culture: "in a Committed Culture, executives involve themselves personally, managers and supervisors drive excellent execution every day, and all employees maintain a sense of vigilance and vulnerability". The value that is created by a committed culture is in both safe and reliable operations by doing "the right thing, the right way" every time. People have the feeling and motivation that they can improve and they are in the starting position to take the steps for improvement. Therefore, developing the process safety competencies of the people in the organization is an essential role of leadership.

The base of the model are the different stakeholders in the organization like customers, suppliers, employees, owners, and external stakeholders with their formal interests and goals. In the end, it is about creating value for the environment in which the organization is acting. The value created in delighting the customer, profit for the organization, developing the talent of the individual's and producing in a safe way.

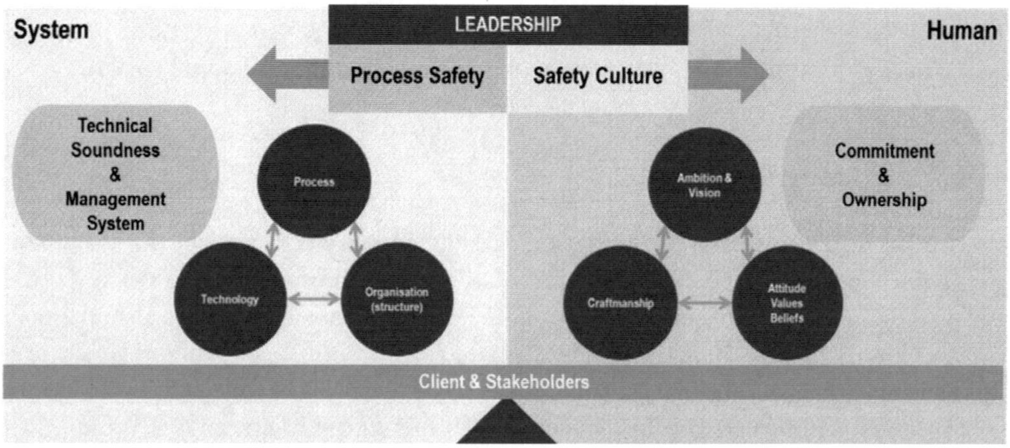

Figure 5. System and human related elements. Both are mutually dependent. If one fails, the other might not be able to compensate. Leadership must ensure the right balance.

The dimensions of balanced leadership that makes an organization successful (5):

- Creating a shared ambition level
  - Long term relationship with all stakeholder (employees, external, contractors)
  - Precondition is to create clear objectives which are shared by the internal and external stakeholders
- Establishing clear but not too narrow boundary conditions
  - Boundary management with a shared management system and technical conditions of the assets
  - Effective working performance management, continuous improvement based on risks in the DNA
- Clear expectation management and discipline in operational execution
  - Open and action driven culture – integrity
  - Wanting to close the loop in performance (action), communication, dialog, environment where people can become successful
- Craftsmanship – adequate mix of (technical) competencies
  - Competency complementary, taking responsibility, flexible, sharp

    - Exchange of knowledge to create a high level of craftsmanship by widely supported awareness that all technical conditions must be there, but if this is met the human factor the quality and level determines
- Sustainable value creation
    - Creating a positive footprint for the environment in reducing risk, financial benefits and development of the stakeholders
    - Knowing the internal and external stakeholders and their needs – the ambition of the company is directly related to the needs of the different stakeholders

The role for leadership is to understand that the road to success is to keep the balance between a system driven and human driven performance improvement to realize the results. Leadership adapts to facilitative leadership competencies, like results focus, managing the boundaries, motivational learning, challenging, etc., that enables them to realize the results by continuously facilitating the transformation to a high-performance organization with the adequate mix of competencies.

4.2 Facilitating continuous transformation

Stagnation is decline for organizations. An organization falls asleep having mindset that they are the best and having the perception that you have the most advance technical installation and excellent organization. Then competition is lurking to take over your position as the leader in the market. Or severe incidents happen because leadership and people in the organization are not anymore sharp to question current situation. There is no sense of urgency to innovate and to work as a team to continuously improve. Based on Kotlers (9) managing change we see the pitfalls for success of change as: no sense of urgency, not having a power coalition, lack of a vision, lost the empowerment of others to act on the vision, no plan and short-term wins, declaring victory too soon, and lack of embedding new way of working.

The key to success of a high-performance organization is to set on an experience driven transformation journey using competency development as the enabler. Establishing a platform for innovation with competency development by continuously creating new knowledge and transforming the organization. The good news for good leaders is that using the balanced leadership approach will facilitate the transformation journey to a high-performance organization. Leadership in transformation in the words of the American president Dwight Eisenhower: "the art to getting someone else to do something you want done because he wants to do it."

The base for transformation is the essence of life and in the words of Darwin: "It is not the strongest or the most intelligent that will survive but the ones most adaptive to change" (Darwin 1808-1882).

The first step in an experience driven transformation journey is that people are aware of their improvement potential in a way that they admit themselves that they can improve. Something like an "ah" moment and becoming aware of the blind spots. I understand now where we can improve in system, culture, technical status and competencies and believe in it that we can improve.

How to get to that moment? Together with the people in the organization visualizing fact based the improvement potential by looking at the situation from different perspectives like the technical soundness, management system and the behavior/culture point of view. Different tools, from brown paper workshops about the processes to case studies support the process in realizing the visualization of mutually understanding the gaps in the system and the individual usage of the system by those people involved indicate the process safety culture within a company.

To understand the culture, it is necessary to get an understanding of the attitudes, values and beliefs of the people in the organization from leadership to the shop floor. This information can be gathered by standard methods such as questionnaires but also with advanced tools like Katalyst, anonymous online workshops in which the approach unveils the attitudes, values and beliefs behind the façade people tend to show.

All the perspectives at the end give a realistic picture of the current safety culture by visualizing the blind spots and creating an environment in which it is safe to become conscious incompetent (the "AH" moment) as an organization and person.

The second step is in cooperation with leadership in the organization developing a transformation roadmap to realize the vision or in other words the picture of the future. It is a roadmap which is owned by the people and leadership in the organization because they have developed it themselves only facilitated by the consultants. Starting journey with the organization and realizing high performance is a stage by stage process. It is like climbing a mountain to go first to base camp and that camp by camp to reach the summit. Now, you are at the summit the decent is the tricky part of the journey and evenly important to get back to base camp – the goal/objective of the organization.

Getting back to base camp is the moment to take a step back and rethink your vision and objectives. Making the plan for the next stage of the journey by assessing the situation. This also accounts for developing the competencies stage by stage – needing continuously step back to understand the changes, i.e. new people.

How does this work for facilitating the development of the adequate mix of competencies that create a high-performance organization by setting up a transformation process safety competency management platform? The steps to take:

1. Development of the process safety competency platform
    a. Development of the competency framework – policy statement, competencies for each role in the organization, training plan, knowledge development
    b. Link into the HR management – maintaining competency level, training and development processes
2. Facilitating process safety competency improvement
    a. Competency assessment and reassessment
    b. Creating an environment people can become successful – Leadership at all levels
3. Managing performance in process safety competency
    a. Having a structure to visualize the results – KPI's
    b. Facilitating knowledge management – maintaining/storing, developing/ improving, sharing

4. Closing the loop by revisiting the objectives of the process safety platform
    a. Continuously check the objectives based on the changing ambition level of the organization
    b. Closing the loop in continuous improvement – balanced leadership

The success of the experienced driven transformation lies in the hands of balanced leadership at all levels creating an environment in which success is possible.

5 Conclusion – Optimal result for all stakeholders

A high-performance process safety culture is one in which people are actively aware of where they are related to best practices and where there is a continuous challenge of the current procedures and behaviors by trying to identify blind spots. It is a culture in which competency development is not seen as a cost but as core to make the experience driven journey to high-performance cultures a success. This committed culture has risk thinking at its core. Leadership at all levels of the organization should know and respect the risk in which the company operates and have the abilities to continuously question to find the blind spots.

In a high-performance process safety culture, there is a balance established by the leadership at all levels through shared ambition levels and clear boundary condition creating sustainable value for the internal and external stakeholders. The transformation approach for developing competency is a continuous experience driven journey of the individual and organization to create an adequate mix of competencies – craftsmanship in the company. The continuous development of the right mix of competencies (craftsmanship) combined with the core values makes a continuous improvement in process safety a success.

6 References.

[1] https://en.wikipedia.org/wiki/Mont_Blanc

[2] http://www.amanet.org/training/articles/the-essentials-of-high-performance-organizations.aspx

[3] J.R Katzenbach, D.K. Smith, "The Wisdom of teams", October 2015.

[4] Dr Andre de Waal, "High performance Organisation", May 2016

[5] C. van Driel, K. Hoeksema, G. Swarte (redactie), "Talentleidershap – Koersen op duurzame waardeontwikkeling, May 2017.

[6] U.S. Chemical Safety and Hazard Investigation Board. Investigation Report - Refinery Fire and Explosion and Fire. BP Texas City March 23, 2005

[7] P. Dehnbostel, "Competence – conceptual introduction from a pedagogic and scientific point of view", Booklet on the ECCE 8's special session on process and plant safety "Process Safety Competence – European Strength degrading to Weakness?", pp. 24-28, 2012,

[8] IChemE, "Process Safety Competency – a Model", IChemE Safety Centre Guidance, 2015

[9] John Kotler, "Managing change: The Power of Leadership", Balanced Scorecard – Insight, experience & ideas for strategy-focused oragnizations, Harvard Business School Publishing, February 2001.

Corporate Governance and Leadership Related to Major Hazard Scenarios is Leader Responsibility and Each Employee Commitment and Accountability

Américo Diniz Carvalho Neto
RSE CONSULTING
CEO&Principal Consultant

1. Introduction

Process safety management is widely credited for reductions in major accident risk and improved process industry performance. Process safety practices and formal safety management systems have been in place in some companies for many years. Over the past 25 years, government mandates for formal process safety management systems in Europe, the U.S., and elsewhere have prompted widespread implementation of a management systems approach to process safety management.
However, after an initial surge of activity, process safety management activities appear to have stagnated within many organizations. Incident investigations continue to identify inadequate management system performance as a key contributor to the incident. And audits reveal a history of repeat findings indicating chronic problems whose symptoms are fixed again and again without effectively addressing the technical and cultural root causes.
While all of these issues may not have occurred fully in one company, they have all happened to some degree in our and other companies. Left unchecked, such issues can do more than cause stagnation; they can leave organizations susceptible to losing their focus on process safety, resulting in a serious decline in process safety performance or a loss of emphasis on achieving process safety excellence. This is one of the reasons that our company created our system based on the Center for Chemical Process Safety (CCPS) next generation process safety management framework – Risk Based Process Safety (RBPS).
The purpose of RBPS Guidelines is to help organizations design and implement more effective process safety management systems. All These Guidelines provide methods and ideas helped us to identify that we had to focus very hard on major hazard scenarios, considering the severity and potential consequences to people, environmental and actives.

The RBPS approach recognizes and helped our company to understand that all hazards and risks in an operation or facility are not equal; consequently, apportioning resources in a manner that focuses effort on greater hazards and higher risks is appropriate. Using the same high-intensity practices to manage every hazard is an inefficient use of limited resources. A risk-based approach reduces the
potential for assigning an undue amount of resources to managing lower-risk activities, thereby freeing up resources for tasks that address higher-risk activities.

This paper shares our experience in reinforcing the PSM with Major Hazard scenarios, after almost 7 years implementing PSM system. We think that one of the important reasons were the occurrence of some potential major hazard incidents we decided to emphasize the role of corporate governance and leaders in good operation performance and High Hazard Scenarios Barriers identification and management.

While we have seen degrees of fluctuation among the top strategic objectives for improving EHS performance among the manufacturers surveyed, a few things are clear. Improving operational performance continues to be the top objective for bolstering EHS performance. Achieving sustained regulatory compliance is also a key driver, and that aligns well with many of the motivations to improve product compliance and, ultimately, product stewardship.

In Fig 01 below, EPSC show a survey with more than 50 companies. We used a similar methodology and these priorities where validated in one internal HES survey, done with more than 70 leaders of more than 25 sites. This reinforce that leaders continue to focus on operational performance and regulatory compliance

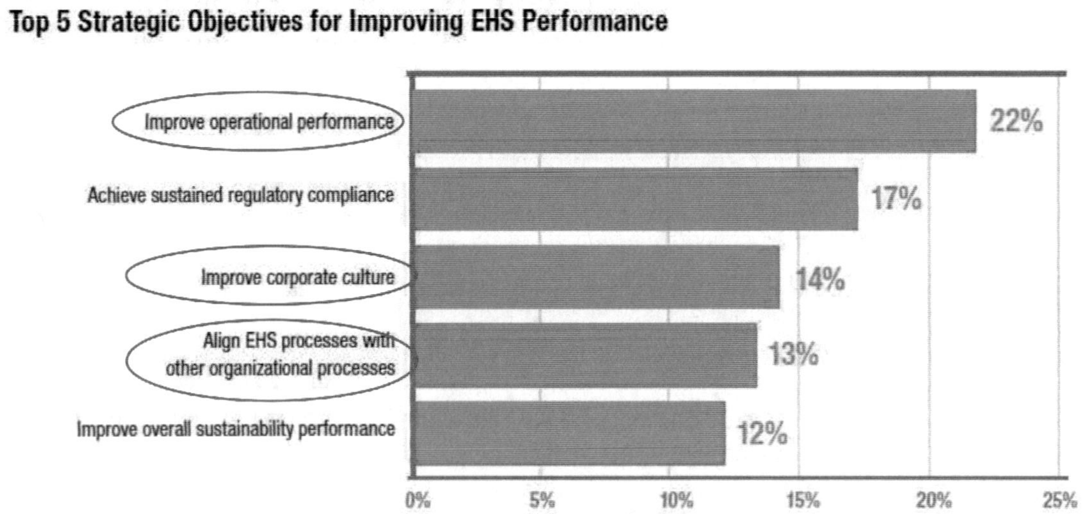

Fig 01 Industry Challenges to Improving EHS Performance (Customized with company information. Based on EPSC survey methodology)

With the results, there is value in integrating EHS and Operational Excellence into enterprise and operational models across departments and establishing value within those approaches. When we decided to invest in Major Hazards scenarios management we got a important benefit that was an integration between Operation, Maintenance, HES, Logistic and others. These were obtained by a clear and free demonstration of all preventive and mitigative barriers and these barriers are related to different areas/processes, and some of them suppose that PSM were not their direct responsibility.

So, analyzing what we mention before and the potential actions that leaders can contribute to process safety culture development we can list below in figure 2:

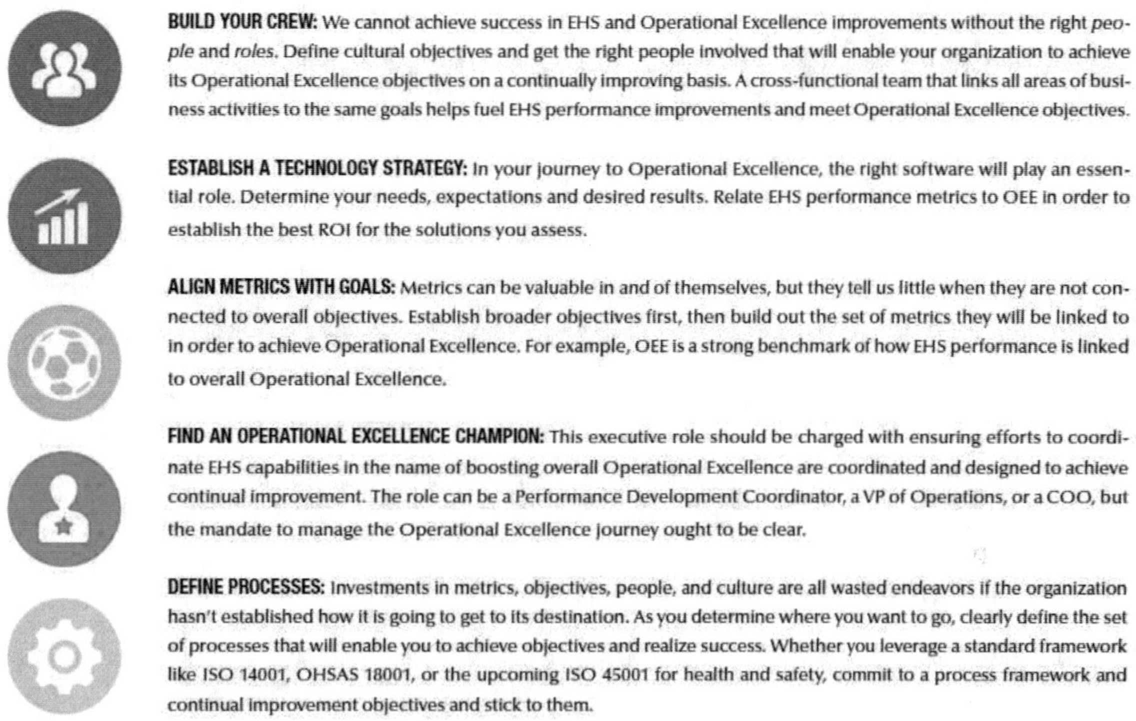

BUILD YOUR CREW: We cannot achieve success in EHS and Operational Excellence improvements without the right *people* and *roles*. Define cultural objectives and get the right people involved that will enable your organization to achieve its Operational Excellence objectives on a continually improving basis. A cross-functional team that links all areas of business activities to the same goals helps fuel EHS performance improvements and meet Operational Excellence objectives.

ESTABLISH A TECHNOLOGY STRATEGY: In your journey to Operational Excellence, the right software will play an essential role. Determine your needs, expectations and desired results. Relate EHS performance metrics to OEE in order to establish the best ROI for the solutions you assess.

ALIGN METRICS WITH GOALS: Metrics can be valuable in and of themselves, but they tell us little when they are not connected to overall objectives. Establish broader objectives first, then build out the set of metrics they will be linked to in order to achieve Operational Excellence. For example, OEE is a strong benchmark of how EHS performance is linked to overall Operational Excellence.

FIND AN OPERATIONAL EXCELLENCE CHAMPION: This executive role should be charged with ensuring efforts to coordinate EHS capabilities in the name of boosting overall Operational Excellence are coordinated and designed to achieve continual improvement. The role can be a Performance Development Coordinator, a VP of Operations, or a COO, but the mandate to manage the Operational Excellence journey ought to be clear.

DEFINE PROCESSES: Investments in metrics, objectives, people, and culture are all wasted endeavors if the organization hasn't established how it is going to get to its destination. As you determine where you want to go, clearly define the set of processes that will enable you to achieve objectives and realize success. Whether you leverage a standard framework like ISO 14001, OHSAS 18001, or the upcoming ISO 45001 for health and safety, commit to a process framework and continual improvement objectives and stick to them.

Figure 2: Leaders Action that support culture development (EPSC Leaders Survey)

2. Communication across the organization

Governance of Process Safety Who needs to know what?

Fig. 03

Analyzing the recommendations above related to information and what has to be manager by leaders and other functions, stimulate HES leaders to propose the HES management process be done by strategic, tactic and operational committees in order to have a systemic approach for these themes.

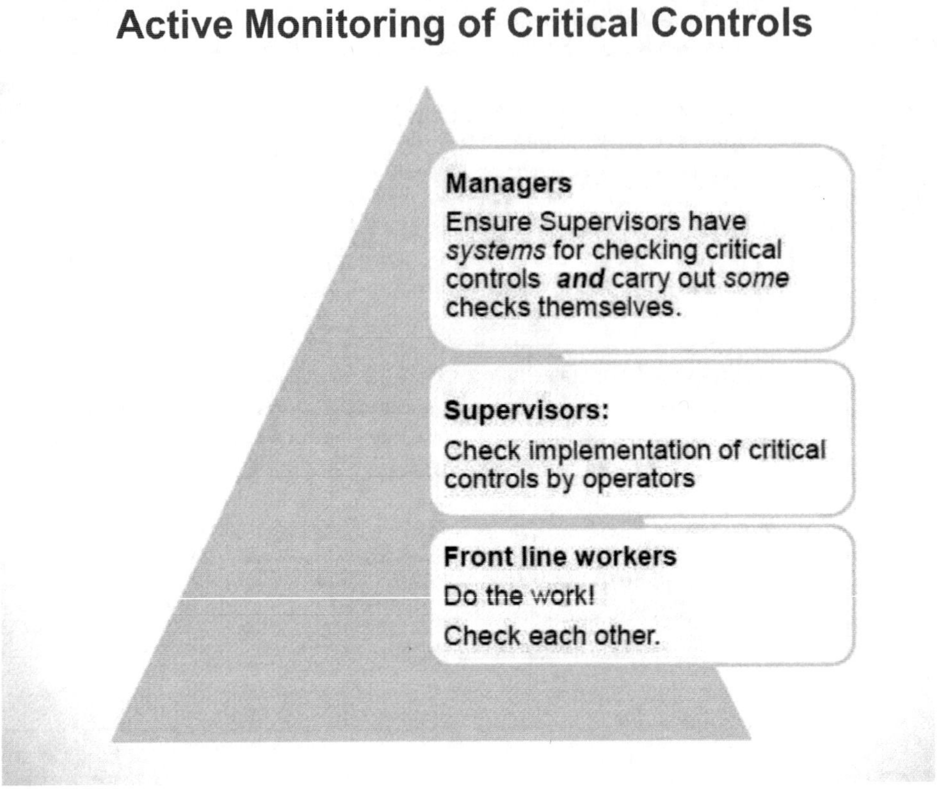

Fig. 04

Simple and objective controls that will assure better results for leaders, workers and people environmental. The objective in each is always evaluate

3. Real Case to develop Major Hazard Management and Culture in the company.

These 02 objectives listed below were identified by audit process in different levels in company facilities and were selected to be developed.
It was based on a company evaluation after a few years implementing PSM and some PSM audits realized annually.

3.1 Improving Personal Accountability

<u>Definition:</u>

- The acknowledgement and assumption of responsibility for ones actions and decisions;
- Being answerable for the results of ones' actions and decisions related to major hazards;
- Doing the right thing to assure high performance in major hazard risk management.

<u>Critical Components:</u>

- Beliefs and Values don't focus Major hazards (MH);
- Clarity of Purpose in MH management;
- Defined Roles & Responsibilities for different leaders;
- Empowerment & Ownership;
- Measurement of right process safety metrics;
- Communication across all levels. Leaders and operation level;
- Consistency of actions and decisions related to major hazard.

<u>Improving Personal Accountability:</u>

- Procedures and Programs Alone are not Sufficient;
- Employees must Possess the Skills to Safeguard Themselves and Fellow Employees;
- Each Employee must Continuously Challenge and Test Work and the Work Environment.

<u>Accountability Problems identified:</u>

- Inconsistent Actions and Messages;
- Value Placed on Safety Fluctuates;
- Limited Involvement "Only a few rowing the boat";
- Performance Evaluation is not Aligned to Desired Outcomes;
- Fear and/or Incentives;
- Excessive Reliance on HSE, Committees and Employee Task Forces;
- Limited or Non-existent Measurement related to major hazard scenarios;
- Roles and Responsibilities are not Clearly Defined;
- Limited Coaching and Mentoring;
- Poor culture and discipline to MOC process.

3.2 Acceptance of Personal Accountability

Definitions of Discipline:

We defined Instruction, to train or develop OD in the company. This subject that is taught Training that corrects or perfects Control through obedience or order. Prescribed conduct or pattern of behavior that we expect.
Company defined Gold Rules that govern activity Reward or administrative punishment related to fails in OD.

Types of Discipline:

Positive:
- Positive reinforcement, acknowledgement of a job well done, etc.

Negative:
- With holding of positive, lack of acknowledgement, etc.

Punitive/Punishment:
- Progressive discipline process, seldom learning specific, etc.

4.0 Project related to develop Major Hazard Management Culture and Management. Strategy and implementation phases

We conducted a project at a major chemical company to assess level of hazards present and the effectiveness of barriers.

The overall objective was to provide management quantitative information so that investment could be directed where it would produce the greatest benefits from a risk reduction standpoint.

The motivation for this project is the identification of conditions that indicates the need of a more in deep knowledge of the Major Hazard scenarios related to process safety from company units. Among this factor, some others as indicated below:

Many of 36 existed sites were in different levels of knowledge of its scenarios with major hazard, their risk levels and barriers reliability and integrity;

Accelerated non-organic growth, incorporating assets with different knowledge and culture related to process safety and knowledge of their risks and barriers integrity;

Large number of scenarios with level of consequences critical or catastrophic, with potential to cause important loss to facilities and reach externally of the units (residential population or other companies);

Occurrence of accidental scenarios over the last months, indicating the need of a better understanding of those scenarios and their barriers;

Need to increase visibility of major hazards to the high levels of leadership in the company corporate level an also sites leaders to show their importance for the decisions related to the sustainability of the business.

The Project was divided into 4 Phases. Follow the description of the 2 first ones.

a. The first phase was the selection of a sampling of Major Hazard Scenarios were identified for each site, logistics installations, terminals, and transportation facilities.
b. After the first phase, the second phase we did for each scenario, the preventive and mitigating barriers were identified and assessed for integrity resulting in a % conformance.

Identification of Existing Barriers in each major hazard scenario.

For each scenario defined on First Phase (The critical ones in each site), the preventive and mitigative barriers are identified based on the risk assessment studies available for each scenario. In the case of non-existence of previous studies, the barriers are identified based on the analysis of technical documentation, as P&IDs and cause and effect matrix.

All identified barriers are validated by the responsible teams from each site. Besides that, to facilitate the traceability of the barriers and identify the responsible for the management of each one, it is determined the area from the site responsible for each barrier.

Responsible Area	Barrier
Inspection	Inspection program of lines, equipment and accessories; Relief devices from equipment and pressurized systems
Instrumentation	Test program for instrumentation and safety equipment Safety alarms Safety Instrumented System Control System (Pressure, Temperature, Level) Remote valves Manual valves
Operation	Training programs for critical procedures Gas detection system Video cameras Emergency shutdown buttons
Civil	Soil protection Dikes Channels systems
Safety	Firefighting system Emergency plan
Maintenance	Maintenance program related to rotating equipment

Develop a Site Action Plan for the Recommendations of the Phase 2 Report, considering the informations:

- Owner: Site Leader
- Site Lead: TBD by Site Leader
- Required Site Resources: Safety Lead, Process Safety Engineer, DCS Engineers, Maintenance & Reliability Engineers, Process Engineers
- Required BU Support:
- Timing: Complete Action Plan
- Deliverable: Communicate Action Plan to Industrials Team

Phases of Major Hazard Culture and Management Development

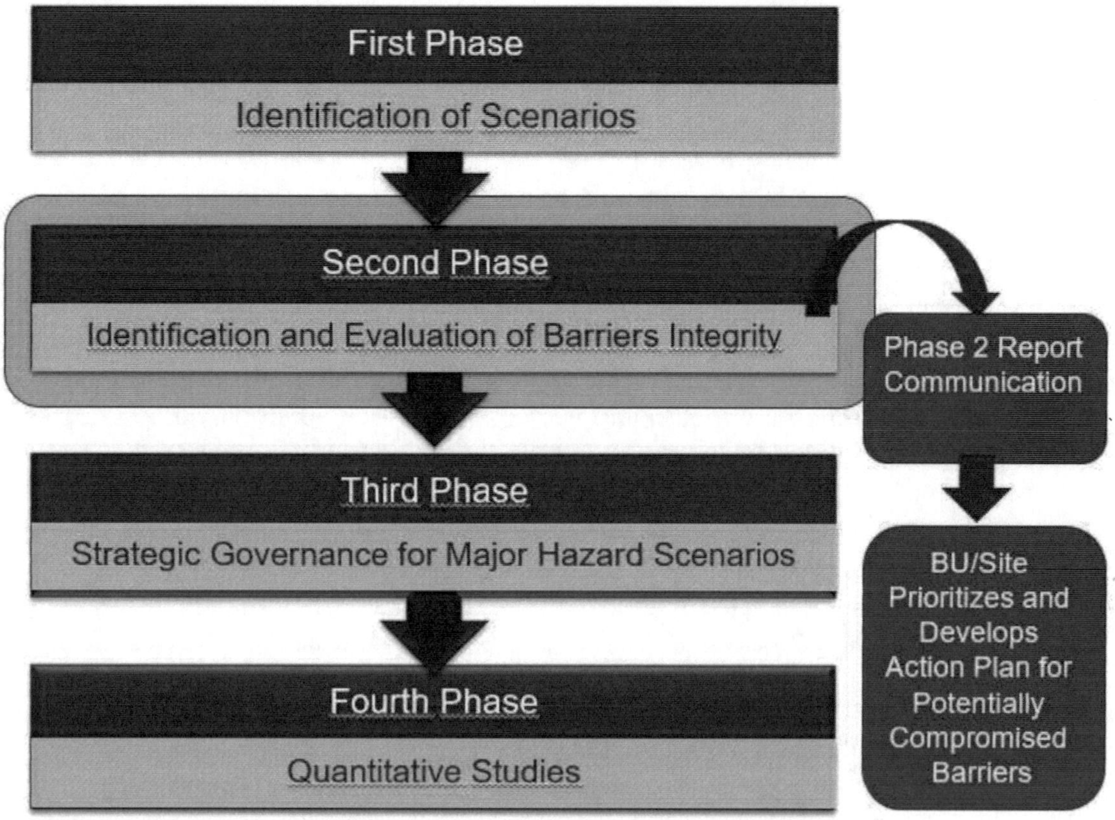

Fig 05 Project Phases (Developed in all 36 facilities)

In phase 1, were asked the following main information from each site:

- Risk Studies (qualitative and quantitative) in the units in the last 5 years
- If there are no studies in the last 5 years, previous studies
- Process flow diagram of the plant
- Inventory list of products
- Layout of the units
- Aerial photos with the region around the unit
- PMP (Probable Maximum Loss scenario)

Note: With all these information, we select the process scenarios that will be evaluated in the project.

In Second Phase- Identification and Evaluation of Barriers Integrity

- For each scenario defined previously, the preventive and mitigative barriers will be identified, through risk analysis information and validation with the teams of the sites involved;

- The Barriers Audit methodology applied will be used for monitoring the condition of each barrier, proposing recommendations to improve the barriers reliability and integrity condition, if necessary;
- It will set a frequency for monitoring of these barriers, for keeping the integrity of them. If necessary LOPA and SIL studies can be applied to better support these settings;
- In the monitoring phase, the recommendations generated in risk studies will also be evaluated, regarding the status of implementation;
- Construction of Bowtie diagram for major hazard scenarios selected;
- For the scenarios which have their barriers audited in less than one year, only the status of recommendations will be evaluated.

For each scenario selected, considering the barriers listed, an audit process is performed based on the methodology of Barriers Audit, to identify the status of each barriers, related to its management, searching for evidences of the process applied by the unit to manage the barrier and guarantee that it will be working accordingly to its specification, when needed. It is important to say that this methodology was applied in Brazil for the first time in a Petrochemical company, in December 2006 and was conducted by engineer Americo Diniz Carvalho Neto leadership.

The Barrier audit is a systematic and independent evaluation that verifies the potential integrity of the item and its compliance with the established requirements in the protocols, based on standards and technical directives.

The methodology of Barriers Audit has the main objective to evaluate the potential degradation of the barriers associated with the control the main accidental scenarios on the site. There are some other important objectives, listed below:

- ✓ The development of a systemic Process Safety Culture;
- ✓ An integrated vision of the management system and the factor that can reduce losses;
- ✓ A support to leadership to define and follow process safety indicators;
- ✓ The evaluation of the leadership and other workers commitment with process;

In the process of identification of the status of the barriers (audit), specialists from process engineer, operation, maintenance, instrumentation and safety participate. Each accidental risk scenario is reviewed, evaluating the preventive and mitigating barriers from the risk assessment. Also new barriers, that were not identified in the risk assessment existed in the unit, are identified and added to the analysis.

The next step is to evaluate the integrity and reliability management of each barrier, considering the protocol established by company team for each type of barrier (emergency plan, firefighting system, safety instrumented system, safety alarm, operational procedures and others).

The standard protocols used on this step of the audit were defined by company and technical team and was adopted in similar studies developed over the last few years.

The barriers audit is a fundamental step for the effective process safety management system (PSMS – Process Safety Management System), as it has the objective of evaluating if the PSM is working. A proper developed PSMS has policies, procedures and planned practices to guarantee that the process safety barriers are in place, being used and effectives, but not always the defined elements of the PSMS are being followed. The protocol applied during the audit will identify the gaps and propose corrective actions.

The auditor must verify how the management of the integrity of the barrier is conducted in the company, considering programs (for maintenance, test and inspection in place, state condition and procedures associated).

As indicated, to support the barriers audit process, standard protocols are applied. For each type of barrier, a different protocol was developed, corresponding to its function and questions/criteria used to collect evidences of needed to evaluate the status of each barrier. As example, some of the protocols are:

- ✓ Inspection program of pipes, equipment and accessories;
- ✓ Operating procedures
- ✓ Safety alarms
- ✓ Safety Instrumented System
- ✓ Relief systems
- ✓ Firefighting system
- ✓ Emergency plan

Each protocol has a series of questions related to the systematics of managing the barrier, where the audit people must answer using the evidences from the site documentation and system as well. As based on the interview that is conducted with the specialists. As each item is evaluated, there are three possible answers:

- ✓ Yes – This answer is only used when the requirement indicated in the question is completely followed;
- ✓ No – This answer is for when the requirements are not being partially or totally followed;
- ✓ Not applicable – This is for the cases where the question is not applicable for the item being analyzed.

To provide evidence that the item is being followed documentation can be used as: design book and assumptions list, records from inspection, maintenance and test plans, inspection reports, tests reports and training records.

In the cases where the barrier has at least one item evaluated on the protocol as not accomplished (answer No on the question), the barrier is considered "Potentially Compromised". If all items are answer with Yes with no gap is identified, the barrier is considered "Not Potentially Compromised".

For the barriers identified as "Potentially Compromised", recommendations are established to guarantee that the identified gap will be treated. The recommendation can be for an isolated item or a systemic recommendation that can be applied to similar barriers or to the whole site.

As example, the flowchart bellow indicated how the protocol for "Inspection Program" is applied. Note that after the evaluation of the barriers for each scenario, the ones that have 70% or more of its preventive barriers classified as "Potentially Compromised" or 90% or more of its mitigating barriers, are classified as Potential High Risks – PHR and will have its recommendations prioritized.

Below the inspection plan barrier process:

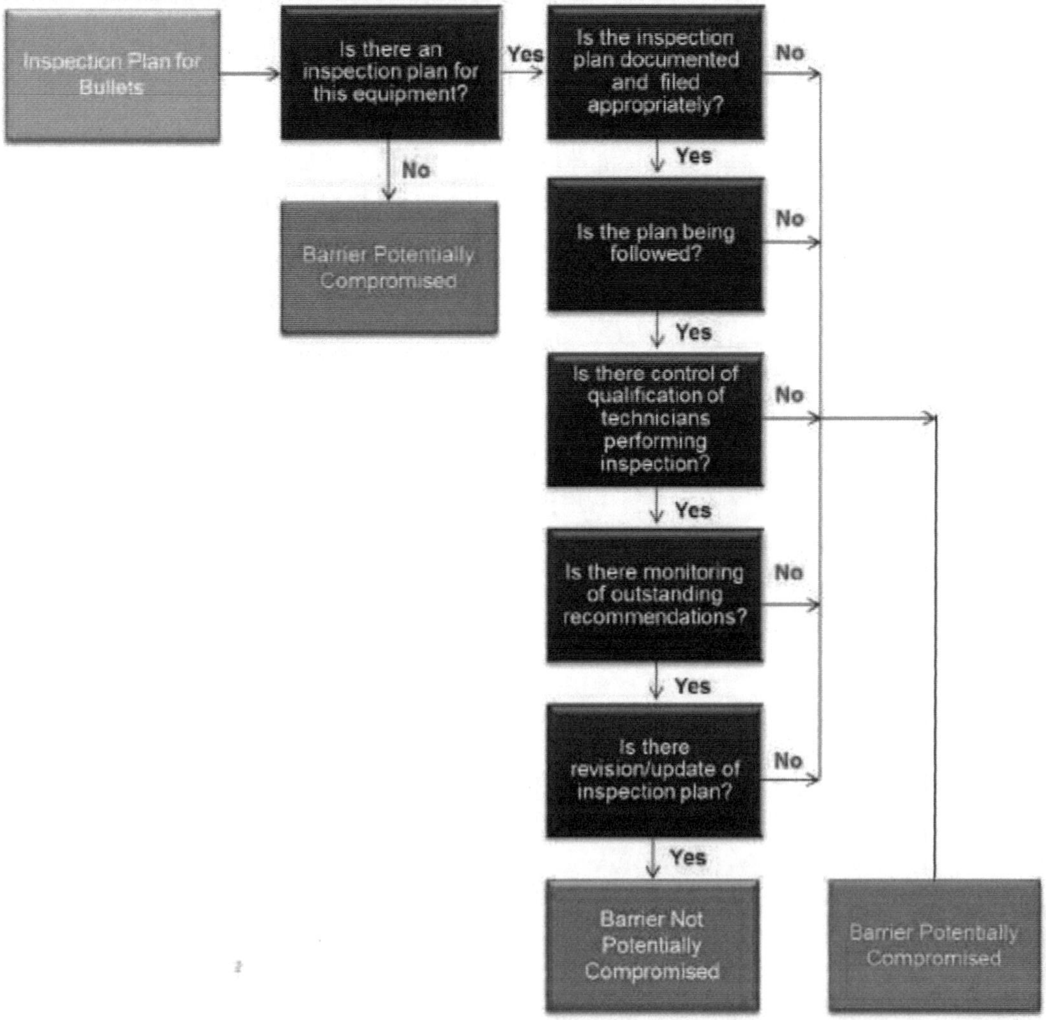

In Third Phase of the project, PSM structure and strategic communication and governance for Major Hazard Scenarios was evaluated and proposed a strategy to assure good performance in all company levels.

- Information and indicators were established to be used for monitoring the condition of the scenarios, considering the different processes of governance, defined in the element 02, the organizational synergy (plant, the industrial unit and the Executive Committee);
- The frequency of monitoring, form of governance and information flow will be set;
- Defined a new PSM structure to support Leaders and different committees;
- Consolidated the career of PSM Specialist and PSM Leader in the company;
- Consolidated the corporate standards to assure the same requirements for PSM in all 4 different countries.

In Fourth Phase of the Project – Quantitative risk analysis were done for selected scenarios, considering probability to achieve outside the site

- Prepare, or complement, quantitative studies of the units, so for the units in which there is no complete or recent study, in the selection of scenarios included in the previous steps will be validated / added the new scenarios identified.
    - For road transport: the calculation will be done for up to 30 scenarios, each one defined by product and location of leakage.
    - For pipelines: the calculation will be done for up to 10 different points of leakage.
    - For processing plants and terminals: accidental scenarios which have potential to achieve residential or industrial areas, outside fans, with effect of 1% fatality or lower flammable limit, will be included.

- Initially the LOPA study must be applied for all PHRs (Potential High Risk) identified in the plant, complementing the other scenarios, as requirements of Risk Matrix and Five-Year Plan.

After this project, the process safety and HES committees changes the way of managing and redesign leaders and employee's responsibilities related to process safety and preventive culture.

Below follow the results for a resin facility. This is one of 36 facilities.

In the first stage of this project in this site, the selection of scenarios of highest consequence was based in previous studies. The list of selected scenarios is totaling 10 scenarios.

In the second stage, which concerns the barriers audit, the scenarios selected were audited and as result 66 barriers were identified by audit team, 46 preventive barriers (70%) and 20 mitigating barriers (30% of total).

About the areas responsible for the management of the protection barriers, it was found that among 66 protection barriers identified, 19 of them belong to the Safety area, 6 to the Inspection, 32 to the Instrumentation and 9 to the Operation.

Also, as a result of the audit, 64% of the protection barriers were classified as potentially compromised and the Safety, Inspection and Instrumentation Barriers are responsible for most of them.

In the analysis, it is briefly mentioned the main points to the management of each area with respect to the barrier compromising, they are:

◆ Safety	◆ **19 barriers identified,** 53% potentially compromised.. ◆ For all scenarios the Emergency Action Plan is Potentially Compromised. ◆ The reason for the potential compromising of Emergency Action Plan is the necessity of review the emergency plan, including the most critical scenarios based on the process hazard analysis.
◆ **Inspection**	◆ **6 barriers identified,** 67% potendially compromised. ◆ The compromised barrier is Inspection Program for Pressure Vessels and it is associated with scenarios 1, 2, 3 and 5. ◆ The barrier potential compromising is due to the necessity of implementing the corporative procedure for critical equipment and pipelines.
◆ **Instrumentation**	◆ **32 barriers identified,** 81% potentially compromised. ◆ The compromised barriers are Blowdown valve, Basic Control System, Shutdown Valve and High/ Low alarm. ◆ As the inspection program is specific to the transmitters, the reason for the potential compromising of Basic Control System and High/ Low alarms is the necessity to test all instruments that are included in the loop control and also is necessary update the calibration plan to some instruments.
	◆ For Blowdown and Shutdown valves it is necessary establish a criteria to do Safety Integrity Level (SIL) studies.
◆ **Operation**	◆ **9 barriers identified,** 22% potentially compromised. ◆ The compromised barriers are Gas detectors and Operating Procedures, associated with scenarios 6 and 8, respectively. ◆ The potential compromising of Gas detectors is due to the absence of systematic control to register failures. ◆ It is necessary check if there is a procedure to Housekeeping program.
◆ **Civil**	◆ There is no barriers identified for Civil area
◆ **Maintenance**	◆ There is no barriers identified for Maintenance area

The audit recommendations proposed were 15 at all and the most representative is the Instrumentation area, with 73% of them. In this audit, the recommendations of risk analysis studies existing in the plant were also evaluated about their status on the date of auditing. (Completed, Ongoing, Delayed, Without Treatment/Not Implemented, Rejected. Most of the recommendations status was classified as Completed (57%).

As a strong aspect in the site, the good technical capacity of the professional involved in the audits was clear. Also, the engagement of leadership and good organizational climate found during the two days of Audit were another positive aspects. Is clearly observed in site the dedication of the teams and focus related to process safety regarding aspects, which was reinforced by American requirement of process safety legal compliance. The management of the safety barriers of the plant, from the studies of risk analysis developed in the site, is in advanced stage of implementation, with priority for medium risk scenarios.

It Is important that the plant follows the evolution of risk analysis studies in accordance with the five-years plan define by corporate procedure, also associated to element of PSM system. It is the same need for complementary risk studies, including Vulnerability (consequence studies), QRA, SIL and LOPA., also in accordance with corporate procedure criteria and methodologic approach.

For this study achieve its final objective, it is necessary that the proposed recommendations are implemented, providing a proper improvement in the management of barriers and consequently a higher rate of non-compromised barriers.

Finally, the audit result of barriers does not aim to demonstrate the real and specific integrity of each barrier, b type questions independent if it is physical or procedural, once the focus is to evaluate the management of the barrier along its life cycle, to assure its integrity within the safety standards expected.

4. Conclusion and Highlights

The strategy designed by this project and applied in more than 33 sites, promote more than technical and mechanical integrity benefits. Below follow some important ones that fully changed the PSM culture and leader's behavior related to Major Hazard scenarios.

- Leadership and all other people directly involved with clear focus on PSM;
- Emergency system well maintained with clear maintenance and operation routines;
- Identified clear routines and procedures to develop risk assessment studies, considering the use of qualitative, semi quantitative (LOPA) and quantitative studies;
- Implementation of maintenance and reliability manual, considering as an important step the criticality pre-classification for equipment's and safety barriers. This practice helps the units to prioritize activities and systems;
- Special attention to be given, regarding possible accidents effects outside company limits, to storage areas and railcars.

We developed this strategy in the company to develop the culture and the risk perception that a black swan can happen. Let's do a brief review in this.

In 16th century London, a statement that described impossibility was referred to as a black swan because all historical records of swans up to that time had reported that swans only had white feathers....

...the extreme impact of rare and unpredictable events, as applied to chemical process risk assessment. 

Our recommendation is: Focus on High Consequence Scenarios

- Understanding of what they are
- Understanding of the barriers/safeguards
- Verification of the barriers/safeguards

Because black swan events continue to occur, we must search more aggressively and continue to audit process safety management systems to ensure their adequacy. These catastrophic events often have management system failures as root causes. Strong process safety management systems may lessen the chance of black swan events even if we can't predict the exact scenarios.

Major process safety incidents always have warning signs and if we can identify and protect against the minor events, perhaps we can ward off the black swan. We should conduct process hazard analysis while also being more open to suggested hazard scenarios, especially multiple failure (double jeopardy) scenarios. If the scenario has not happened and the potential consequences are severe, perhaps safeguards should be implemented anyway unless the hazard can be reduced or avoided.

Referencies:

CCPS. Guidelines for Risk Based Process Safety. Center for Chemical Process Safety, American Institute of Chemical Engineers;

EPSC. European Process Safety Center. Guide Developing process safety indicators. A step-by-step guide for chemical and major hazard industries;

Control of Major Accident Hazards Regulations 1999 SI 1999/743 The Stationery Office 1999 ISBN 0 11 082192 0 (COMAH), as amended by the Control of Major Accident Hazards (Amendment) Regulations 2005 SI 2005/1088 The Stationery Office 2005 ISBN 0 11 072766 5, which implements the Seveso II Directive (96/82/EC) in Great Britain;

James Reason Managing the Risks of Organizational Accidents Ashgate Publishing Limited 1997 ISBN 1 84014 104 2;

Leadership for the major hazard industries Leaflet INDG277(rev1) HSE Books 2004 (single copy free or priced packs of 15 ISBN 0 7176 2905 8)

C331 Carvalho Neto, Américo Diniz.The occurrence of industry accidents at work and their correlation with the human factors: case study: Braskem - Basic Unit - Bahia /A hard tackle from Américo Diniz Carvalho Neto. - 2006.148p. yl.

Project Strategy developed and implemented in a major Petrochemical Company. Results of more than 200 Audits realized in more than 10 years. And the result of Major Hazard Project.

Case Study in Organizational Risk Profiling – Bridging the Gap Between Risk and Decision Making

Nathan Turner, Senior Manager
DuPont Sustainable Solutions
Wilmington, DE, USA
nathan.a.turner@dupont.com

Ehsan Akhavan, Principal
DuPont Sustainable Solutions
Dubai, UAE
ehsan.akhavan@dupont.com

Khaled Metwally, Senior Manager
DuPont Sustainable Solutions
Dubai, UAE
khaled.metwally@dupont.com

Mark Hause
DuPont
Wilmington, DE, USA
Mark.C.Hause-2@dupont.com

Prepared for Presentation at American Institute of Chemical Engineers
2018 Spring Meeting and 14th Global Congress on Process Safety
Orlando, Florida, April 22 – 25, 2018

Keywords: Process Safety Competency, Management Review and Continuous Improvement, Measurement and Metrics, Risk Profiling

Abstract

Many organizations today struggle with bridging the gap between the asset-level process safety risks known by the process safety professionals and operations leaders and the actions taken in relation to these risks by executives. Organizations have increasingly used tools such as risk registers and audit scores to shape management direction but too often these are not compelling enough on their own. In this paper, the author will share an experience working with multi-national high hazard manufacturing organizations in the creation of a corporate-wide risk profiling tool to enable more real-time and informed decision making by the corporate safety team and executives. The author will describe the selection process of inputs into the tool, such as kpi's and existing work process outputs, in addition to new inputs and the impact of local plant culture on the results. Finally, the author will describe how this tool is used by organizations to drive higher levels of risk-based decision making and asset prioritization.

1.0 Introduction - Managing Risk in High Hazard Industries

Industry is constantly changing. Commodity prices change daily, business models are changing at an accelerating rate, and companies are changing organizational structure, consolidating, or even changing ownership continuously. Another area of change in industry is around HSE. As we know, there are a number of famous incidents related to all facets of operating and maintaining high risk assets, with many valuable lessons to be gleaned from each. One key learning from incidents in the past decade has been the tendency of top leadership to focus on employee safety vs. process safety.[1]

DuPont Sustainable Solutions (DSS) recently conducted a global survey of high hazard industry executives to determine their perception of how risks are managed in their company. Responders included representatives from oil and gas, chemical/petrochemical, utilities, metals/mining, other manufacturing, building and construction, and transportation industries. The results were quite revealing. Two major learnings are:

1) Executives recognize that they are not devoting enough resources and capabilities to effectively manage risks within their companies, which threatens their continued right to operate.[2]

2) Executives acknowledge that there is a significant organizational disconnect and misalignment among leadership and employees with respect to risk management, which greatly contributes to the likelihood of a catastrophic event.[2]

Furthermore, executives noted in this survey other topics which they believe are critical to appropriately manage organizational level risks. The executives included their corresponding perceptions of their organization's strength on these topics.

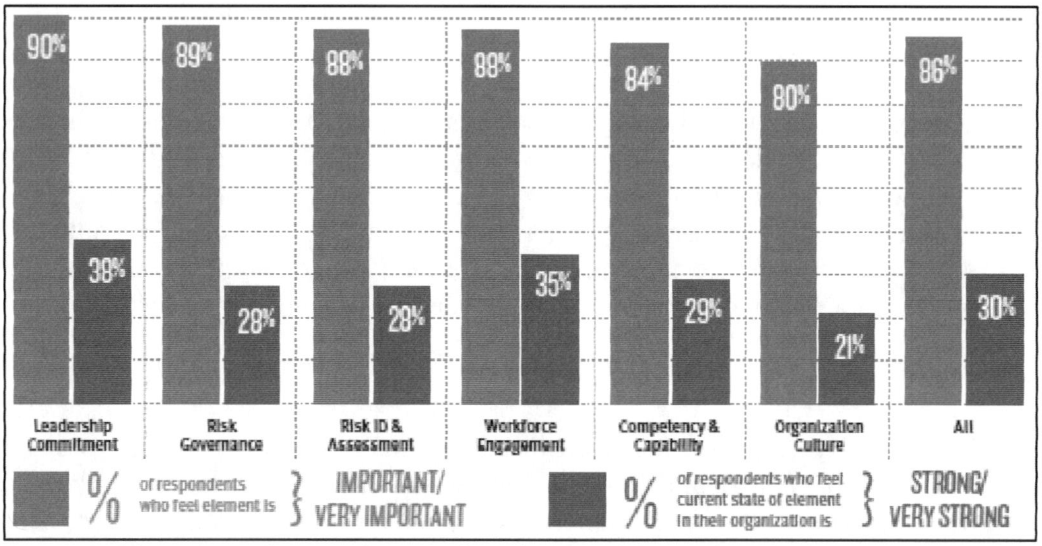

Figure 1 – A DSS survey of executives on key facets of Operational Risk Management (ORM)[1]
© 2018 DuPont

The results of this survey suggest that executives are generally aligned on what it takes for an organization to manage its risks in a robust and sustainable way. However, these same executives responded with a much lower degree of confidence when asked about their own organization's strength on this topic. Why would there be such large deltas on topics such as leadership commitment, risk governance, and risk identification and assessment?

Perhaps the disparity identified above can be partially explained by another statistic from the same report. Executives were asked to identify the topics most frequently and infrequently discussed in Board meetings. The results are shown below.

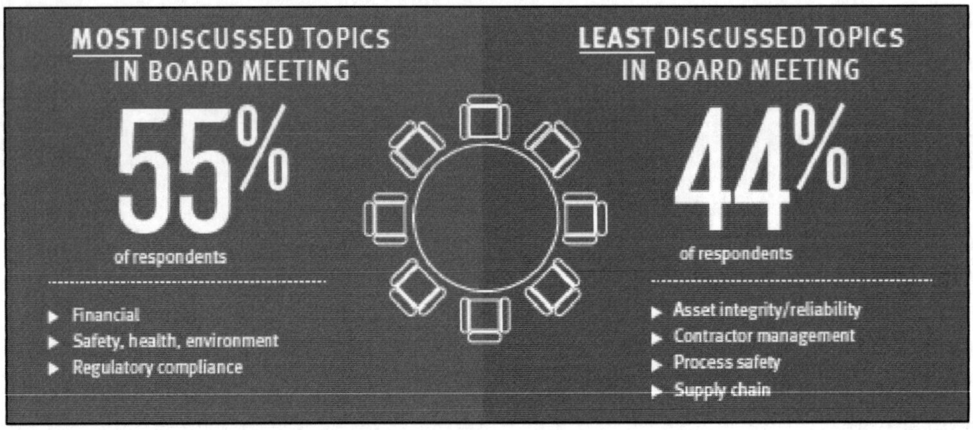

Figure 2 - DSS ORM Survey highlighting topics most and least discussed in board meetings.[1]
© 2018 DuPont

These results show that most Boards are discussing financial results, regulatory compliance, and HSE and Boards are discussing items such as asset integrity/reliability, contractor management, process safety, and supply chain less frequently. On the surface, this is not altogether surprising. There is an argument to be made that executives should not be functioning "in the weeds" of an organization; their role is to strategically guide the organization toward prosperity, results and, of course, company sustainability. However, when coupled together, these data sets suggest that there is a disconnect between what the Board discusses (results) and the elements which together contribute to these results, including technical systems such process safety and asset integrity, and softer components which make these systems function in practice, such as leadership commitment, risk governance, and organizational culture identified earlier. Even at the Board level, HSE tends to be performance and compliance focused with an emphasis on lagging metrics, as opposed to forward looking discussions about the organization's risk portfolio.

1.1 Leadership Response

As indicated in the study, executives know the basic building blocks necessary for managing risk and generally express concern about their organization's level of readiness. Executives and managers tend to respond to identified deficiencies with action. But the actions, even with the best of intentions, have not always resulted in improved performance or met the intent.

For instance, the authors of this paper have worked with executives who desired to demonstrate increased levels of ownership and interest in risk related topics, but inadvertently usurped control of the decisions and work processes from the company experts, effectively marginalizing them. Others have responded to poor risk management results with a "shoot the messenger" approach. This consequently leads to hiding of problems and non-transparent communications. There is, of course, a population that will remain apathetic and view risk management as strictly an HSE function or think that HSE and risk management improvements can be accomplished in a matter of weeks.

The most common approach observed by the authors is a rapid-firing of initiatives and blind resourcing to address the issues highlighted. Typically, there is a push to do more frequent risk management audits at site level, and a desire to speed up gap closure by supplying additional resources to address perceived problems. This is not to say that change and initiatives are inherently bad for organizations; in fact underperforming companies often need a shock or stimulus to change versus just incremental improvements. However, constant change initiatives can have a detrimental impact on organizational capabilities, knowledge absorption and morale, and often don't lead to the desired effects.

1.2 The Executive Advisor Role

Many of the risk management components of an organization are typically owned by HSE, process engineering, and maintenance management professionals reporting into one or two executives. Many readers might be wondering what their role is in this process. After all, shouldn't HSE personnel assist the executive team in identifying strategic direction and executing? Is this survey suggesting that there are widespread capability gaps in these functional roles across all surveyed clients? Conversely, are there capability gaps in managing risk amongst executives?

Far more likely explanation for the findings identified in the survey is related to alignment between the various roles in an organization. It is the executive's role to demonstrate value and provide resources for key risk management processes and to serve as a visible, felt leader for his/her organization. It is the advisor's role to assist the executives in determining when their involvement is needed and in what capacity, which key decisions are needed, and what else is needed to maintain this leadership position.

1.3 Typical Governance Practices

Numerous tools have been created and applied to better manage risk in organizations. A non-exhaustive list includes risk registers, process hazard analysis (pha) and bowtie reports, and enterprise risk management practices. These tools and practices are very powerful and are based on sound methods and logic. There is a soothing feeling associated with simply listing all the top hazards that we as leaders have to contend with. Once these are listed, it's simply a matter of managing our resources in a way that addresses these items, isn't it?

Unfortunately it's not this simple. These techniques help to identify and catalogue our sources of risk. If we as leaders want to understand how well we are managing risk, we still need to understand the controls that are in place and how well those controls are maintained. Traditionally we have constructed elaborate KPI dashboards to indicate when a site, or region, or business line is struggling. The top quartile companies have moved away from a sole focus on lagging metrics and shifted the focus toward leading metrics, even at the corporate levels. The best companies have avoided creating dashboards containing dozens of metrics that overwhelm lower levels of the organization but are still getting detailed enough to enable decision making.

KPI's are a vital part of any robust governance process. However, they cannot serve as a substitute for on-site auditing. Both activities are needed in the governance process. Auditing is invaluable to executives because it helps to cut through management layers and fill in the gaps about actual site practices. It helps top executives to feel confident that a site is complying with required practices. Audits can give insight into the local culture, and can ensure that actions and findings are closed. Of course, this assumes that the audits are conducted by highly skilled people with a consistent approach and a heavy focus toward implementation versus management system design and documentation. And of course, the results are inherently biased toward what is witnessed or examined during the exercise.

1.4 Putting it all Together

All of the techniques and tools described above are necessary and helpful in connecting executives with their risks. Each have their limitations. But it's important to remember that executives come from a variety of backgrounds and don't all have a rich operating or safety experience to draw upon. Even those with a deeper subject matter expertise can struggle as these are simply exhausting jobs which demand attention to be split in multiple directions and come with overwhelming pressure to perform. Fundamentally, when it comes to risk governance, many executives are struggling with a basic set of questions:

1) Where are my biggest risks?
2) Am I confident that we're addressing them adequately?
3) What else should we be doing?

As advisors to our executives, can we honestly say that we've clearly articulated the answers to these questions? On the surface it's easy for us to point to the tactics above and say that we have done everything possible. But let's discuss a method that starts bringing these elements together more holistically – risk profiling. Traditional tactics such as KPI dashboards and audit results are inputs into the overall risk profile, instead of a singular message to executives.

2.0 Risk Profiling – A Case Study

Many of the conditions described so far apply to a number of our clients. Our clients have complex and growing portfolios, operate in areas with increasing regulatory complexity, are facing significant business pressures for production growth and operational efficiency and deal with high levels of organizational change and capability challenges.

The risk profiles of these companies have changed over time. Their approach to risk management continued to grow as well, employing all the traditional risk management tactics described above. However, while these traditional methods were useful to members of the organization, they were not helping **top management, the decision makers,** to understand their risk profiles and address the top risks.

Over time, a number of these companies, particularly multi-site and multi-regional clients, developed gaps between their risk management approach and their risk profiles. Typical symptoms included:

- Inability of comprehensive HSE programs to form consolidated view of risks. Singular concerns were expressed to top management and getting lost in the ocean of topics to be addressed.
- Corporate decisions were based on indicator performance and not consistently risk-based.
- Growing gaps between top management decisions and site realities.
- Risk-based thinking used in multiple ways, but often fragmented.
- Lack of alignment on initiatives and priorities among sites.
- Overwhelming audit programs creating redundancies and wasted efforts. Focus shifted from addressing risks to addressing findings and actions before next audit.
- The top audit resources were stretched thin and others were recruited to audit with less attention to training and capability development, making for highly variable audit results
- Corporate support to sites often top down and compliance driven, challenging the relationship between corporate functions and sites.

The symptoms described above are a result of traditional governance processes. As one client described it – "If we had a site that performed well on an audit, we would plan to re-audit in 2-3 years. If a site performed poorly or experienced a few incidents, we would plan to re-audit in 2-3 years." We realized that we had been thinking about risk in a linear fashion and without prioritization. Rather than simply thinking of risk in terms of identification, quality of control and resulting likelihood of occurrence, we needed to start looking at aggregate and individual risks relatively, which means more focus on higher risks and risky assets and less attention to others – risk profiling.

Risk Profiling is a is a means to collect and assess relative risk level based on a set of criteria across competency areas and sites (units and assets) to enable differential risk treatment decisions. This informs and drive priorities, EHS actions, capability development needs, clarity on risks and investments at the site level, and audit plans, investments, leadership action, corporate EHS programs and asset prioritization at the corporate level. It is not intended as a replacement of a company risk matrix (likelihood vs. consequence), risk register, or an enterprise risk management (ERM) program. The techniques described in this paper can be extrapolated to include business risks, similar to an ERM program. However, risk profiling per our method is focused on health, safety, and environmental risks. This method places risks in a real-time prioritization tool for management and integrates inherent hazards and dynamic factors (such as frequency of high hazard operations activities and personnel changeover) to help management focus efforts and attention on highest risk sites and singular risks. For example, if a site experiences high employee turnover in critical roles or increases in high voltage electrical lock outs, the risk profile of that site will increase and visibly show to management the need for action or at least increased oversight.

We worked through a five step process to develop a tailored risk profiling tool. It's important to keep in mind that each company has different work processes and contributors to its risk profile, which means that there is not a one size fits all result.

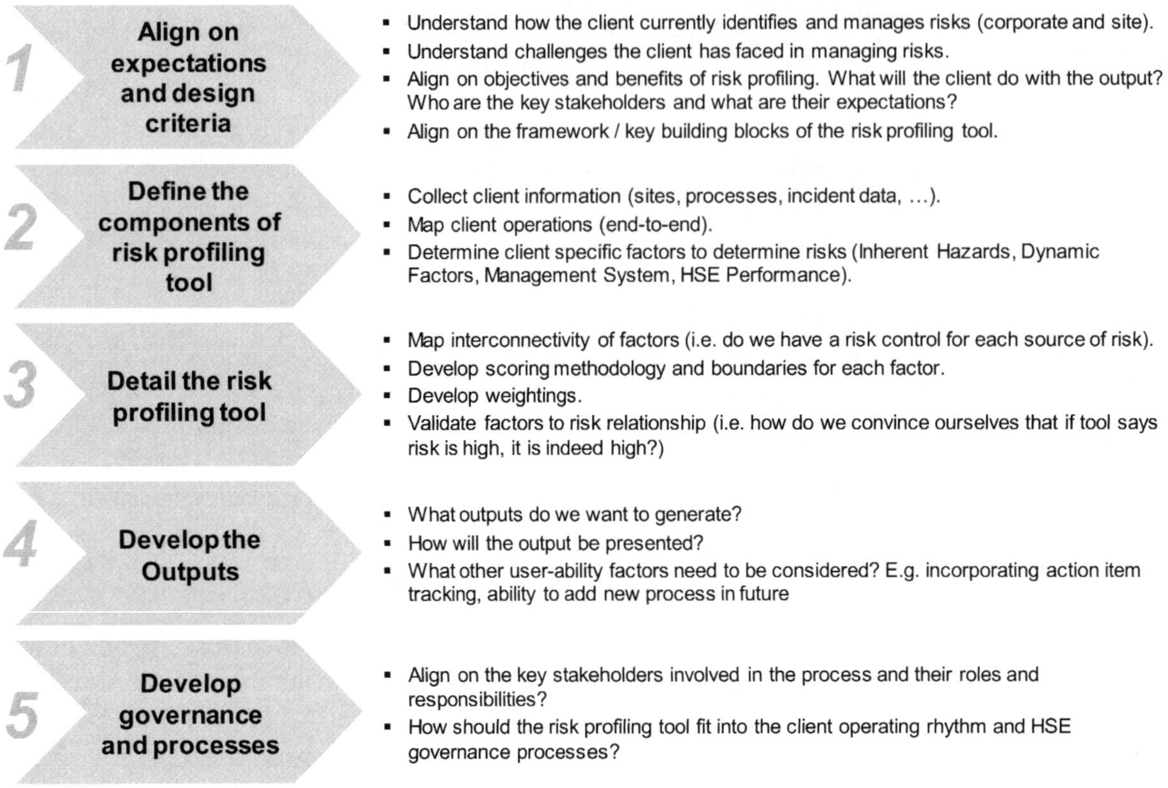

Figure 3
© 2018 DuPont

Within step 1, the team had to define and select the important building blocks for each of these dimensions. These dimensions are tailored to their operations and represent high risk areas and perceived challenges by the organization. The key building blocks consisted of the following:

- Inherent Hazards – hazards or complexity that are integral to the process or system that create potential for injury, illness, incident, or non-compliance. Includes, chemical, physical, regulatory complexity, etc.

- Dynamic Risk (Internal and External) – Internal changes to the activity that create the potential for additional hazards or complexity or likelihood of injury, illness, incident, or non-compliance. Examples - MOC – Physical, Chemical, Technical, Regulatory, Personnel, Process, etc. External changes to the process or system that create potential for additional complexity or negative impact from the process or system. This could include proximity to community, natural receptors, regulatory enforcement, natural resource shortages, transportation risk, community relations, etc.

- Management Systems – Leading metrics and controls that measure the strength of the system or process in managing the risk – Certifications, time to close audit findings, overdues, survey results, near misses, repeat audit findings, capability of personnel vs. requirements, training, etc.

- Performance – Lagging metrics that measure the performance of the system or process in managing the risk – Injuries, illnesses, deviations, incidents, audit findings, non-conformances, etc.

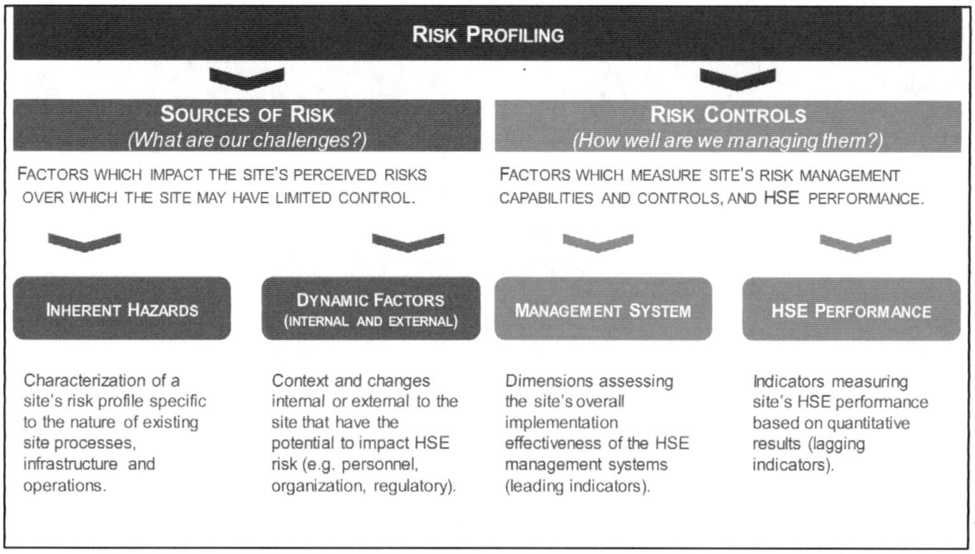

Figure 4 – A Risk Profile Approach.
© 2018 DuPont

The process to define the building blocks and factors involved a careful review of all existing governance items and selection of the most important for inclusion in the process. From there, operational definitions and boundaries were established in order to promote consistency of reporting and application across the asset base. These boundaries were based on industry benchmarks and existing data from the organization. Finally, the team aligned on factor weighting and challenged the results to ensure that the things that mattered most rose to the top of the list.

With the parameters defined, a data collection phase begins. This involves establishing an iterative process where the assets provide annually a selected set of metrics along with other gathered data for loading into the risk profiling tool. From that point, the resulting profiles are examined and regular action planning begins. This process is repeated on a regular basis and over time, as risk sources and risk controls (including action plans) change, the prioritization of the assets changes correspondingly.

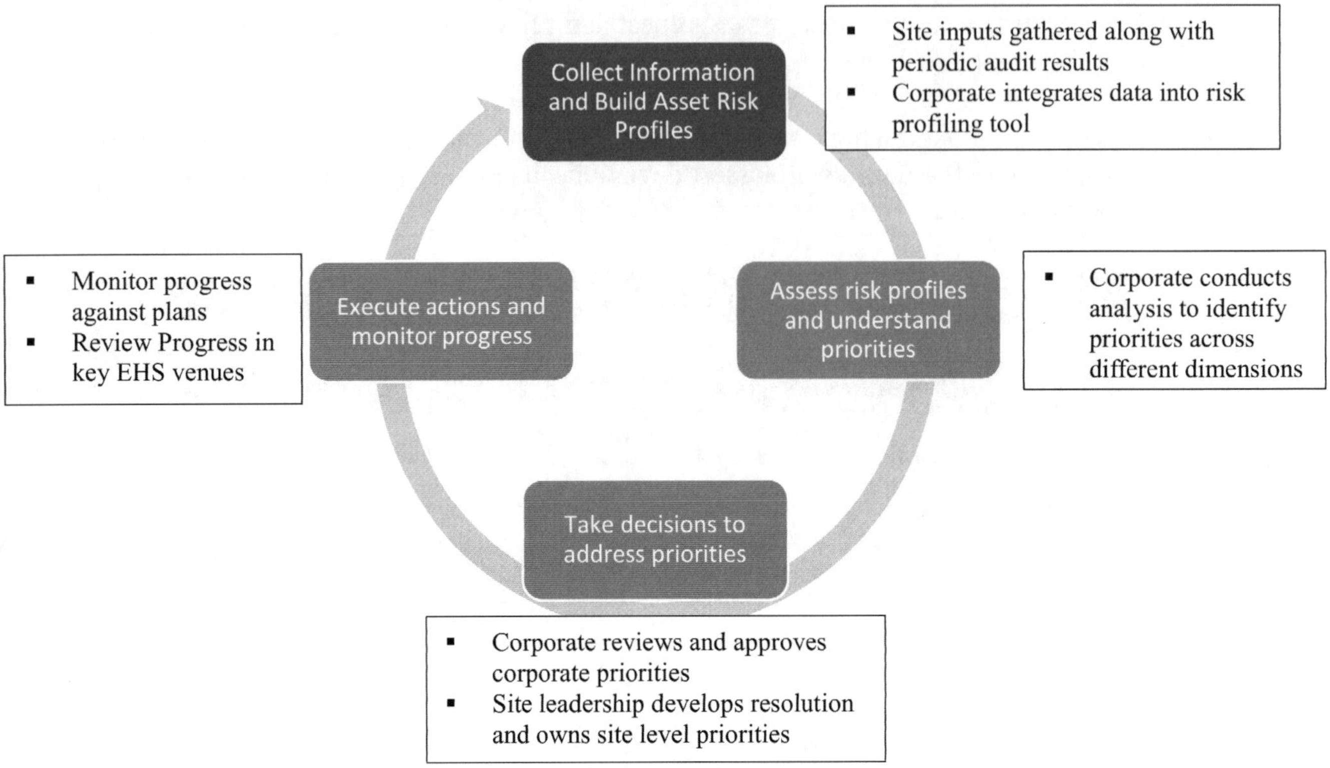

Figure 5 - The Profiling Process
© 2018 DuPont

2.1 Results and Conclusions

It is early to determine the long term impact of this profiling method in terms of client risk governance impact. However, there are some preliminary learnings and results to share:

- Clients are reporting better visibility of key risks and priorities at corporate level, enabling cost-effective allocation of (limited) resources.

- As part of one client data collection process, we discovered that similar assets experienced a high variability of high risk activities conducted for the time period in question. Upon review, we observed that sites had different understanding levels of some of the corporate requirements and metrics related to these activities. Identifying this issue and addressing it was already a big learning for the organization.

- HSE Corporate functions are using the data to evolve from auditors and policy makers to internal consulting and support teams for the assets. Over time, this will reduce tension between the functions and introduce a collaborative problem-solving climate.

- It is anticipated that over time the targeted risk treatments will result in day-to-day site risk reduction and focused initiatives, with the added benefit of easing the burden on sites.

- The corporation has learned in more detail how assets compare in terms of challenges in managing hazards and complexity.

- One company has decreased its audits by 75% and experienced greater consistency of results with focused auditors executing the balance.

- As Figure 6 shows for the corporation, each dot represents a different asset with a composite risk profile (including occupational safety, environmental risks, health, process safety, etc). In one simple view, leadership can decide efficiently how to prioritize its resources against the assets in need. This treatment may include focused audits or corporate support or investment, or the results may suggest that some assets need no corporate initiative at all. Similarly, within each of these dots each asset splits the data into individual risks, which can be used at the site level for prioritization.

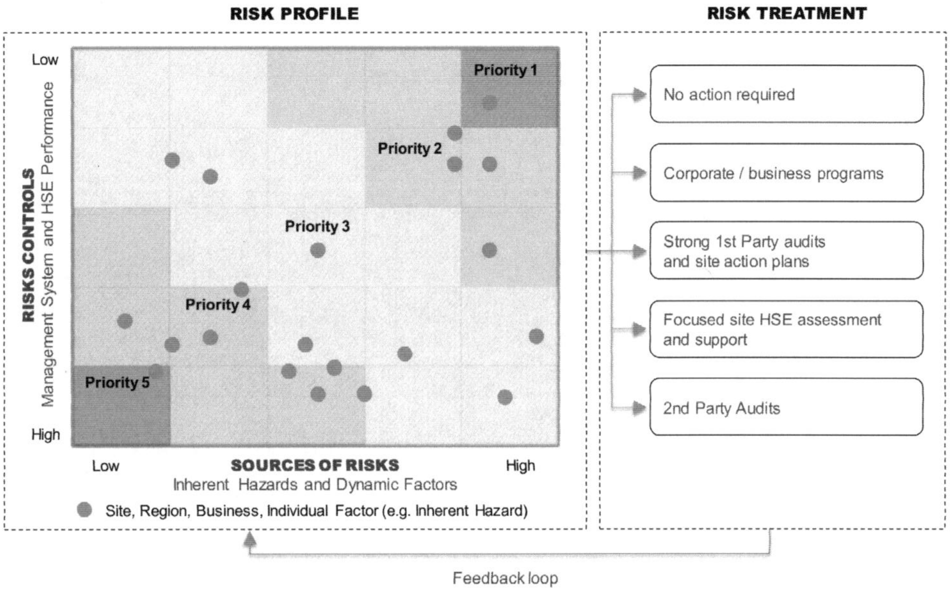

Figure 6 – Processing the Profile Results at Corporate Level
(Note: this is for illustrative purposes only and is non-exhaustive, site level prioritization is not shown)
© 2018 DuPont

As described above, for many companies there is a disconnect between what we as HSE advisors are telling our leaders to be concerned about and the fundamental questions they are trying to address. Traditional governance processes, while helpful, tend to focus on lagging HSE results and often exclude topics such as process safety and asset integrity. There is a need to raise the profile of key topics to executive levels but it must be done in a way that does not overwhelm the organization – either in terms of data collection for the assets or key messages presented to the decision makers. In this paper, the authors have introduced a risk profiling concept which is a method for meeting these needs, in addition to helping organizations to focus their initiatives and resources toward the assets which need it most. Furthermore, it helps sites to prioritize and address their risks systematically, provided that there is ongoing strong buy in and support from each level of management. This method will inevitably require adjustment and maintenance as organizations continue to evolve, but the initial results are promising in advancing risk governance processes and effectively bridging the gap between executives, HSE professionals, and asset leadership.

3. References

[1] **Baker, J, Leveson, N, Bowman, F, Priest, S, Erwin, G, Rosenthal, I, Gorton, S, Tebo, P, Hendershot, D, Wiegmann, D, and Wilson, L, The Report of the BP U.S. Refineries Independent Safety Review Panel, Available at** http://www.csb.gov/assets/1/19/Baker_panel_report1.pdf, **Accessed on March 15, 2018.**

[2] **Bahr, N and Ariawan, A, DSS Study – Lack of Internal Alignment and Commitment of Resources to Manage Risk Threaten Corporate Business Performance, Available at** http://consultingservices.dupont.com/ORM?cid=Social-ON-Lnk&utm_source=Social-ON-LnK, **Accessed on March 1, 2018.**

509732 Digitalization in the Process Industries – How and Where to Implement

Abstract:

Digitalization... It's THE buzz word in the industry. But what is it? And how can you go about implementing digitalization in your plant and achieve the benefits that it promises? The goals of the presentation are to provide attendees with an overview of digitalization, including the expected benefits, and explain some practical examples of how it can be implemented. Specific focus will be on the vast array of field devices in your plant that include an increasing amount of intelligence and methods you can use to capitalize on the information they provide from the following perspectives... engineering, operations, and maintenance. Also discussed will be ways to use digital fieldbus technology to provide effective communications between smart field devices and the process automation system.

Specific questions to be answered include...

- *What is digitalization and what are the benefits?*
- *What information is available and how can it be effectively transmitted?*
- *How can I use this information to more productively operate and maintain my plant?*
- *How can this technology be used to more efficiently engineer my automation system?*

514096 Operational Analytics: Transforming Operational Data into Actionable Insights

Abstract:

The advent of digital revolution has now enabled Chemical Industry with numerous tools that could be leveraged to transform our operational data into actionable insights. Key opportunities with digitization include better visualization, integrated planning and execution for value-chain optimization, that results in smarter production, intelligent response to changes in raw materials, process and equipment conditions, reduce energy and waste along with prevention of asset breakdown, safety and environmental issues.

The goal is to link processing plant performance to operational insights, integrating traditionally siloed data and KPIâ€™s for a more holistic approach to optimizing Supply Chain Plant Operations. Integrating operational targets with actual operational insights. The operational events become the new transactions in a process plant.

This paper discusses strategic data hierarchy methodology, findings, and challenges in the ongoing journey of implementing an analytics platform that evolves from being retrospectively descriptive to anticipatively prescriptive.

511688 Empirical Models for Analyzing BIG Data. What's the Difference?

Abstract:

Many issues around "BIG Data" (eg. collection, warehousing, integration, cloud services) are just infrastructure issues that need to be improved in order to ultimately use the data to extract actionable information. If the data analysis step is not performed well, then most of the infrastructure effort will have been wasted. To analyze historical data, one needs to make use of models, usually empirical – such as regression, data mining (deep learning neural networks, decision trees, etc.) or latent variable models. My PhD supervisor in Statistics (G.E.P. Box) used to often say **"All models are wrong, but some are useful"**. The problem is that most people lump all empirical models into one category – "empirical models" - as though any of these models are interchangeable irrespective of the nature of the data or the objectives of the problem.

But, whether a model is useful depends upon three factors:

1. The objectives of the model
2. The nature of the data used for the modeling
3. The regression method used to build the model

From an objective point of view, there are basically two major classes of models – those to be used for **passive** use and those to be used for **active** use. Models for **passive** use are intended to be used just to passively observe the process in the future. Such passive applications include classification, inferentials or soft sensors, and process monitoring (MSPC). For such **passive** uses one does not need or even want causal models, rather one wants to just model the normal variations common to the operating process. Historical data is ideal for building such models. Models for **active** use are intended to be used to actively alter the process. Such active applications include using the models to optimize or control the process or to trouble-shoot process problems or gain causal information from the data. For such active use one needs **causal** models. Causality implies that for any active changes in the adjustable or manipulatable variables in the process, the model will reliably predict the changes in the output of interest.

The problem is that most of the data available in industry is historical operating data and such data almost never contain causal information on individual variables. This poses a major problem if we want causal models. The question then becomes what analysis methods are useful for obtaining **active** models from historical operating data?

Machine Learning (ML) methods are currently the rage in "BIG Data" communities. These include deep learning neural networks, and massive decision trees. These new ML approaches are improvements on the older "shallow" neural networks (a few layers connecting all variables to all nodes) and single large decision trees, both of which led to overfitting of the data and to large variances in the predictions. The newer ML approaches are aimed at overcoming some of these deficiencies. Deep learning NN's use many simplified layers, regularization and averaging to reduce the effective number of parameters and the overfitting. New decision trees involve building many decision trees based on fewer randomly selected variable and then averaging or voting on the results to effectively reduce the variance and bias of the results.

These newer ML methods can be very good for passive uses. But they cannot be used for extracting interpretable or causal models from historical data for active use. With historical data, there are an infinite number of models that can arise from any of these machine learning methods, all of which might provide good predictions of the outputs, but none of which is unique or causal. This does not allow for meaningful interpretations, even more so if the results come from averaging or voting on many (eg often 1000 or more) models.

Nevertheless, these ML models have proven to be very powerful in **passive** applications – eg. deep learning NN's for image analysis, and Random Forests for medical diagnosis, both of which are passive applications (where there is no interest in altering the image or the patient).

Latent Variable models such as PLS (Partial Least Squares or Projection to Latent Variables) were developed specifically to handle "BIG Data" where the real number of things affecting the process is much smaller than the number of measured variables. Typically, the number of latent variables needed to extract useable information from hundreds of process variables is more in the order of 3 to 10, implying that the true number of degrees of freedom that affect the process is often quite small.

These latent variable models such as PLS or PCR (Principal Component Regression) are a total break from the classic statistical regression of machine learning models in that they assume that the input or regressor (X) space and the output (Y) space are not of full statistical rank and so they provide models for both the X and Y spaces rather than just the Y space. Without simultaneous models for both the X and Y spaces, there can be no model uniqueness nor any model interpretabilty or causality when using historical data. LV models do provide causality, but only in the reduced dimension space of the latent variables as this is the only space within which the historical process has varied. By moving the latent variables one can reliably predict the outputs (Y) – implying causality for the latent variables. However, to move the LV's one cannot just adjust individual x variables, rather combinations of the X variables that define the LV's as defined by the X space model.

ProSensus has used this uniqueness and causality provided by LV models for many years to trouble-shoot and understand processes and to optimize both processes and products based on actionable information extracted from "BIG" historical data.

513189 Holistic Analysis of Emissions from a Small, Modular Fast Pyrolysis System for Conversion of Biomass and Mixed Waste

Abstract:

Total emissions for a small-scale (45 kg/hr, nominally 1 ton/day), transportable pyrolysis system, including on-site utilization of the fuels and byproducts, were determined from measured burner emissions using pyrolysis bio-oil, char, and gas fuels derived from mixed solid waste (MSW) and biomass. Fast pyrolysis provides an alternative to open burn pits that have been used at remote bases, and emit harmful gaseous emissions and particulate matter. By comparison, fast pyrolysis can convert MSW into useful heat and power while producing sterile ash and relatively clean combustion products. Combustion of pyrolysis byproducts (i.e., char and gas) was required to generate process heat for drying the feedstock to the required moisture content (<10%) and to achieve the desired volumetric waste reduction target (>95%). Utilization of byproducts was especially important for these small-scale systems where heat losses and parasitic power draws became more challenging as the surface-to-volume ratio of the unit operations increased. Emissions were measured from a process gas burner, bio-oil burner, and char burner that compose the fuel utilization section of a pilot-scale pyrolytic solid waste remediation system (*py*â €'SWRS), and the emissions were compared to the relevant standard set by the U.S. Environmental Protection Agency (EPA). The gas burner was evaluated using propane (LHV of 47 MJ/kg) and a surrogate process gas (LHV of 8 MJ/kg) with a molar composition of 16% CH_4, 39% CO_2, and 45% CO. Several 80% (m/m) ethanol and 20% (m/m) bio-oil fuel blends were prepared to measure the CO_2, CO, NO_x, THC, and PM emissions of a heated, air-atomizing bio-oil burner. Fast pyrolysis char produced from pine feedstock was combusted with a pilot-scale fluidized bed reactor to demonstrate the feasibility of process heat generation. In the char burner, the fluidizing gas flow rate, gas inlet temperature, excess air, and reactor temperature were varied to determine the effects on emissions and combustion efficiency. Based on the emissions of each burner, the total emissions for CO, NO_x, SO_2, and total PM were determined for the pilot-scale *py*â€'SWRS and compared to the EPA standard for other solid waste incinerators (OSWI), which regulates nine pollutants. Of the four common pollutants, CO, NO_x, SO_2 met the EPA OSWI standard, while total PM (32.6 mg/dscm) was slightly over the 30 mg/dscm limit. The bio-oil burner contributed the most to the total PM emissions, based on the tested burner conditions. We postulate that the PM could be reduced below the limit by tailoring the fuel-air equivalence ratio of the bio-oil burner. The other five pollutants included in the OSWI standard (dioxins, cadmium, lead, mercury, and HCl), which depend primarily on the feedstock composition, were also considered. The total emissions data were fed into a decision support model for waste remediation at remote locations, which considered environmental compliance as one of the performance metrics.

512939 Modular System for Converting Lignocellulose into Pyrolytic Sugars and Phenolic Oil

Abstract:

Easy Energy Systems and Iowa State University are collaborating on the design, construction, and demonstration of a 50 ton per day pyrolysis system based on modular manufacturing principles. The goal is to commercialize a system suitable for processing distributed resources of lignocellulosic biomass into pyrolytic sugars and phenolic oil. Success is premised on intensification of several processes and designing major unit operations and subsystems as modules that fit into standard ISO shipping containers. The modules are field assembled into a complete biomass processing plant at scales much smaller than once imagined for commercial production of cellulosic biofuels and biobased chemicals.

Processes to be intensified include: provision of energy to the pyrolysis process through autothermal operation; production of sugars through thermal deconstruction of plant polysaccharides; and anaerobic digestion of the aqueous phase produced during pyrolysis. In contrast to conventional pyrolysis in which oxygen is excluded from the reactor, autothermal pyrolysis adds air at equivalence ratio less than 10% (considerably lower than autothermal gasification, for example) to provide the enthalpy of pyrolysis without the need for external heat transfer equipment, greatly simplifying system design. Autothermal operation removes the heat transfer bottleneck in fast pyrolysis, increasing biomass throughput by several fold compared to conventional pyrolysis in a similarly sized reactor. Pyrolysis sugars are levoglucosan (a dehydrated form of glucose) from cellulose and principally xylose from hemicellulose. Although sugar yields from pyrolysis of lignocellulosic biomass is normally very low, a simple pretreatment prior to pyrolysis enhances sugar production several fold. The sugar is recovered as a syrup contaminated with phenolic monomers. The syrup is cleaned with a resin adsorption filter, which allows its use as fermentation substrate. The phenolic oil, derived from lignin in the biomass feedstock, has several applications. It can be cured to a solid known as Lignocol, which has applications as a renewable fuel substitute for coal in electric power generation. It can also be emulsified with diesel fuel or dissolved in ethanol to produce a low-viscosity fuel oil suitable for small-scale power generation. With suitable hydroprocessing the phenolic oil can also be converted into renewable gasoline and diesel fuel. The aqueous phase is anaerobically digested to bio-methane after removal of fermentation inhibitors.

This presentation will describe pilot-scale tests and design activities to support construction of the 50 ton per day demonstration plant in 2018.

513197 Anaerobic Treatment of High Strength Industrial Wastes By Moving Bed Bioreactors (MBBR)

Abstract:

Anaerobic Treatment of High Strength Industrial Wastes by Moving Bed Bioreactors (MBBR)

Somnath Basu
Headworks International
11000 Brittmoore Park Drive, Houston, TX 77014
sbasu@headworksintl.com

ABSTRACT
Effluents from many industrial processes contain high concentrations of organic wastes (COD). These include petrochemical, food & beverage, and breweries & distilleries among others. Treatment of high COD wastes pose major challenges to the operating cost and environment traditional aerobic activated sludge process due to their:
a) high energy cost of aeration as every kilogram of COD requires one kilogram of oxygen; and
b) high cost of transportation and disposal of residual solids resulting from biochemical reactions.
These two activities constitute the largest operating cost of a wastewater treatment plant.
Anaerobic processes are net energy positive and sustainable for high strength waste treatment compared to their aerobic counterpart as:
a) they do not consume any energy for aeration as these processes take place in absence of air;
b) very low generation of residual biosolids requiring much lower hauling and disposal costs.
On the other hand, anaerobic processes produce biogas that can be utilized as a source of energy for heating, steam and power generation on-site. From energy viewpoint, the production far outweighs the consumption to maintain the bioreactor well mixed by agitators and at 35 to 38 C. Thus, anaerobic processes are sustainable as biogas is a source of 'Green Energy'.
Despite these advantages, anaerobic treatment faces a significant operating challenge in solid-liquid-gas separation can be challenging because:
a) trapped gas impedes solids settling
b) waste sludge does not flocculate well
This difficulty leads to occasional washout of biomass from the reactor. Restart and stabilization of reactor can be time consuming and can affect the smooth operation of production processes.
An anaerobic MBBR reactor overcomes this constraint as the biomass in the reactor is attached to media surfaces while the media themselves are freely suspended and uniformly distributed within the reactor. Other advantages of this type of reactor configurations, as listed below.
• Robustness against toxic or shock loads;
• Resiliency to peak flows and loads; and
• Easy recovery from washout situations as attached growth provides an inventory of biomass
This paper will present the fundamentals of anaerobic MBBR as a sustainable process for high strength waste treatment with the results of a case study of a pilot scale brewery wastewater treatment.

The Roundabout Way to Disaster:
Recognizing and Responding to Normalization of Deviance

Jennifer F. Mize
Engineering Associate
Process Safety Design, Worldwide Engineering and Construction
Eastman Chemical Company
P.O. Box 511, B-54D
Kingsport, TN 37662
jfmize@eastman.com

Prepared for Presentation at
American Institute of Chemical Engineers
2018 Spring Meeting and 14th Global Congress on Process Safety
Orlando, Florida
April 22 – 25, 2018

The Roundabout Way to Disaster: *Recognizing and Responding to Normalization of Deviance*

Jennifer F. Mize
Engineering Associate
Process Safety Design, Worldwide Engineering and Construction
Eastman Chemical Company
P.O. Box 511, B-54D
Kingsport, TN 37662
jfmize@eastman.com

Keywords: Normalization of deviation (deviance); process safety; near misses; conduct of operations; operational discipline; safety culture

Abstract

Humans develop processes to structure their activities to set and reach goals. Wherever a process exists, there will be always be deviations from that process. Skipped steps, workarounds, shortcuts, improvements, and innovations can be incorporated over time, gradually resulting in new practices that are accepted as the norm. Small changes that seem insignificant can easily be overlooked and therefore not managed. To the extent that individuals, teams, and organizations tolerate non-conformance, normalization of deviation can be established. The cumulative effect of these deviations may be unmitigated hazards that can lead to a catastrophic process safety event.

A CCPS project subcommittee (Project 269) was established to develop a text to address the initial and long-term effects of normalization of deviations. The finished book, *Recognizing and Responding to Normalization of Deviance*, will be published in early 2018. This paper will provide examples of the depth and breadth of normalization of deviation throughout processes, procedures, and organizations, and will illustrate techniques to address its potential impact.

1 Introduction

During an average day, humans utilize numerous processes and procedures to complete tasks at home, at school, and in the workplace. Many of these are so engrained into the everyday routine that they are initiated and completed without conscious thought. Who hasn't arrived in the parking lot at work, gathered their things, and walked into the office without a clear recollection of the drive to work? These processes and procedures provide structure to activities and allow people to set and reach goals and to complete complicated tasks. Where a process exists, there will be deviations from that process. These deviations may be intentional and deliberate, or they may be small changes that seem insignificant, allowing them to be easily overlooked. Skipped steps, workarounds, shortcuts, improvements, and innovations can be incorporated into procedures and processes over time, gradually resulting in new practices that are accepted as the norm. As individuals, teams, and organizations tolerate non-conformance, normalization of deviation can be established. The cumulative effect of these deviations may be unmitigated hazards that can lead to a catastrophic process safety event.

2 CCPS Project 269 Subcommittee

A CCPS project subcommittee (Project 269) was established to develop a text to address the initial and long-term effects of normalization of deviations. The finished book, *Recognizing and Responding to Normalization of Deviance*, is anticipated to be published in early 2018. This paper will provide examples of the depth and breadth of normalization of deviation throughout processes, procedures, and organizations, and will illustrate techniques to address its potential impact.

2.1 Subcommittee Members

- Dan Sliva - CCPS Staff Consultant
- Jennifer Mize, Chair – Eastman Chemical Company
- Steve Arendt – ABS Group
- Michelle Brown –FMC Corporation
- James Caudill – Marathon Petroleum Company LP
- Kelly-Ann Charles – Methanex Corporation
- Joey Cranston – Albemarle Corporation
- Scott Haney – Marathon Petroleum Company LP
- David Hill – Occidental Chemical Corporation (OxyChem)
- Greg Horton – SABIC
- Patti Jones, Praxair, Incorporated
- Jai Karia – Chevron Corporation
- Neil Maxson – Covestro
- Mikelle Moore – Buckman International, Incorporated
- Louisa Nara – CCPS
- Mark Paradies - System Improvements Incorporated
- Michael Pelupessy – Akzo Nobel
- Sara Saxena - BP
- Joan Schork – Baker Engineering and Risk Consultants, Inc.
- Adrian Sepeda – AIChE Emeritus
- Tony Strawhun – Afton Chemical Corporation
- Ken Tague – Archer Daniels Midland Co.
- Karen Tancredi – Chevron
- Daniel Wilczynski – Marathon Petroleum Company LP
- Elliot Wolf – Syngenta

AntiEntropics, Inc. of New Market, Maryland, was selected the contractor for this project, with Robert J. Walter as the principal author and project manager, and Sandra A. Baker as co-author and editor. After Bob Walter passed away in 2016, Albert Ness and Bruce Vaughn with AIChE stepped into assist with completion of the book.

2.2 Peer Reviewers

- Salvador Avila, CCPS
- Pedro A. Bonilla, CCPS
- Jeffrey S. Caudill, Marathon Petroleum Company
- Erin P. Collins, Jensen Hughes
- Scott A. Haney, Marathon Petroleum Company LP
- John W. Herber, CCPS
- David E. Herrick, Universidad de los Andes (Colombia)
- Casey Johnson, Covestro LLC
- John Kusowski, Engineering and Technical Associates, Inc.
- Peter N. Lodal, Eastman Chemical Company
- Michell L. LaFond, Dow Corning
- William Mosier, Syngenta
- Juliana Schmitz, Praxair
- David Thaman, PPG

Many thanks to these peer reviewers for their insights, comments, and suggestions, which helped to ensure a balanced perspective for the book.

3 Defining "Normalization of Deviation"

Deviation is "the action of departing from an established course or accepted standard."[1] As defined in the CCPS Process Safety Glossary, normalization of deviance is "a gradual erosion of standards of performance as a result of increased tolerance of nonconformance."[2] The subcommittee identified three characteristics to identify normalization of deviation:

- It is a human-based deviation.
- The deviation occurs repeatedly, over time.
- The deviation does not cause an immediate incident or noticeable process effect.

This does not necessarily mean the exact same deviation occurring over and over, but the same *types* of deviations occurring repeatedly. Normalization of deviation certainly occurs when everyone on the team adopts the same change in the way things are done. In this situation, the deviation may be readily identified by observing the operation or task in comparison with the written procedure, such as during Behavior Based Safety observations, since every practitioner will demonstrate the same deviation. However, normalization of deviation also occurs when individuals and teams fail to challenge undocumented change such as skipped steps, workarounds, shortcuts, improvements, and innovations. This is much more difficult to identify due to its subtlety; each deviation is unique to the individual committing it. The person committing the deviation may not even be aware that they have deviated from the expected procedure, especially for routine tasks that are conducted without the use of a formal checklist. As deviations persist over time, a new norm is established.

4 Examples of Normalization of Deviation

4.1 Misuse of a Safety Instrumented System

A tank containing flammable solvent is batch-filled. The operating procedure specifies that the tank is to be filled to 90% level. There is no history of overfilling incidents in 10 years of operation. During the most recent Process Hazard Analysis (PHA) revalidation, the PHA team evaluates the overfill scenario and determines that there are insufficient layers of protection to meet the company's risk tolerance. The team recommends installation of a Safety Instrumented System (SIS) to close a block valve in the feed line and stop flow into the tank on high level. The SIS is installed utilizing the existing level transmitter as the initiator and the existing block valve as the final element with a trip point of 95%. The change is properly managed using the company's MOC process. No other changes are made. Two years after the SIS is installed, the tank overfills resulting in a large release of flammable solvent to the diked area. The release is a Loss of Primary Containment (LOPC) event. [3]

Figure 1

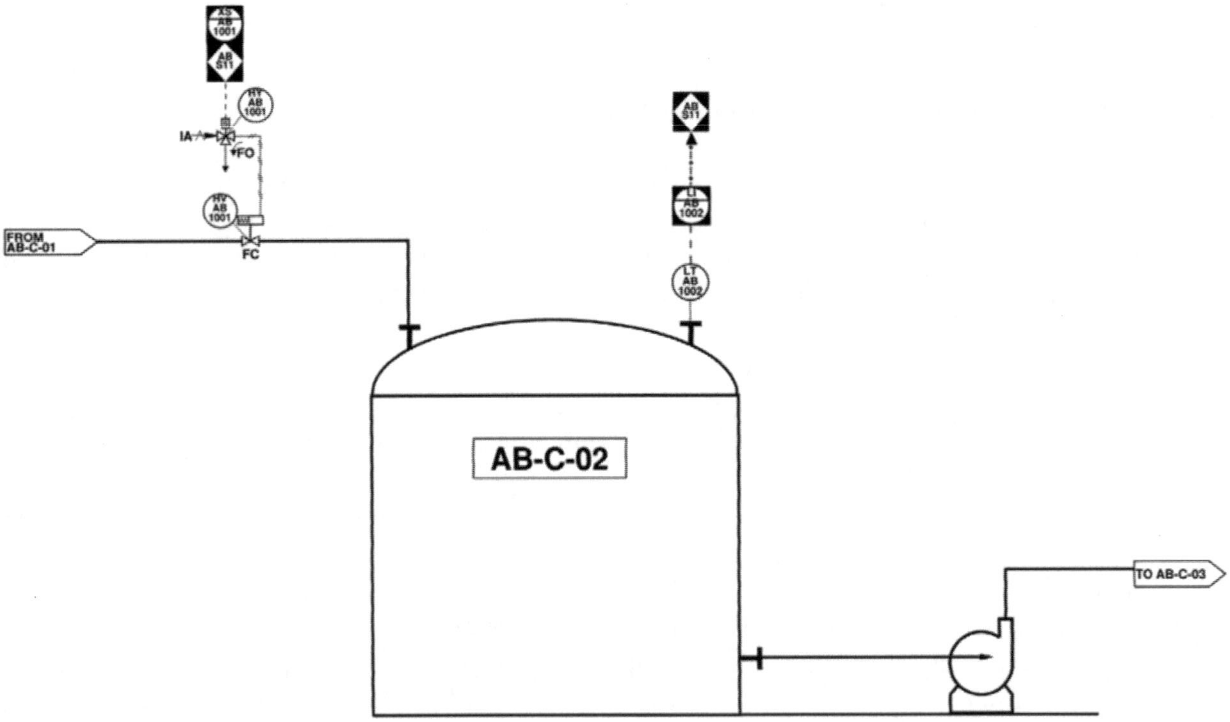

4.1.1 Results of the Incident Investigation

The incident investigation revealed that the single level transmitter on the tank had malfunctioned and was not providing correct data to the operators. The SIS did not trip when the tank reached and exceeded 95% level due to the failure of the level transmitter. The investigation also revealed that normalization of deviation was also a significant factor in the release event. Over time, the operators began using the SIS as the normal means of level control, allowing the tank to fill to its trip point of 95% level and then letting the SIS trip to shut off the feed. In addition to putting the SIS in high demand mode with this practice, which was not reflected in the reliability calculations, the operators stopped routine monitoring of the tank level during filling. They had confidence that the interlock would shut off the feed. If they had been routinely monitoring the filling operations, it is possible that the failure of the level transmitter could have been detected and addressed before an incident occurred.

Allowing the tank to fill to 95% level resulted in fewer fills per shift. The operators were able to attend to other tasks with the time they saved by not monitoring the tank level and using the SIS to stop feeds. The operators became overconfident in the reliability of the interlock, the installation of which was viewed as a significant safety improvement, and lost their sense of vulnerability with this routine task. The operators did not understand the intent of the SIS to function as an Independent Protection Layer (IPL) to be utilized only in the event that the other safeguards (which existed prior to the installation of the SIS) failed to prevent the tank from overfilling.

The incident investigation report also noted that improvements should be made to the installation of the SIS. Installation of a dedicated level transmitter and block valve for use in the SIS, while allowing operators to use the existing block valve and level transmitter in their filling operations, will greatly reduce the likelihood that the SIS will be used as normal level control. Use of external comparison between the data on the two level transmitters on the tank can reveal some types of failures in the transmitters.

4.2 "Доверяй, но проверяй"—An Old Russian Proverb ("Trust but Verify")

A multi-purpose batch production facility has a number of identical reaction units, consisting of a reactor, condenser, and receiver. The facility produces many different specialty chemicals, with campaigns scheduled based on production forecasting. This flexibility in production capability is ideal for sales and marketing, but presents challenges to operational discipline. Operators have to be trained on multiple operating procedures, understand a variety of chemistry concerns and chemical hazards, and routinely manage transient operations including start-up, shutdown, and equipment cleaning, each with their own equipment and instrumentation configurations.

Two operators are running a batch of Specialty Chemical ABC in Unit 1. Following the batch recipe, they prepare to charge a solid reactant to the mixture already in the reactor. The solid reactant is packaged in bags. Earlier that morning, an operator assigned to the chemical warehouse delivered several bags to the staging area near the reactor as ordered by the operators in the control room through the inventory system. Several bags of solid materials were already sitting in the staging area when the delivery was made.

The operators charge four bags of the solid reactant to Unit 1 and then start the agitator. Within 15 minutes of completion of the charge, the control room operator reports that the temperature and pressure in the reactor are rapidly increasing. The relief system on the reactor activates and relieves to the effluent handling system. However, this doesn't prevent the overpressure scenario and the vessel ruptures, releasing the contents into the operating area resulting in a large fire and damage to nearby equipment with additional release of ignitable chemicals.

4.2.1 Results of the Incident Investigation

The incident investigation revealed that one of the four bags of solid reactant charged to Unit 1 was an incompatible material. Once the incompatible material was incorporated into the batch mixture, the contents began to react uncontrollably. The relief system was not designed for this runaway reaction and was not able to prevent the vessel from exceeding its Maximum Allowable Working Pressure (MAWP) and building pressure to the failure point.

The operating procedure and PHA for the process listed several administrative controls to address hazards associated with charging of materials to the batch. The primary administrative control for charging is a checklist requiring signature by two operators at each step. This checklist requires operators to positively verify each bag by checking the label prior to charging. Additionally, the chemical inventory system was intended to serve as a strong administrative control. Operators were instructed to order only the number of bags or drums needed for a single batch and to remove unused or partially used bags or drums from the operating area after charging tasks are completed. Managing the inventory in the operating area in this manner mitigates the potential for over or under charging the batch with the intended chemicals or for charging the wrong chemicals to the reactor.

The operators in this unit had been working together for several years. They were under increasing pressure to improve productivity, which was proving to be a difficult mandate in the highly manual production facility. Over time, they deviated from the procedure, ordering enough bags or drums for several batches and using the staging area to manage that inventory on the operating floor. As production targets changed and the number of batches completed varied from the planned production, they often had excess bags or drums sitting in the area. The experienced operators managed this by carefully checking the labels before making charges. In many cases, they could recognize the bags by their color and design without needing to actually check the label.

On the day of the incident, several bags of solid chemicals were present in the staging area. One operator moved four bags to the charging area after glancing at the labels. He initialed the checklist and handed it to the second operator. Trusting his partner, the second operator initialed the checklist without checking the labels. The four bags all looked very similar. However, the first operator had made a mistake. One of the four bags was not the same as the other three. The vendor had changed packaging and the visual verification of the bag design was no longer sufficient to distinguish between the intended chemical and the incompatible chemical.

4.3 Multi-tasking, Bad Timing, and Partially Continuous Attention

A crew of operators is responsible for managing a neutralization process. This part of the process is run as a batch operation. Acidic product is collected in a neutralization tank until the desired operating level is reached. There are two identical neutralization tanks; one is being filled with product via gravity drain while the contents of the other tank are being neutralized. Caustic pumped from a storage tank is added to neutralize the product. The tank contents are pumped through an external cooler to remove the heat of neutralization. A neutralization is conducted every 8 hours.

Two field operators, one experienced and one new to the operating area, handle the manual tasks required to switch product feed between the tanks and enable caustic to be fed to start the neutralization. The caustic header enters the building near the top of the neutralization tanks. The product header is two floors above the tanks and there is no elevator in the building. A checklist is available for the switchover tasks, but experienced operators don't carry checklists with them when performing tasks. The new operator is being mentored and trained by the experienced operator, mimics the demonstrated behavior, and therefore also doesn't use the checklist.

Figure 2

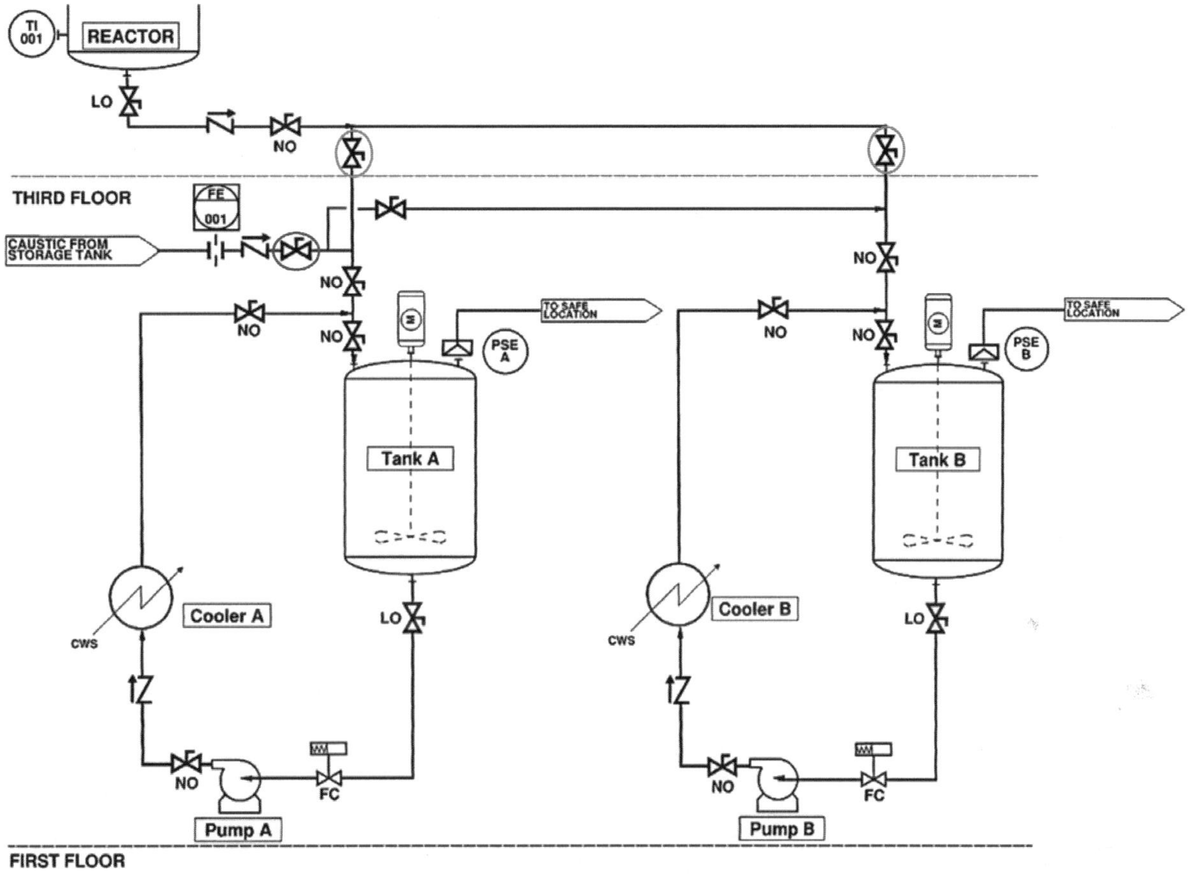

The two operators enter the building to switch the product flow from Tank A to Tank B (green valves) and valve in the caustic flow to Tank A (blue valve) to start the neutralization process. The experienced operator heads to the third floor to switch the valves on the product header, and the new operator stays on the first floor to open the valve on the caustic header to Tank A. When he arrives on the third floor, the experienced operator decides to walk over to inspect a temperature gauge since the instrument connected to the Basic Process Control Systems (BPCS) seems to be malfunctioning.

While the operators are still in the building, the control room operator calls over the radio that high temperature and pressure are being experienced in the reactor. The control room operator shuts off the steam to the reactor that feeds the neutralization tanks, but the temperature and pressure continue to increase. The field operators evacuate the building. The emergency pressure relief system activates and discharges the contents of the reactor to the blowdown tank. Although the relief system prevents the reactor from failing due to overpressure, the reactor is damaged due to pressure and temperature well above its MAWP and MAWT.

4.3.1 Results of the Incident Investigation

The incident investigation revealed that the caustic header valve was opened to allow flow into Tank A before the product header valves were switched. The discharge pressure from the caustic pump was significantly higher than the reactor operating pressure. Caustic was pumped into the reactor through the product header, leading to an exothermic reaction between acid and caustic with significant increase in temperature and pressure as well as gas generation.

A review of the procedure and associated checklist for the neutralization steps listed the tasks associated with the product and caustic header valves in the following order:

1. Switch product header valves to direct flow into the empty tank.
2. Open caustic header valve to start the flow of caustic into the full tank.

The procedure did not explicitly state that a single operator was to perform these tanks, nor did it prohibit reordering the steps or conducting them simultaneously rather than in series. A number of deviations had occurred in the operating unit over a period of several years. As more tasks in the facility were automated, the size of the crew was reduced by two operators. Over time, the operators optimized the performance of all their tasks to adjust to the reduced manpower. When highly manual operations were conducted, the crew was short-handed and by necessity became accustomed to looking for ways to be more efficient. Efficiency was gained by conducting multiple unrelated tasks during single visits to the operating area. It became the new norm for operators to conduct routine walkthroughs and collect samples while in the building to conduct the switchover. It also became common for two operators to split the tasks on the first floor and third floor of the building so that one operator would not have to go up and down the stairs repeatedly.

On the day of the incident, the two operators divided up the tasks on the first and third floors as they had become accustomed to doing. However, on this particular day, their timing did not match up; the caustic path was open and pressurized before the process flow could be isolated from the full tank and directed to the empty tank. The unrevealed failure of a check valve in the product header allowed caustic to backflow into the reactor leading to the incident.

5 Why deviate?

These three examples demonstrate that the operators became accustomed to deviating from expected practices for several reasons:

1. Relying only on experience and training / overconfidence

   The operators relied on experience rather than using checklists to perform tasks. They were overconfident in their abilities and viewed the written checklist to be a paperwork exercise rather than a tool to ensure operational excellence.

In the batch unit, the operators relied on their familiarity with visual identification of the bags rather than checking the labels.

In the neutralization unit, the operators incorporated multiple separate tasks into their field visit, which distracted them from the switchover task. They failed to coordinate their efforts on the first and third floors, which resulted in improper valve sequencing and backflow of caustic. It is likely that this situation had occurred in the past, but the check valve prevented the caustic from reaching the reactor. The operators did not realize that the check valve was the only effective safeguard against backflow with this mode of operation.

Having knowledge and skills alone are not sufficient by themselves to ensure that manual operations are properly performed; conduct of operations practices should include the expectation that all procedures and safe work practices are followed to the letter. Procedures should be followed or changed. There are certainly improvements that could be made to the procedure for the neutralization unit, including explicit instructions on the order of operations and specifying that the tasks are to be performed by a single operator. Checklists should be used when performing operations, not filled out after those tasks have been completed. When a checklist requires signature verification, that signature must be a certification that the task was performed in its entirety.

2. Inadequate Housekeeping Inspections

In the batch unit, the practice of staging sufficient materials for several batches rather than ordering just enough for the needs of one batch should have been identified during housekeeping inspections. The presence of multiple bags and drums of different chemicals in an area not intended for inventory should have been easy to spot. Personnel in the area became used to using this area for inventory and no longer considered the practice to be a deviation.

3. Physical layout of the building or equipment

While the two neutralization tanks are located on the first floor, the two sets of valves that need to be manipulated during the switchover are not both located near the tanks. The lack of an elevator in the building contributed to the decision of the operators to "divide and conquer". The difficulty of access to equipment, such as manual valves and field instruments, can contribute to a higher likelihood that operators may deviate from instructions to check valve positions or instrument data as part of routine walkthroughs. If historic data shows that the valve position is always the same or the instrument always reads the same, paperwork may be filled in based on previous readings rather than a true independent verification.

4. Inadequate Management of Organizational Change (MOOC)

Permanent reductions in staffing can have significant implications for work practices. Expecting the same amount of work to be done with fewer people sets the stage for deviations. Onboarding apprentice operators without making staffing decisions such that the apprentice is always accompanied by an experienced trainer until they achieve sufficient competency to operate alone can introduce deviations as the apprentice performs tasks without observation and opportunity for correction.

5. Doing More with Less

   Productivity initiatives can incentivize personnel to deviate to save time and money. Exceeding the safe operating limit for high level, reducing the number of fills made in a tank, saves time. Relying on a SIS to stop flow rather than monitoring the tank level allows the operators to spend time on other tasks, as well as spending less on instrumentation. Staging more inventory in the operating area saves times as operators do not have to spend as much time ordering materials and waiting for delivery.

6 Recognizing Normalization of Deviation

Normalization of deviation can exist even in high performing organizations. Initially, these deviations do not have a noticeable impact on the operations and can even lead to increased productivity. However, the presence of these deviations can lead to compromising safeguards and layers of protection. Recognizing normalization of deviation is the first step to mitigating its presence and potential impact. Proactively searching for normalization of deviation allows the organization to evaluate their process safety culture and operational discipline.

6.1 We Do What We Say

Procedures should either be followed or changed--period. Annual procedure reviews should include discussion with operators to get feedback on how relevant the written procedure is to the activities they conduct in the operation of the process. When reviewing procedures, look for the use of "should" versus "shall" and "must" versus "may". Instructions that are written with "should" and "may" provide room for interpretation whereas the use of "shall" and "must" reinforce that the instructions must be followed exactly as written. Ensure that procedures clearly specify the order for steps and tasks and identify those steps that must be done by the same operator and those that require independent verification. Leadership should make resources available to update procedures as changes are made and set and reinforce the expectation that updating procedures is a critical step for adequate completion of an MOC.

Another clue for the presence of normalization of deviation is comments made during crew meetings or written in log books.

- We never do that step.
- That was good enough
- We made it work.
- I always bypass that.
- Wait for the alarm before you close the valve.
- We don't have to follow that procedure

These statements indicate that operators are in the habit of making deviations from written procedures or expectations. These types of comments may also be heard during a Process Hazard Analysis (PHA) as the team reviews the procedures and checklists. Operators that rely on experience, training, and institutional knowledge to perform tasks rather than referring to the written procedures are paving the way to a culture that fosters normalization of deviation. The PHA team may also find that the available Process Safety Information (PSI) is deficient, inaccurate, or simply unavailable. Failure to maintain accurate and up-to-date PSI is often an example of normalized deviance and can trigger additional deviations due to lack of information and well-intended shortcuts or work arounds taken by employees to address the deficiencies.

6.2 Use Behavioral Safety Techniques and Processes

Behavior-based safety programs encourage all personnel to identify unsafe practices and provide immediate, direct, and caring feedback to their co-workers when unsafe practices are observed. An organization with a robust behavior-based safety program develops the habit of "see something, say something" and often has a database with numerous examples of observations and outcomes to share across the entire organization. Many of these observations are those that are easy to spot, such as those that violate the Safety Musts ingrained in the employees including failure to hold the handrail, failure to wear basic PPE, or texting while walking or driving. Build upon this foundation to dig deeper to identify normalization of deviation. Observations should include a comparison of the task as performed to the expectations in the written procedure or checklist. The same task should be observed with operators from different crews to look for inconsistencies and deviations. These observations can identify opportunities for additional training or updating procedures to address inconsistencies. Immediate feedback to employees when deviations are identified can prevent them from becoming normalized. A strong behavioral safety program supports high levels of operational discipline. Maintaining a culture of operational excellence where all employees strive to perform every task correctly, every time, results in increased accuracy in performing day-to-day operations, which leads to fewer deviations.

6.3 Use Routine Walk-throughs and Inspections

Walk-throughs and inspections can reveal deviations in the operating area, such as inventory control issues, housekeeping concerns, and maintenance issues. Look for markers of situations that are tolerated and not promptly addressed, such as Do Not Operate tags that are obviously faded and tattered Caution tape that indicates that equipment has been bypassed or out-of-service for an extended period of time. Poor labeling, product build-up on equipment and piping, poor lighting, clutter, and other housekeeping issues can be an indication of a lack of operational discipline. Develop a checklist specifically for these walk-throughs and inspections so that auditors can focus on deviations from the operating area's expectations.

7 Responding to Normalization of Deviation

Many of the tools and techniques for responding to normalization of deviation are already part of a company's process safety management program. Leadership must set expectations for strong operational discipline and commit to preventing normalized deviance from persisting in the organization's culture.

7.1 Operator Training, Procedures, and Process Safety Information (PSI)

Operator training and procedures must clearly identify safe operating limits, consequences of deviation, and steps to correct. Safety-critical steps should be clearly identified. Management systems must establish clear expectations for every operations activity, including following procedures, adhering to safe operating limits, and correcting deviations immediately. Deviations outside of safe operating limits should be considered as near misses and should be investigated to identify learnings that can be shared with all personnel.

Installation of a SIS is a significant safety improvement when it is maintained as an effective Independent Protection Layer. The PHA documentation and associated PSI should include an itemized list of ALL safeguards and description of their part in the layers of protection for the hazard scenario. Operators and engineers supporting the area should be familiar with the hazards in the area and understand the layers of protection that are essential to mitigating those hazards.

Engaging the entire workforce is critical for reducing normalized deviance. Leadership must set the expectation that process safety is everyone's responsibility. Employees must be trained to have sufficient process safety competency for their role. Train everyone on normalization of deviation and to raise concerns when they identify it within their crew and operating area.

7.2 Rigorous Management of Change

Formal, rigorous management of change is essential to combat normalization of deviation. How do personnel view the MOC process – as a paperwork exercise only, or as a tool to evaluate and manage change? Leadership must establish the expectation of excellence in execution of each MOC and work to ensure that employees internalize this expectation. Appointing a dedicated MOC coordinator for each operating unit who can act as a gatekeeper in the process ensures that MOCs meet the expected standards before processing. Feedback from the gatekeeper to the MOC initiators when the process has not been properly followed helps to change the culture to one where appropriate rigor is applied to the MOC process.

Changes proposed as part of productivity initiatives need to be managed as true changes, through the established, formal MOC process. Deviations for the sake of increased productivity should not be tolerated. Organizational changes must also be properly managed. When personnel changes are considered, adequate time for transitioning responsibilities to other employees must be provided. Time should be provided for training, mentoring, and onboarding activities for employees who are moving into a new role or operating area before assigning them as a fully functioning member of the crew. When proposing a permanent reduction in staffing, procedures and tasks should be thoroughly evaluated to determine what impact will be experienced with fewer people to perform the necessary operations. Procedures and processes should be evaluated based to determine if and how they should be changed to be safely accomplished and managed by fewer people. This should also be done when personnel changes result in loss of experience, such as replacing a retiree with a new hire.

7.3 Provide Adequate Resources

Lack of adequate resources can allow normalization of deviation to persist in an organization. Human beings are extremely innovative when forced to find ways to accomplish tasks with inadequate resources. Maintaining a strong process safety culture is not possible without leadership commitment to provide sufficient resources for the tasks required to support process safety management programs. Rigorous use of MOC requires that people have the time and skills needed to update P&IDs, operating procedures, and other documentation. Operator training requires significant time and expense to develop, deliver, and maintain. Maintain focus and support for day-to-day operations rather than diverting all attention to capital initiatives. Evaluate productivity initiatives to ensure they send the correct message rather than encouraging workarounds or shortcuts to meet the stated goals and ask "how lean is too lean?". Ensure staffing levels are appropriate to allow the expected tasks to be completed as written. Practice effective Management of Organization Change (MOOC) and evaluate all proposed staffing and organizational changes for their impact on process safety performance. Leadership must set expectations for operational discipline and adherence to procedures and standards and follow up with adequate resources that allow employees to meet those expectations. Additionally, employees should be rewarded for adherence to procedures and standards rather than praised for innovations that result in improperly managed change.

7.4 Investigate Safety System Activations

All safety system activations should be investigated as incidents or near misses. Exceeding safe operating limits and reaching the trip point of a SIS is a Tier 3 Process Safety Event (PSE) [3]. A Tier 3 event typically represents a challenge to a barrier or safeguard. An SIS that functions as designed when challenged prevents an incident from progressing to a Loss of Primary Containment (LOPC) event. Rigorously investigating challenges to safeguards such as SIS activations can be a valuable tool for identifying when normalization of deviation has resulted in safeguards being used as basic process control. Investigation of the activations of the SIS in the example in section 4.1 and recording those activations as Tier 3 events would have highlighted the deviation from use of the established procedure to reliance on the SIS to control the tank level before it could become normalized.

8 Conclusion

Deviations that persist in an organization can lead to failure to properly manage hazards. Safeguards can be compromised or defeated by shortcuts or workarounds. Normalized deviance is truly a roundabout way to disaster. Its effects are not immediately noticeable, but are revealed with sudden clarity in hindsight when a catastrophic process safety event occurs. Leadership must set expectations for strong operational discipline and commit to preventing normalized deviance from persisting in the organization's culture. Through dedication to operational discipline and excellence in execution of operating tasks--every task, every time--a safer, and ultimately more reliable and productive facility will result.

9 References

[1] CCPS Process Safety Glossary, Normalization of deviance, Available at http://www.aiche.org/ccps/resources/glossary?title_op=contains&title=normalization+of+deviance&=Apply, Accessed on February 26, 2018.

[2] OxfordDictionaries.com, Deviation, OxfordDictionaries.com, 2018, Available at http://www.oxforddictionaries.com/definition/english/deviation, Accessed on February 26, 2018.

[3] American Petroleum Institute, API RP 754: Process Safety Performance Indicators for the Refining and Petrochemical Industries, Second Edition, June 2017.

511477 CCPS Ever Green Process Safety Metric Project

Abstract:

In 2006, the CCPS Technical Steering Committee authorized a project to develop a Guideline book for the development and use of Leading and Lagging Process Safety Metrics. In the initial meeting, the committee identified that the key breakthrough opportunity for industry was the development of an industry metrics that would become the benchmark across the chemical and petroleum industries for measuring process safety performance. To achieve this objective, CCPS extended the committee composition beyond CCPS membership to involve representatives and members from each of the major chemical and petroleum trade associations as well as other key stakeholders.

Since then, there has been several other metrics systems based upon the original CCPS work that continues to evolve and adapt to new interpretations, clarification of intent, and most importantly, learnings from process safety incidents. Whereas this process is continuous, guidelines take considerable time to complete the review and revision process and once completed, initiate subsequent revision and harmonization of other guidelines to maintain consistency within global industries.

To provide a source of current information and updates to the various metrics activities as well as a place for ongoing research, publications, and new techniques; the CCPS has started an evergreen webpage for process safety metrics. This paper will describe what is included in this evergreen project and its key attributes, specifically, the current status of consensus metrics, leading metrics, and a new CCPS survey.

514178 Essential Practices for Developing, Strengthening and Implementing Process Safety Culture and Leadership

Abstract:

In June 2018, CCPS will publish a new Concept Book that addresses developing, strengthening and implementing Process Safety Culture. This book also provides guidance for leaders at all levels to assess and develop a culture improvement plan. This paper will discuss the lessons presented in this new book, and also discuss its relationship to overall process safety leadership, as described in the forthcoming CCPS Process Safety Leadership book which will be completed later in 2018.

513716 Justifying Energy Savings in Ethylene Crackers in a Low Energy Price Environment

Abstract:

Global energy prices have dropped to their lowest level in a decade and have taken with them the days of easy justification for energy improvements. In specific countries where energy prices are still subsidised, this is a familiar situation in everyday energy management. The subject of this paper is the different ways that have been used to improve the economics of energy projects in low price situations in order to make viable project investments.

Case studies will be used to illustrate such methods as:

- Debottlenecking energy critical equipment to increase throughput,
- Overcoming an energy cap to improve severity and product yield,
- Making energy savings to meet legislative targets, and
- Using energy reduction to justify changing catalysts to improve yield.

511653 Retrospective Health Impact Assessment for Pm 2.5 Pollution in Mexico City 1990-2015

Abstract:

Fine particulate matter has a well-documented association to mortality. We have conducted a retrospective health impact assessment to quantify the lives saved in Mexico City from 1990-2015 due to PM 2.5 reduction resulting from pollution control policies, utilizing the Integrated Exposure Response model developed for the Global Burden of Disease study. In addition, we monetized the health benefits using city-specific value of statistical life (VSL). Our analysis suggests that 21,000 lives saved during that period and benefits in the order of 18 billion 2015 USD. We also developed a novel model to retrospectively decompose the health benefits and attribute them to one policy utilizing sulfur reduction in fuel policy as an example. Our model indicates that this policy has resulted in 1,600 lives saved. This suggests that Mexico City air pollution policies could serve as a model for developing countries and the industry with environmental regulations at infancy stage.

512814 Sustainability Analytics - the Critical Link between Sustainability and Business Strategy for Actionable Insights

Abstract:

Advances in Enterprise Systems Are Making It Feasible for Corporations to Track and Transform Sustainability Performance. the Materiality of These Seemingly Non-Economic Impacts Is the Critical Link between Sustainability and Business Strategy. Business Analytics Enable Organizations to Convert Raw Data into Actionable Insights to Achieve Their Sustainability Goals. Sustainability Analytics Analyze the Data and Provide Insights That Help Transform Sustainability Information into Action across the Value Chain and Life Cycle. This paper will Provide Business Leaders with Insight into How to Determine Which Sustainability Metrics Are Material to Them and Relevant to Their Business, and How to Incorporate Them into Corporate Strategy, Resource Allocation Prioritization and Decision Making Process.

Based on My Book Introduction to Sustainability Analytics Under Publication By CRC Press

512839 The Reaction Mechanism and Kinetics of Magnesium Oxide for CO_2 Immobilization

Abstract:

The CO_2 immobilization is significant for the sustainable energy and chemistry industries. And the magnesium oxide is a clean and economic agent using for the CO_2 immobilization. Therefore, the reaction of CO_2 with magnesium oxide suspended in water was investigated, the mechanism proposed, and the kinetics regressed as well. The CO_2 pressures from 1 to 5 (MPa), as well as ratios of solid to liquid, stirring speeds and temperatures, significantly influenced the dissolution of magnesium oxide. And as high as 26.63 g/L MgO was dissolved in the solution under the condition of CO_2 pressure 5 MPa at 50 °C.

In the process, the magnesium oxide became magnesium hydroxide through hydration reaction, and then the hydroxide dissolved by carbonated water. In the carried out reaction of CO_2 with magnesium hydroxide suspension, no solid product was found covering the solid $Mg(OH)_2$. Therefore, a reaction mechanism based on the hydration of Mg^{2+} was proposed, where the concept of â€˜progressive liquid filmâ€™ was introduced to the reaction system. And the derived kinetic model for the dissolution was fitted the experiment data well. The pressurized magnesium hydroxide dissolution was found to be rate-controlled by the diffusion in the progressive liquid film around the solid material. The apparent activation energy, obtained from 17.1 to 25.4 (kJ/mol) in the experiments, indicated also that the dissolution inclined to be diffusion rate controlled. Moreover, the apparent diffusion coefficients of H^+ in the liquid film were calculated and an empirical expression was obtained for the first time. The kinetic model might be applicable to other dissolution reactions similarly with just soluble product.

513948 Moving Towards the Smart Factory Leveraging Big Data Advancements

Abstract:

Moving Towards the Smart Factory

Leveraging Big Data Advancements

James Moyne, Michael Armacost, Mingwei Li and Amos Dor

Applied Materials—Applied Global Services

Santa Clara, CA, USA

Key Words—Big Data, Smart Manufacturing, Industry 4.0, Analytics, Prediction Methods, Predictive Maintenance, Virtual Metrology, Digital Twin, Cyber-Physical Systems Abstract Introduction

Smart manufacturing (SM) is a term "generally applied to a movement in manufacturing practices towards integration up and down the supply chain, integration of physical and cyber capabilities, and taking advantage of advanced information for increased flexibility and adaptability" [1,2,3]. It is often equated with "Industry 4.0" (I4.0), a term that originated from a project in the German government that promotes a 4th generation of manufacturing that uses concepts such as cyber-physical systems, virtual copies of real equipment and processes, and decentralized decision making to create a smart factory [4,5].

As "big data" requirements and capabilities in data volumes, velocity (rates), veracity (quality), variety (merging) and value (analytics) increase rapidly, manufacturers across industries are beginning to see opportunities to leverage big data advancements and enable the many facets of SM to improve quality, capability and throughput, and reduce cost [6,7]. The facets of SM that are aggressively being pursued include (1) leveraging advanced analytics especially for predictive technologies; (2) tightening integration up and down the supply chain network; (3) improving the use of cyber-physical system (a term used to refer to the "tight conjoining of and coordination between computational and physical resources" [8]) to improve qualities such as adaptability, reconfigurability, functionality, and reliability; (4) improving the use of real-time simulation-like tools to begin to realize the "digital twin", which refers to a digital replica of the physical system at multiple levels creating the "virtual factory" [9]; and (5) relying more heavily on a knowledge network to incorporate subject matter expertise (SME) in analytical solutions [10]. Determining the State-of-the-Art of Smart Manufacturing

While advancements are being made across these facets of SM, the progress varies from industry-to-industry and from facet-to-facet, based in large part on the pace of evolving needs in the industry and the progress of research and development in a particular facet. With respect to industry progress, the microelectronics manufacturing industry (which includes semiconductor and display manufacturing) has seen significant progress on the SM front because it is needed to address device production requirements coming from the constant innovation in areas such as mobile devices, Internet of Things (IoT) and even big data, as well as the pressures to produce high volumes, flexibly at low cost. During microelectronics manufacturing precision processes are used to create "chips" or "panels with multiple layers of complex circuitry. Each layer contains features that must be controlled to the nanometer or even molecular level. A typical microelectronics manufacturing process has hundreds of these precision control steps.

Figure 1 illustrates the SM vision for the microelectronics industry, incorporating the previously mentioned SM facets [11]. One SM area where the microelectronics manufacturing industry has excelled is in the application of analytics. As an example the dynamic quality control of these steps in this industry is facilitated with a set of capabilities collectively termed "Advanced Process Control" or APC, which includes dynamic model-based process control, Fault Detection and Classification or "FDC", and more recently predictive technologies including Predictive Maintenance (PdM), Virtual Metrology (VM) [1]. As an example of the use of cyber-physical systems, high-mix product production requiring flexibility and high throughput is addressed with real-time scheduling and dispatch capabilities. These capabilities incorporate characteristics such as product priority, production process capability, transport times and queue length, and maintenance schedules in scheduling and dispatch decisions. These and other SM capabilities make semiconductor manufacturing an idea arena with which to benchmark the current state of the art in SM.

Figure 1: A Smart Factory vision for the microelectronics industry.

In determining the progress in SM facets across the manufacturing base, it is important to understand that some SM capabilities can be implemented today, while others are still in the research phase. The latter might best be implemented today with existing leading-edge capabilities that are effective forerunners to the ultimate SM vision, rather than focusing on the ultimate solution. As an example of SM capabilities that can be implemented today, PdM is a technology that has been shown to be effective in semiconductor manufacturing. While the challenges of wide-spread PdM deployment are significant, interim solutions for PdM such as extrapolation on diagnostics data can often be effective as a complement to more complete and complex solutions, as long as prediction horizon and prediction confidence are still conveyed with the maintenance prediction event. As an example of capabilities that are effective precursors to the ultimate SM vision, the concept of the digital twin across the enterprise is a vision of factory integration in semiconductor manufacturing [6], however it is generally accepted that achieving the final vision in the industry won't happen for at least five years. In the meantime the aforementioned rules-based real-time scheduling and dispatch is an effective precursor to the digital twin. It utilizes configurable rules and workflows to capture processing and product behavior, thereby providing real-time reconfigurability decisions in order to optimize throughput in the face of quality, production schedule and product priority constraints. Further the capability is structured such at the rule base can be supported by or replaced with real-time simulation/emulation capabilities as these capabilities evolve [12].

Summary

Effective adoption of SM technology requires an understanding of the current state-of-the-art of the tenants of SM as well as the path towards the ultimate SM vision. In this tutorial the SM vision will be described including a detailed description of the individual SM tenants. The current across-industry state-of-the art of each tenant will then be detailed, identifying current solution approaches that can provide a path to the ultimate vision. State-of-the-art solution approaches and case studies in microelectronics manufacturing will be used to illustrate the SM evolution approach.

References:

[1] J. Moyne and J. Iskandar, "Big Data Analytics for Smart Manufacturing: Case Studies in Semiconductor Manufacturing," *Processes Journal*, Vol. 5, No. 3, July 2017. Available on-line: http://www.mdpi.com/2227-9717/5/3/39/htm.

[2] Wikipedia: Smart Manufacturing. Available online: https://en.wikipedia.org/wiki/Smart_manufacturing.

[3] Davis, J., Edgar, T, Porter, J., Bernaden, J., and Sarli, M. "Smart manufacturing, manufacturing intelligence and demand-dynamic performance," *Computers & Chemical Engineering*, 2012, vol. 47, pp. 145–156.

[4] Project of the Future: Industry 4.0, Germany Ministry of Education and Research, Available online: http://www.bmbf.de/en/19955.php.

[5] Kagermann, H.; Wahlster, "W. INDUSTRIE 4.0 Smart Manufacturing for the Future", *Germany Trade and Invest*, 2016.

[6] *International Roadmap for Devices and Systems (IRDS): Factory Integration White Paper, 2016 edition*. Available online: http://irds.ieee.org/images/files/pdf/2016_FI.pdf.

[7] Moyne, J., Samantaray, J., and Armacost, M., "Big Data Capabilities Applied to Semiconductor Manufacturing Advanced Process Control", *IEEE Transactions on Semiconductor Manufacturing*, 2016, Vol. 29, No. 4, pp. 283-291.

[8] *Cyber-Physical Systems (CPS) Program Solicitation NSF 10-515*, Available online: https://www.nsf.gov/pubs/2010/nsf10515/nsf10515.htm.

[9] Wikipedia: Digital Twin. Available online: https://en.wikipedia.org/wiki/Digital_twin.

[10]Armacost, M. and Moyne, J. "Moving towards the 'smart factory' in microelectronics manufacturing", *Nanochip*, vol. 12, N. 2, 2017.

[11]Hasserjian, K. "Emerging Trends in IC Manufacturing Analytics and Decision Making", Keynote, *Advanced Process Control Conference XXVIII, October 2016*. Available online via: http://apcconference.com.

[12]Lopez, F., Moyne, J., Barton, K., and Tilbury, D., "Process-capability-aware scheduling/dispatching in wafer fabs", *Advanced Process Control Conference XXIX, October 2017*. Available online via: http://apcconference.com.

513921 Future of Latent Variable Methods for Big Data Analytics in the Process Industry 4.0

Abstract:

In modern industries adopting the Industry 4.0 paradigm, the result of Industrial Internet of Things (IIoT) connecting intelligent physical entities to each other allows complex equipment units to have embedded sensors and special modules (agents) providing connection to the monitoring center. This is leading to the so-called Big Data problem in Industry 4.0.

Big data exhibit high volume and correlation, rank deficiency, low signal-to-noise ratio, changing structure, and missing values. Data in process/manufacturing industries are especially complex in structure and the useful information is often subtle.

Apart from the infrastructure (hardware) needed to manage these Big Data streams, the key point is how to analyze them to effectively extract information to give organizations new insights about their products, customers and services and steer the decision-making process. This can be particularly valuable when it is critical to maintain quality and uptime, such as in process monitoring applications, by quickly detecting and diagnosing abnormal activities, predicting the time-to-failure of equipment units or when rapid new products development is critical for company survival.

In this talk we illustrate the potential of latent variable-based multivariate statistical methods for Big Data Analytics to analyze and visualize extracted information in a way that is easily interpreted and that is useful for process understanding, real time process monitoring, fault detection & identification, process improvement, predictive maintenance, and optimization.

A discussion on the pros/cons of latent variable models with classical statistical and machine learning methods in Big Data environments will also be addressed.

Applying an Enterprise Risk Management Framework to Process Safety

Muhammad Mushahid R. Qureshi, Ph.D.
Loss Prevention Engineer
Saudi Aramco
C-06, Building No. 3150, (STAR Building), Dhahran, Saudi Arabia 31311
Muhammad.Qureshi.4@aramco.com

Prepared for presentation at
the American Institute of Chemical Engineers
2018 Spring Meeting and the 14th Global Congress on Process Safety
Orlando, Florida
April 22- 25, 2018

Applying an Enterprise Risk Management Framework to Process Safety

Muhammad Mushahid R. Qureshi, Ph.D.
Loss Prevention Engineer
Saudi Aramco
C-06, Building No. 3150, (STAR Building), Dhahran, Saudi Arabia 31311
Muhammad.Qureshi.4@aramco.com

Keywords: Enterprise Risk Management, Process Safety, ERM

Abstract

Process safety incidents remain a major challenge to all industries involved in the handling, processing, manufacturing, and transport of hazardous materials, such as hydrocarbons. Over the past decade, a number of incidents have led to significant changes in the manner in which major industrial incidents are being perceived and handled. As a result, many organizations have adopted a risk-based approach when managing potential hazards associated with major process safety incidents. Larger organizations benefit from an Enterprise Risk Management (ERM) framework when assessing potential corporate risk. This approach can prove valuable when evaluating risks associated with major process safety incidents at a corporate level. This presentation outlines the six-step ERM approach applied when assessing, analyzing, and recommending treatment actions to bridge potential gaps related to major process safety risks. This approach also takes into account the development and utilization of performance metrics — used for monitoring risk at a corporate level — on an ongoing basis.

1 Introduction

Enterprise Risk Management (ERM) includes the methods and processes used by organizations to manage risks and seize opportunities related to achieving their business objectives [1]. ERM provides a framework for risk management, which typically involves identifying particular events or circumstances relevant to an organization's objectives, assessing them in terms of likelihood and magnitude of impact, determining a response strategy, and monitoring progress. By identifying and proactively addressing risks and opportunities, business enterprises protect and create value for their stakeholders, including owners, employees, customers, regulators, and society overall. This ERM framework can be applied at a corporate level to address the risk of a major process safety incident occurring in the operating facilities handling hazardous materials. A typical application could be for corporations involved in various business lines (Upstream, Downstream, Chemicals, Petrochemicals, etc.) having multiple facilities across various geographic locations around the globe. The application of an ERM framework, as presented in this paper, is aimed at addressing the corporate view of risk related to a major process safety incident. The deliverables from this are useful to board members of a company that intends to have a high-level understanding of the risks associated with their companies, and those interested in managing and monitoring these risks on an ongoing basis. The key information used in this process is derived from front line employees in processing facilities and this framework provides the necessary link between the facilities management and corporate leadership.

2. History

There have been numerous process safety incidents in the past decade, which have made a significant impact on a wide range of industries and regulations. For example, the blowout at the Macondo well in 2010 resulted in a loss of over U.S. $42 billion to date [2]. This incident is regarded as the largest environmental disaster in modern history, and forced the U.S. government to reassess its management of the oil and gas industry. New regulations and oversight bodies have been formed to increase the scrutiny of similar activities. While these are reactionary measures, if applied effectively, the lessons learned from this incident could effectively be used to prevent similar accidents. In addition to regulatory changes, this incident also served as a wakeup call for many companies, including those within the oil and gas sector, to take a deeper look at their individual prevention and protection measures and safety programs.

Although an ERM framework had been established long before the above-referenced incident [3], larger corporations have recently adopted this approach more religiously to gain an in-depth understanding of business risks and its impact on stakeholders. Recently, this approach has also been promoted by large insurance companies when evaluating the corporate risks of their clients [4]. The ERM framework provides a structured approach to understanding overall risks and taking proactive measures to prevent companies from catastrophic losses. In the end, the results help in boosting shareholder confidence and position the company for sustainable growth.

3. ERM 6-Step Process

Enterprise risk management provides a 6-step framework to conduct risk evaluation in a structured manner. The steps of this framework are:

1. Context
2. Identify
3. Analyze
4. Evaluate
5. Treat
6. Monitor and Review

Figure 1 – The 6 Steps of ERM Framework Application

These steps are applied in a cyclical manner to ensure that the process is ongoing. Despite risk level fluctuations within an organization, following this process will ensure that risk level monitoring continues.

For application to process safety risk, this framework can be used to assess, analyze, and address any risks associated with a major industrial incident (e.g., significant process safety incident). The subsequent sections below describe the application of the ERM process to manage risk of a process safety incident at a corporate level.

3.1 Context

Establishing context is the most important step in this process to achieve accurate outcomes effectively and efficiently. This step should establish the following aspects:

- Highest level sponsorship of the risk management program.
- Alignment with business objectives.
- Roles and responsibilities associated with risk management activities.

When managing risks associated with process safety incidents at a corporate level, this first step ensures that the company's highest level of Health, Safety and Environment (HSE) is overseeing the program and monitoring its application. It is important to secure buy-in from the HSE committee and have this documented in the form of a policy statement. Aligning context of process safety risk to business objectives is essential so that people associated with this work are able to understand the link between this work and business outcomes.

Involving the correct stakeholders and other participants in the risk management process is key to the success of the entire process. It is important to establish a multidisciplinary corporate level taskforce team from central business support organizations and to engage stakeholder organizations within operating facilities. Generally, the taskforce should be made up of members responsible for process safety, engineering, maintenance, inspection, operations, and environmental protection.

Documenting the scope will ensure a more focused risk assessment effort. Defining the key roles and responsibilities of various stakeholders involved in this risk management process is also essential as it establishes the key expectations, as well as accountability. The taskforce should also clearly establish high level deliverables and outcomes. This activity will ensure that everyone involved in the process is aligned to achieve the desired outcomes.

3.2. Identify

The purpose of risk identification is to generate an inventory of process safety risks that could significantly affect achieving safety and business objectives for a relevant organization. The taskforce should conduct workshops focused on engaging field organizations to generate a list of safety risks in the field. This input is fundamental to generating a meaningful and practical list of potential risks.

The taskforce should then consolidate this input to generate more high-level risk events. In addition to workshops, additional information should be gathered through document reviews, incident history reports, interviews, survey reports, and regulatory audit findings.

For a corporate overview of process safety risks, this step helps identify high level sub-risks that could potentially lead to a major process safety incident. The list below outlines a few examples of identified sub-risks:

- Major loss of primary containment of highly flammable materials.
- Major gas leak resulting in a fire or explosion.
- Major tank fire.
- Loss of a critical asset due to a power outage.
- Major toxic material release.
- Major dust explosion.

3.3 Analyze

Risk analysis involves assessing each sub-risk's potential consequence, its likelihood, and its potential overall severity. It also includes identification and assessment of controls to address these risks.

3.3.1 Determine the "credible" worst case scenario

It is important to identify credible worst case scenarios. These should not be the most extreme or unlikely outcomes, nor the most expected. They should be somewhere in between, but still represent a pessimistic outcome with a severe consequence. By describing the worst case scenario, a validation tool for the assessment of consequence and likelihood of an event can be created as part of this step in the risk management process.

3.3.2 Develop a Risk Assessment Matrix (RAM)

Developing a risk assessment matrix (RAM) is a fundamental tool to ensure that analysis is performed in a standardized and consistent manner. Companies can develop a RAM based on size, geographic locations, complexity of operations, business environments, etc. The RAM is a tool used to apply risk-based process safety to address identified sub-risks, as a corporation may not be able to apply the same resources to address all levels of risk.

It is important to note that the risk assessment matrix does not suggest the level of risk that an organization is willing to accept, but it is merely a tool used to support the consistent prioritization of actions in eliminating or mitigating safety risks. Similar to risk identification, risk analysis should involve workshops and should ideally be a collective and consensus-based exercise aimed at reducing individual bias and providing alternative insights.

An example of the RAM's consequence categories - with varying thresholds - is provided in Table 1 for reference only. In addition, companies may choose to analyze consequences related to additional categories, such as "Compliance" and "Business Interruptions," in which case it will be necessary to expand Table 1 to include these categories and develop applicable thresholds accordingly.

Table 1 - An example of the consequence catageories and thresholds for a risk assessment matrix adopted from [5].

Level	Health and Safety: Onsite Workers	Health and Safety: Public	Environmental Impact	Financial Loss	Reputation
6	>200 acute or chronic fatalities	>50 acute or chronic fatalities	>100,000 bbl of oil in sensitive coastal waters. >1,000,000 bbl of oil in other coastal waters. Prolonged regional/global contaminations.	>$10 billion	Global outrage. Global brand damage and/or affecting international legislation.
5	>50 acute or chronic fatalities	>10 acute or chronic fatalities	>10,000 bbl of oil in sensitive coastal waters. >100,000 bbl of oil in other coastal waters. Short-term damage at a regional level. Prolonged contamination affecting extensive nature conservation or residential areas.	$1 billion – $10 billion	International media coverage. Regional outrage, regional brand damage and change of regulations at a regional level.
4	>10 acute or chronic fatalities	1 or more acute or chronic fatalities, multiple permanent injuries, or irreversible effects.	>10,000 bbl of oil. >1,000 bbl of oil in a sensitive area. Long-term damage affecting an extensive area.	$1 billion – $10 billion	Regional media coverage or severe national outrage. Threat of or loss of license to operate for affected business/site. Likely to lead to change of regulations at the national level.
3	1 or more acute or chronic fatalities. Multiple permanent injuries or irreversible health effects.	Permanent injury or irreversible health effect affecting a single person. Non-permanent injuries or short-term health effects affecting multiple people.	Uncontained release of reportable quantity, e.g., >100 bbl of oil, less if in a sensitive location. Extensive short-term pollution/contamination. Prolonged pollution/contamination affecting a limited area.	$1 million – $10 million	National media attention or severe local outrage. Prosecution by regulators.
2	Multiple moderate injuries, reversible health effects	Minor injury or short-term health effects to a single person.	< 100 bbl of oil, no short-term pollution/contamination.	< $100K - $1 million	Limited to internal stakeholders.
1	First aid injuries	No impact on public.	Immaterial environmental contamination.	<$100K	Immaterial reputational damage.

Source: Adopted from *John Atherton and Frederic Gill, "Incidents that Define Process Safety" Center for Chemical Process Safety: a John Wiley & Sons Inc. Publication 2008.*

3.3.3 Analyze Consequence

The consequences of process safety events should be assessed before the likelihood is assessed. It should be noted that a process safety risk event may have multiple consequences in different categories, as identified in Table **1**. Once the consequences across all categories are determined, the worst credible consequence should be attributed as the final consequence of the sub-risk. For example, a release of toxic substance may not have severe environmental consequences; it could have severe health and safety (onsite worker) consequences. Similarly, a blowout at an offshore location could result in a significant environmental damage, as compared to health and safety consequences. Therefore, the consequences of each sub-risk need to be carefully evaluated and documented.

3.3.4 Analyze Likelihood

The next step is assessing the likelihood, which could be expressed as being in one of five possible ranges of probabilities.

1. Highly likely
2. Likely
3. Possible
4. Remotely possible
5. Unlikely

Assessing the probability of a particular sub-risk event is usually subjective and can be difficult to apply in practice. Testing the likelihood against different time periods (such as 1 year, 5 years, or 10 years) can provide a useful "sense check." The team should conduct this exercise as a workshop and obtain the consensus and agreement of all participants involved, as well as management. For process safety risk management methodologies, it is quite possible to derive the likelihood from historical incident data.

Frequency of past incidents (especially the high-impact risks that ERM focuses on) may be a useful input in the analysis, but is not a substitute for applying judgments about the likelihood of a future sub-risk event.

3.3.5 Determine Risk Severity

Risk severity is derived from the consequences and the likelihood, and used in prioritizing and reporting risk. It is obtained by reference to a heat map, such as the one outlined below in Figure 2.

Neither the color depicted, nor the severity level, should be confused with risk appetite. Severity is merely a means of identifying prioritization to ensure that adequate focus, in terms of treatment and management, is placed on the most significant sub-risks.

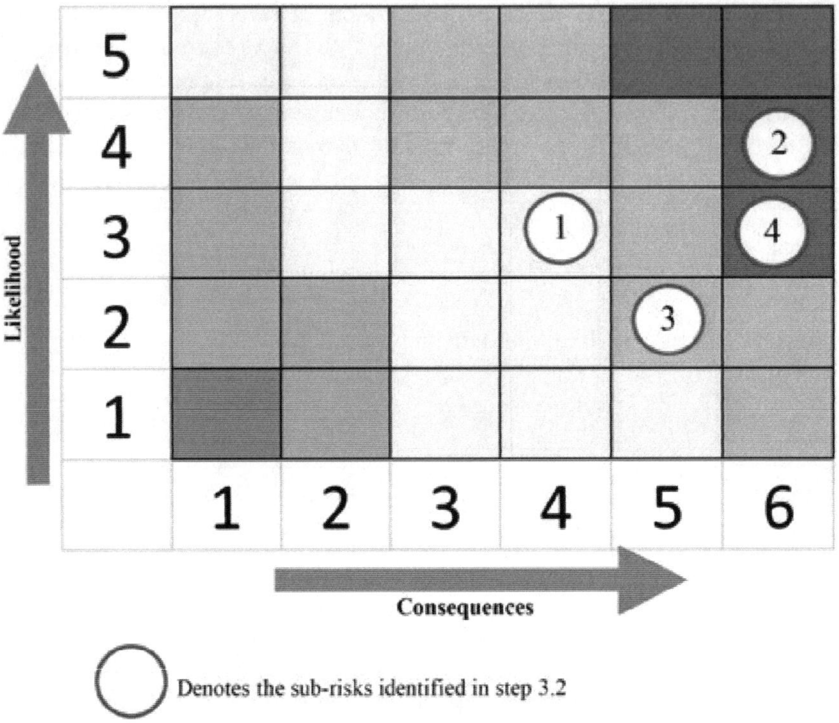

Figure 2 - An example risk assessment matrix (RAM)

3.3.6 Assess Controls

Controls, including safeguards and layers of protections are measures applied by a company that are intended to reduce risk. Controls are associated with the process safety sub-risks, and therefore need to be identified and assessed to analyze a sub-risk's likelihood and potential consequence. Generally, a bow-tie diagram is helpful in understanding corporate level controls used for reducing the risk of a major industrial incident. This helps in defining the type of control, whether it is preventive (trying to prevent an incident from occurring) or reactive (trying to reduce consequence levels).

The controls for a specific process safety sub-risk should be defined at various levels, such as management system level, design and technology, facility processes, administrative, and human aspects. The health of a company's safeguards should be evaluated to define what is working and what might require attention. This is solely based on the knowledge and experience of the team regarding the practices at various sites. This evaluation can be supported by specific findings from various audit and compliance review activities.

3.4 Evaluate

This step involves measuring the level of assessed risks against the willingness of the company to accept the risk or deciding to address the risk based on the control effectiveness, possibility of risk manageability, and cost-benefit analysis of managing the risk. This step also includes the process of determining the treatment objective of each sub-risk and developing a broader strategy for achieving that objective. Having understood how risks are currently being controlled and the effectiveness of the controls, it should now be possible to determine the risk management objective and the strategy to achieve it.

Applying this step to the process safety sub-risks could mean that each sub-risk is evaluated and a strategy is applied to reduce the likelihood and/or consequence of the sub-risk. The taskforce should document the evaluation and be in agreement with corporate management on the treatment strategy for each sub-risk. This will ensure that each sub-risk is evaluated and an informed and collective decision is made by corporate management on how to deal with the risk.

3.5 Treat

Risk treatment involves selecting one or more options for adjusting risks and applying those options to achieve the strategic objectives as agreed in Step 3.4 (Evaluate) above. The "Four-Ts" approach is a model that can be used in risk management to support the formulation of a risk management strategy.

1. Terminate: The activity that gives rise to a risk is ended. In this option, the objective is to eliminate or avoid a risk by removing its source or deciding not to start or continue with an activity that gives rise to it. This could translate into closing an aging facility or opting not to build a plant that could generate a high level of risk for the company.
2. Treat: Reduce the risk by introducing new controls or improving the existing ones. In this option, the strategy is to apply additional controls and treatments or improve existing controls that manage the likelihood and/or the consequence of the risk to a more desirable level. This option is most often applied to process safety and generally identifies areas of improvements related to the process safety management system including enhancing field implementation practices, developing additional control processes, enhancing competencies related to process safety, and improving process safety governance in the field. One of the ways to achieve this is through the application of recognized and generally accepted good engineering practices (RAGAGEP).
3. Tolerate: Accepting risk. Some organizations could decide to accept risk at its face value as it is low enough or inherent to the nature of a business. This option is also utilized when the cost of treatment is greater than the risk consequence or when there are no effective treatments within management control such as "external factors" impacting its likelihood. This option is also applied in combination with the "treat" option (above), where the organization could decide to tolerate residual risk after treatments have been implemented and the risk is brought down to as low as reasonably practicable (ALARP).
4. Transfer: Sharing risk with another party or parties through insurance, outsourcing, or other contractual arrangements. This involves transferring potential risk and its consequences to a third party that is more capable of managing it.

Risk treatment also includes developing and implementing tasks to achieve the treatment objectives and strategies as identified above. These tasks are documented in a treatment plan that specifies tasks to be taken in support of the strategy, how they will be implemented, when, and by who. Selecting the most appropriate treatments involves balancing the costs and efforts of implementation against its benefits, and the treatment plan should identify priorities for treatment implementation. Components of the treatment plan should be specific, measurable, achievable, realistic, and time-bound. Once these treatment plans are approved, they should be executed and undergo periodical review and monitoring. The taskforce should ensure that:

- Treatment objectives and strategies are documented.
- Task owners are identified, made accountable, and have sufficient authority and resources.
- Completion dates are identified and agreed.
- Necessary authorizations are granted.

When applied to process safety, this step would involve the development of the details for an identified treatment strategy and how to execute tasks. The taskforce should work with assigned action owners to ensure a comprehensive treatment plan with clearly defined gaps, expected benefits, stakeholders, and specific tasks, as well as an estimated time of completion. Once these are documented and agreed upon, the taskforce should work with assigned action owners to ensure a successful launch of the treatment execution program.

Some treatments related to field implementation of existing process safety controls, such as the improving management of change (MOC) practices, upgrading aging equipment, or rolling out new leak detection technology, should be given special emphasis as these can be seen as "low-hanging fruit" and can bring about quick changes at the ground level. Treatments addressing enhancements in management systems, improvements in safety culture, and behavioral changes can take time to be fully implemented and executed. Organizations should not lose focus on these long-term areas of improvements, as they will bring about long-term sustainable enhancements to reduce the overall process safety risk.

It is also important to ensure that treatment impact and effectiveness in the field is measured and reported as part of the treatment execution.

3.6 Monitor and Review

Without proper monitoring and review, risk management processes and appropriate information and responses could eventually become obsolete. Monitoring and review provides assurance that:

- Major risks have been identified and remain relevant.
- Controls continue to be effective and efficient in design and operation.
- Risk management objectives remain appropriate and are supported by effective treatment activities.

A monitoring and review plan should include:

- Identifying appropriate and specific indicators of risk trends, including severity, application and effectiveness of controls, and implementation of treatment plans.

These are commonly referred to as Key Risk Indicators (KRIs) [6] and should be collected and reported on a regular predefined basis.

- Defining responsibilities, mechanisms, and schedules for collecting, analyzing, and reporting KRI data.

When applied to process safety, the KRI development should involve a review of industry-recognized key performance indicators related to process safety. There are industry guidelines [7] and survey results [8] that can be used as a starting point. In addition, a company should consider existing process safety metrics at various levels within organizations. A list of all potential metrics should be generated; all potential metrics cannot be monitored at a corporate level. This is not practical; therefore, a selection criterion should be used to filter relevant process safety metrics. The selection criterion could be as follows:

- Industry practice for adopting the metric at a corporate level.
- Credibility of the data.
- Ease of collection of data.
- Frequency of collection.

The criteria could be adjusted to fit the organizational need and applicability of the KRIs. The list of potential metrics is then filtered, based on the finalized criteria, to select only a few (4-6) KRIs that can then be used for monitoring process safety risk at a corporate level. These KRIs could be a combination of lagging and leading indicators; however, more focus should be given to leading indicators due to their predictive nature. Lagging indicators could be used for benchmarking an organization's performance against the industry and to keep an eye on the reactive controls. Some examples of the potential corporate level KRIs are highlighted in Table 2.

Table 2 - Example of Potential Corporate Level Key Risk Indicators (KRIs)

No.	Indicator
1.	Process Safety Event Rate
2.	Overdue (Process Safety) Audit Action Items
3.	Process Safety Competency Assessment
4.	Overdue Incident Investigation Items
5.	Primary Containment Inspection or Testing Results outside Acceptable Limits
6.	Other Loss of Containment Events (Tier 3)
7.	Overdue Inspection Items
8.	Safe Operating Limit Excursions
9.	Demand on Safety Systems
10.	Safety Systems that Failed on Demand

4. Discussion

ERM's 6-step framework provides a unique approach for corporate management to assess and manage process safety risks at a corporate level. However, there is the possibility that organizations may encounter some initial challenges when applying this framework.

Table 3 below provides a list of potential key challenges for each step and recommendations to overcome those issues.

Table 3 - List of potential key challenges and recommendations

Step	Potential Challenges	Recommendations
Context	Securing buy-in from corporate management	• Ensure that purpose of establishing this program is linked to business objectives. • Conduct awareness sessions for executive management
	Multidisciplinary team formation	• Establish strong technical leadership for the taskforce. Develop clear roles and responsibilities for members of the team to avoid any possible conflicts.
Identify	Multiples sub-risks identified	• Focus on events that could lead to certain high potential consequences with certain level of likelihood. Avoid listing all hazards (similar to Process Hazard Analyses). • Consolidate a list of only a few sub-risks by merging similar sub-risks together. Some sub-risks can be effectively handled by the facility management and do not require escalation to corporate management.
	Prioritization of sub-risks	• Document why the sub-risks are a potential threat to business objectives. Justify what was considered during this step and why certain sub-risks have been delegated/parked? This information is also helpful during the subsequent cycle of review.
Analyze	Exaggerating the credible worst-case scenario	• Utilize a balanced view of the hypothetical worst-case scenario and the credible worst-case scenario. Also, note what is credible for some organizations may not be the same for others and vice versa.
Evaluate	Under-evaluation of impact on sub-risks	• Conduct and document the qualitative evaluation (high, medium, low) of the impact of each treatment on the relevant sub-risk to determine if the desired impact on the risk can be achieved. Even though it is a subjective exercise, it is quite useful when presenting the expected sub-risk levels after treatment implementation to the executive management.
Treat	Generic Objectives	• Be as specific as possible in the treatment objectives. Avoid using generic verbiage.

	Impact and effectiveness measurements	• Establish the baseline data before executing the treatments to help gauge the impact and effectiveness of treatment after implementation.
Monitor & Review	Limitation of KRIs	• Develop only a few number of KRIs, to be monitored at corporate level, that provide a good preview of overall process safety health. Additional KRIs could be monitored at site level/business line levels as appropriate.
	Complexity of KRIs	• Ensure that KRIs objectives are clear and calculations are simple and easy to be understood by corporate management. A complex KRI may not be of interest to corporate management and could potentially be misinterpreted.
	Lagging vs leading indicators	• Ensure a dual assurance model when selecting the lagging and leading indicators. KRIs should be strategically selected to complement each other.

5. Conclusions

Board members of large corporate organizations require an understanding of potential major risks associated with their company. The ERM's 6-step process can be utilized by corporate management to conduct a structured analysis aimed at managing any major risks faced by a company, including the risk of a major process safety incident. The identification of process safety sub-risks, evaluation of existing controls, and addressing the gaps in those controls will help ensure that corporate management remains aware of any potential process safety risks. Specific treatments are developed to enhance controls, and a set of Key Risk Indicators can help alleviate risk monitoring at a corporate level. The process is cyclical and repeatable to ensure that corporate management is aware of any emerging risks and can provide support to manage these risks in a timely fashion.

6. References

[1.] ISO 31000:2009 "Risk Management-Principles and guidelines" International Organization for Standardization, First Edition, published 2009.

[2.] Fontevecchia, Agustino "BP Fighting A Two Front War As Macondo Continues To Bite And Production Drops – Forbes." Forbes. https://www.forbes.com/sites/afontevecchia/2013/02/05/bp-fighting-a-two-front-war-as-macondo-continues-to-bite-and-production-drops/ Accessed on 6 February 2018.

[3.] "Overview of Enterprise Risk Management" Casualty Actuarial Society, May 2003 http://www.casact.org/pubs/forum/03sforum/03sf099.pdf, Accessed on 6 February 2018.

[4.] "Briefing Sheet: Enterprise Risk Management"

http://www.willis.com/Documents/Publications/Services/Enterprise_Risk_Management/BriefingSheet_ERM.pdf Accessed on 6 February 2018.

[5.] John Atherton and Frederic Gill, "Incidents that Define Process Safety" Center for Chemical Process Safety: a John Wiley & Sons Inc. Publication 2008.

[6.] Mark S. Beasley, Bruce C. Branson, Bonnie V. Hancock, "Developing Key Risk Indicators to Strengthen Enterprise Risk Management" https://www.coso.org/Documents/COSO-KRI-Paper-Full-FINAL-for-Web-Posting-Dec110-000.pdf Accessed on 11 February 2018.

[7.] "Guidelines for Process Safety Metrics," Center for Chemical Process Safety, a John Wiley & Sons Inc. Publication 2010.

[8.] "Process Safety – Leading Key Performance Indicators," Supplement to Report 456, International Association of Oil & Gas Producers, July 2016.

What is the Safest Way to Move LNG?

Ryan J. Hart, PhD, PE
Exponent, Inc.
4580 Weaver Parkway, Suite 100
Warrenville, IL 60555
rhart@exponent.com

Delmar "Trey" Morrison, PhD, PE
Miguel Garcia, PhD
Exponent, Inc.

Prepared for Presentation at
American Institute of Chemical Engineers
2018 Spring Meeting and 14th Global Congress on Process Safety
Orlando, Florida
April 22 – 25, 2018

What is the Safest Way to Move LNG?

Ryan J. Hart, PhD, PE
Exponent, Inc.
4580 Weaver Parkway, Suite 100
Warrenville, IL 60555
rhart@exponent.com

Delmar "Trey" Morrison, PhD, PE
Miguel Garcia, PhD
Exponent, Inc.

Keywords: quantitative risk analysis, LNG

Abstract

The abundant supply of domestic natural gas in the US has led to the rapid development of infrastructure to process, handle, store, and transport the commodity. Liquefying the natural gas and handling as the cryogenic liquid LNG has been attractive due to low storage pressures and relatively high energy density. The LNG can then be stored onsite or distributed to end users. The three primary transport methods that can be considered for moving the LNG from the remote liquefaction facility to the domestic end user include: truck, rail, and pipeline. All three of these transport methods have pros and cons associated with them, including: regulatory requirements, deployment time scales, existing infrastructure, cost, and risk to the public. The risk of transporting LNG via any of these three methods is explored in detail here using commercially available software (PHAST Risk) along an example route. The resulting societal and individual risk will be compared for the different transport modes.

1 Introduction

Increasing domestic production of natural gas has quickly outpaced domestic consumption. This has resulted in record levels of underground gas storage capacity and saturation of pipeline capacity; in turn, this has increased the pressure on getting gas from the production facilities to end-users (whether domestic or foreign). The United States is no stranger to accommodating natural gas storage in regions where existing infrastructure does not allow – through the use of LNG peak shaving facilities. Unlike traditional peak shaving strategies, however, the current high demand of natural gas on the gas pipelines and limited gas pipeline infrastructure reaching all sources and users may not necessarily accommodate the field-to-peak shaver delivery of gas, or the LNG-to-gas redelivery when demand spikes. Thus, there has been increasing interest in production of LNG near the well fields (close to the source) coupled with transport of the LNG to the various end users (whether they be peak shaving facilities, ports, or filling stations).

There are several options potentially available to transport the field-produced LNG to the various end users, including: pipeline, over the road, or rail. Note that for this analysis, vacuum insulated double wall pipelines are considered. Each method has various pros and cons, related to logistics, costs, and availability.

Table 1. Summary of potential LNG transport modes.

Consideration	Pipeline	Over the Road	Rail
Ease of access	Limited endpoints, may require additional transport mode to get to end-user	Can deliver directly to end-user	Limited endpoints, may require additional transport mode to get to end-user
Cost	Highest capital expenditure, significant infrastructure and operating costs	Lowest equipment cost, may fluctuate considerably depending on trailer/driver availability	Higher equipment cost than over the road, but lower shipping costs than over the road
Availability	Requires construction, long timeline to use	Utilize existing roads, new access roads can be built quickly	Limited to existing main line infrastructure, additional spurs to reach end-users can be built quickly
Proximity to public	May be routed to avoid high population centers depending on route availability	May go through high population centers, but abundance of routes makes limiting exposure possible	Routes are fixed and may need to pass through high population centers
Reliability	Highest; delivery rates are constant	Generally reliable assuming availability of trailers/drivers; subject to weather disruptions	Generally reliable if on a schedule pickup/delivery route; less subject to weather disruptions
Delivery timing	Quickest delivery rates; capped by size of pipeline	Relatively fast, fleet limitations may limit delivery capacity	Slowest but can accommodate large shipment sizes

Risk inherently influences the selection of transport mode, but the risk may not necessarily be well characterized for a given application. The models and constraints for calculation of risk also differ to an extent between the modes of transport. Oftentimes the question 'What is the safest way to move the LNG?' goes unanswered. To address this question, one may turn to the use of a Quantitative Risk Assessment (QRA) to estimate the risk associated with the various transport modes available. In this paper, we use QRA to compare the three potential transport modes for a hypothetical scenario of LNG production in a gas field with transport to storage/end-users. For this example, the route is 500 miles long (approximately the distance from Lubbock to Houston). The transport capacity considered here is 430,000 gallons/day of LNG.

1.1 Understanding Risk

Risk, simply defined, is the potential to lose something of value. Risk is evaluated by taking the product of event likelihood with the event outcome severity, and then comparing the product to some benchmark risk which is considered by the stakeholders as being acceptable. Unlike qualitative risk assessment methodologies where risk matrices are used to assign the approximate risk of a given process, a QRA assigns a mathematical value to risk. A QRA is comprised of several integrated analyses that culminate in the quantification of risk:

- Identification of systems/hazards – What systems are included in the scope of the QRA and what hazards are represented are first identified.
- Development of failure rates – Event tree methods are routinely used to determine the likelihood that a system or component creates a hazardous environment. These event trees may consider layers of protection or mitigation strategies that are designed to prevent a hazardous event. Many data sets exist which may be used to develop event trees.
- Determination of hazardous consequence – Mathematical models or empirical tables can be used to assess the potential impact from a hazardous event. These calculations consider the potentially affected populations or exposed assets in order to quantify the damage extent, usually as a function of time. Damage criteria, such as toxic or flammable endpoints, shall be defined for the specific systems/hazards.
- Analysis of Receptors – the potential receptors surrounding the proposed activity shall be determined, and may include fractions of time indoors and fluctuations in exposure based on time of day.
- Calculation of risk – The previous three steps are all combined in order to determine the risk of a proposed facility, process, or component. Meteorological data and indoor/outdoor populations are often considered in order to determine the extent of the hazardous event effects. Risk is reported as both individual risk (IR) and societal risk SR).

Here the commercially available software tool PHAST Risk was used to perform the QRA calculations for the three potential LNG transport modes.

1.2 LNG Hazards

LNG poses unique hazards relative to other non-pressurized liquid fuels. LNG has a shipping identification number of UN1972 for refrigerated cryogenic methane. LNG, comprised primarily of methane, has a flammable range when mixed with air in concentrations of approximately 5% to 15%; outside of this range, the fuel will not burn. The liquefaction of natural gas is achieved by cooling the material to its normal boiling point, - 260°F. This is unlike other low molecular weight hydrocarbon fuels, like propane, which can be liquefied by pressurization. At the boiling point temperature, LNG does not need to be stored under pressure but it must be insulated to avoid excessive boiling due to heat transfer. As the liquid boils, it does so at its constant, low boiling point temperature. To avoid excessive pressure buildup under extended duration (e.g., on the order of 30 days) storage conditions, LNG transport containers will typically vent low volumes of natural gas to the atmosphere via a pressure relief valve.

The cryogenic temperatures of LNG pose unique hazards to operating personnel. Due to a large difference in temperature, the rapid transfer of heat from an object into the cryogenic liquid can cause burns if direct contact with skin occurs or if PPE is inadequate to prevent cold-temperature injury due to an exposure. Additionally, large spills of the liquid onto metal structures can cause embrittlement and fracturing. Methane is odorless and LNG contains no odorant (unlike residential natural gas supplies), making detection difficult without a flammable gas detector device.

The behavior of a spill of LNG is unique due to the cryogenic temperature of the liquid. For example, a spill of LNG will vaporize rapidly when it contacts ambient air and even faster when in contact with warm solids such as the ground. The cold vapors may condense humid air, causing fog formation and decreased visibility. After vaporization, the cold vapors are denser than ambient air and they will tend to stay close to the ground as they disperse, getting pushed by prevailing winds. The dense vapors can travel great distances without significant dilution, as the mixing with ambient air is limited near the ground, and the vapor will tend to accumulate in low spots or trenches along the ground.

The operational hazards of handling LNG were not considered in this study; only large scale releases and ignition that could cause fire and explosion events were explored. The specific fire and explosion scenarios, as well as release, ignition, and consequence probabilities are discussed below.

2 Quantitative Risk Assessment

A quantitative risk assessment considers both the likelihood that an event occurs, combined with the hazardous outcomes of the event. The modeling tool PHAST Risk was used to calculate the Societal Risk (SR) and Individual Risk (IR) for the three transport modes. The model inputs used in the calculations are provided below.

2.1 Accident Models

The likelihood of a release of LNG is evaluated through the use of a release model. The model used here considers an initiating event (accident for truck/rail transport and release) coupled with an outflow probability (e.g. hole size). The accident models for rail, truck, and pipeline transport are described below.

2.1.1 Rail Accident Model

For the purposes of this study, it was assumed that the LNG would be transported by rail in 30,000 gallon DOT-113 rail cars. For the basis of 430,000 gallons of LNG per day, it was assumed that the LNG was transported in a unit train (100 cars of LNG) once per week. The likelihood of experiencing a release of LNG is summarized by the successive steps described in Figure 1.

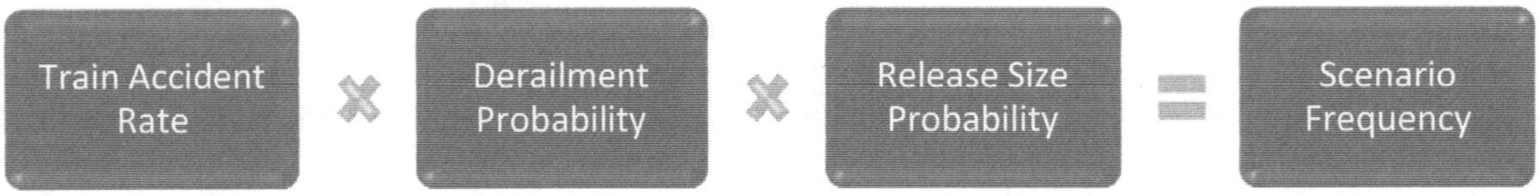

Figure 1. LNG DOT-113 train accident model overview.

LNG shipping by rail is historically uncommon in recent U.S. rail industry history; thus, accident data for the movement of LNG in DOT-113 rail cars do not exist. Publicly-available data from the Federal Railroad Administration (FRA) was used to estimate train accident rates for the QRA. The FRA Office of Safety Analysis maintains an online database that provides historical accident and failure rate data for the rail industry [1]. Accidents in the database include broken equipment, highway grade crossing collisions, train collisions, and derailments. The FRA industry-wide database for train accidents with reportable damage data [2] was first queried and downloaded for all accident reports during the twenty year period from 1997-2016, yielding a total count of 65,940 accidents. The accidents are identified in the database by category and include multiple types of collisions, explosions, fires, other impacts, and other events. These types of accidents are consistent with the events necessary to lead to a release of LNG from a DOT-113 rail car. The FRA data was filtered for all accidents from 1997-2016 (all railroad classes), and the results were analyzed to determine accident frequency for mainline accidents and derailment probability at high speed conditions. The values are summarized in Table 2 \.

Table 2. Train accident rates from FRA data.

	Statistic	**1997-2016**
Mainline, Speed > 25 to ≤ 50 mph	Total Accidents	6,580
	Total Derailments	3,501
	% of All Accidents	9.98%
	Probability that Derailment Occurs	53.2%

The raw accident numbers were then divided by train mileage to develop accident frequency estimates as accidents per train mile for the QRA. Operational data tables provided by the FRA were used to determine the total number of mainline (non-yard) train miles for the period from 1997-2016 for all classes of railroad represented in the data [3]. The operational data tables did not subdivide the mainline train miles according to track speed; thus, only the total accident frequency per mainline train mile for all trains could be calculated. The average accident frequency was found to be 2.42×10^{-6} (accidents/train mile) for the mainline travel. The results are summarized in Table 3.

Table 3. Train accident rates from FRA data.

	Statistic	1997-2016
Mainline	Total Non-Yard (Mainline) Train Miles	12.92×10^{9}
	Non-Yard Accident Rate (/train mile)	2.42×10^{-6}

The mainline accident frequencies from Table 3 were then multiplied by the total number of annual train miles estimated for the example route to arrive at the yearly accident frequency (accidents per year) used in the calculations for the route. A summary of the calculated annual accident rate for the example route is provided in Table 4.

Table 4. Calculated annual accident frequencies for the example mainline route.

Estimated Total Annual Route Length (train miles/yr)	Accident Frequency (accident/train mile)	Calculated Annual Accident Frequency (accident/yr)
26,000	2.42×10^{-6}	6.29×10^{-2}

Not all accidents-leading-to-derailment will involve an LNG DOT-113 car. Several factors are expected to affect the likelihood that an LNG DOT-113 car is derailed including: (1) the position of the LNG DOT-113 car(s) within the train and (2) the number of LNG DOT-113s grouped together. In this analysis, it was conservatively assumed that the unit-trains of LNG DOT-113s started at train position eleven (11) and continued throughout the remainder of the train. The historical FRA accident data was analyzed to develop a model for estimating the probability of derailment of an individual car versus its position in the train.

The probability of derailment for one or more LNG DOT-113 cars is dependent on the position of the first car derailed in the train, the average number of cars derailed during an accident, and the location of LNG DOT-113s in the train. The FRA 20-year accident data from 1997-2016 was first filtered to include only mainline accidents, with derailment, from speeds greater than 25 mph up to 50 mph. The average number of cars derailed was determined to be 11.

Based upon the dynamics of a derailment, it was assumed that in an average derailment, the first car would derail plus the immediately following sequence of n-1 cars would derail, where n is the average number of cars derailed. The data reveal that when a train accident results in a derailment, the first car derailed is usually the head car (position 1). In fact, the first car derailed is one of the first ten cars in nearly a third (28%) of all mainline derailments where train speeds are between 25 mph and 50 mph. Representative probability values of first car derailed versus position are provided in Table 5.

Table 5. Representative probability of position-in-train of first car derailed (1997-2016) given that an accident with derailment has occurred.

Statistic	Car Position in Train			
	1	11	21	31
Mainline Derailment Accident, Speed > 25 to ≤ 50 mph	13.4%	1.20%	0.91%	0.80%

Assuming the train configuration is described by LNG DOT-113 tank cars starting at position 11 and continuing until the end of the train, the probability of having a given number of LNG DOT-113 rail cars involved in the derailment was calculated. These probabilities are provided in Table 6.

Table 6. Probability of having 'X' number of LNG DOT-113 rail cars involved in a train accident with derailment, by train speed.

'X' Number of LNG DOT-113 Cars Involved in Train Accident with Derailment	Probability of 'X' Number of LNG DOT-113 Cars Involved in Train Accident with Derailment
	High Speed (25 - 50 mph)
1	13.4%
2	2.17%
3	3.06%
4	1.54%
5	1.60%
6	1.40%
7	1.49%
8	1.23%
9	0.943%
10	1.40%
11	71.8%

The release probability data for LNG in DOT-113 tank cars does not exist, so general rail industry data was used and reasonable engineering assumptions were made, as necessary. Pressure tank cars and cryogenic tank cars have an extensive history of operation with corresponding accident data, and with some engineering judgement, this type of accident data was applied to shipping LNG DOT-113s. The Pipeline and Hazardous Materials Safety Administration (PHMSA) maintains an online database that provides historical loss of containment data for rail tank cars, among other transportation vessels [4]. The database complements the FRA database in that the PHMSA database records the inventory of HAZMAT cargo released for each accident; whereas, the FRA database only identifies that a release has occurred. The PHMSA database provided relatively scant accident data for DOT-113 rail cars (and did not contain catastrophic release scenarios), but it did list a significant number of pressure tank car release accidents. Thus, pressure tank cars were used as an analog to estimate the probability of a release if a car was derailed.

The PHMSA database listed accident data from 1971 to the present. All rail car data was queried from 1971 to 2017, for incidents including spillage, vapor (gas) dispersion, and no release. The resulting data was then filtered for pressure tank cars only, and incidents where no tank car specification was available were excluded from the analysis. The resulting 5,542 pressure tank car incidents (as of September 20, 2017) were then sorted by amount released (units are either cubic feet (ft^3) or gallons).

The PHMSA data was grouped into four release volume ranges in order to estimate the probability of a certain leak size. The categories were no release (less than 100 gallons), small release (100 to 1,000 gallons), large release (1,000-30,000 gallons), and catastrophic release (30,000+ gallons) [5]. These volumes were chosen as the PHMSA data appeared to reflect mostly 30,000+ gallon tank cars.

Representative hole sizes were chosen for each release category, in line with PHAST calculation results assuming storage at 90 psig. Small releases were modeled using a ½-inch hole while a 2-inch hole was used for large releases. A catastrophic release assumes that the tank shell has been ruptured, leading to an instantaneous spill of the entire tank contents. Catastrophic releases were thus assumed to represent the PHMSA database cases where 30,000 gallons or more of contents were spilled. The resulting release probabilities are provided in Table 7.

Table 7. LNG rail accident release probabilities.

Quantity Released in gallons	Incident Count	Probability	Release Scenario
≤ 100	5,297	0.955	No Release
$100 < x \leq 1,000$	88	0.016	½-inch Leak
$1,000 < x \leq 30,000$	142	0.026	2-inch Leak
$> 30,000$	15	0.003	Catastrophic

2.1.2 Truck Accident Model

For the purposes of this study, it was assumed that the LNG would be transported by truck in 10,000 gallon ISO shipping containers. For the basis of 430,000 gallons of LNG per day, it was assumed that the LNG was transported by 43 trucks per day. The likelihood of experiencing a release of LNG is summarized by the successive steps described in Figure 2.

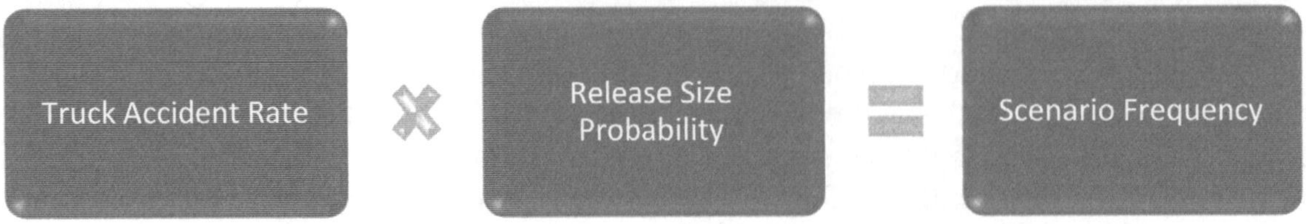

Figure 2. LNG ISO truck accident model overview.

The methodology for developing has been reported previously [6]. The methodology is consistent with the development of the train accident model. The truck accident rate was developed via analysis of the DOT PHMSA online databased for highway accidents of HAZMAT pressure tankers and CFS for the annual mileage. A summary of the road transport accident rates is provided in Table 8.

Table 8. Road transport accident rates for LNG containers.

	LNG (HAZMAT Pressure Tankers)
Annual Miles (CFS)	10,357,300,000
Annual Accidents (PHMSA)	978
Annual Frequency (/mile/yr)	9.4×10^{-8}

The highway accident frequencies from Table 8 were then multiplied by the total number of annual highway miles estimated for the example route to arrive at the yearly accident frequency (accidents per year) used in the calculations for the route. A summary of the calculated annual accident rate for the example route is provided in Table 9.

Table 9. Calculated annual accident frequencies for the example highway route.

Estimated Total Annual Route Length (truck miles/yr)	**Accident Frequency (accident/truck mile)**	**Calculated Annual Accident Frequency (accident/yr)**
7,850,000	9.40×10^{-8}	7.38×10^{-1}

Similar to the rail release probability methodology, the probability of having a release of LNG in the event of a truck accident were determined using the DOT PHMSA database for highway accidents and filtered by release volume. It should be noted that the analysis assumed a catastrophic release as one having larger than 5,000 gallons release (compared to the 10,000 gallon ISO case presented here) as a number of the accidents in the database involved 5,000 gallon bobtail containers. A summary of the truck transport release probability estimates used in this study are provided in Table 10.

Table 10. LNG road accident release probabilities.

Quantity Released in gallons	Incident Count	Probability	Release Scenario
≤ 100	31,514	0.770	No Release
100 < x ≤ 1,000	5,297	0.129	½-inch Leak
1,000 < x ≤ 5,000	2,989	0.073	2-inch Leak
> 5,000	1,127	0.028	Catastrophic

2.1.3 Pipeline Accident Model

For the purposes of this study, it was assumed that the LNG would be transported by pipeline in an 8-inch diameter pipe. Although this is unlikely to be an economically feasible method of transportation, it is evaluated her to demonstrate the theoretical risk. For the basis of 430,000 gallons of LNG per day, it was assumed that the LNG was transported at 300 gpm. The likelihood of experiencing a release of LNG is summarized by the successive steps described in Figure 3.

Figure 3. LNG pipeline accident model overview.

Both the pipeline accident rate and release size probability were based on published data for onshore liquid hydrocarbon (i.e. oil) pipelines [7]. For the 8-inch LNG pipeline, the calculated failure frequency is described in Table 11.

Table 11. Calculated annual release frequencies for the example pipeline route.

Estimated Total Annual Route Length (pipe miles/yr)	Release Frequency (accident/pipe mile)	Calculated Annual Release Frequency (accident/yr)
500	1.3×10^{-3}	6.5×10^{-1}

The hole size probabilities for a release of LNG from the 8-inch pipeline are summarized in Table 12.

Table 12. LNG pipeline release probabilities.

Probability	Release Scenario
0.460	½-inch Leak
0.020	2-inch Leak
0.340	Catastrophic

2.2 Population Density

The population surrounding a release event influences both the probability of ignition (which affects both the IR and SR) and the consequence impact (for SR). Three population densities were considered in the analysis: 100/mile^2, 1,000/mile^2, and 10,000/mile^2. These population densities correspond to transport through a rural area, a suburban area, and a high-density city area, respectively. It was assumed that the example 500-mile route had a composition described in Table 13. It should be noted that although it may be expected for the different transport modes to be exposed to different population densities (e.g. pipeline may be routed away from dense populations), they were assumed to have the same potential exposures here for comparison.

Table 13. Summary of the composition of the example 500-mile route.

Population Type	Population Density (people/mile^2)	Fraction of Total Miles
Urban	10,000	10%
Suburban	1,000	25%
Rural	100	65%

2.3 Flammable Cloud Formation

Several types of computational tools are available that can calculate LNG releases, flashing of the liquids, flammable vapor cloud formation, and cloud dispersion. Depending upon the computational tool, implementing the results into a QRA can provide varying levels of effort. We have found that using a comprehensive modeling package that incorporates the statistical and hazard impact criteria is the most effective. We used PHAST Risk v6.7 to model release conditions, vaporization, cloud formation and dispersion, flammable cloud envelope, and hazard impact. PHAST Risk is a commercial software package developed and distributed by Det Norske Veritas (DNV). PHAST Risk combines a phenomenological release and consequence analysis model with a risk analysis sub-model to evaluate spills, sprays, and gas dispersions and the resulting toxic, fire, and explosion consequences on populations.

PHAST is widely used for the calculation of hazard distances from the release of several hazardous substances, including LNG. The PHAST code uses the Unified Dispersion Model (UDM) as an integral model to calculate the dispersion following a two-phase pressurized release or an unpressurised release. The PHAST UDM is one of only a few computational models approved by the U.S. Piping and Hazardous Materials Safety Administration (PHMSA) for evaluating LNG release exclusion zones. PHAST model calculations assume that the terrain is completely flat and does not account for any obstructions (either natural or nearby equipment) on the dispersion distance of flammable clouds. In many cases, this assumption produces a conservative overestimate of the distance to hazardous outcomes.

2.4 Ignition of a Flammable Cloud

Given a release of LNG and the formation of a flammable cloud, the hazardous outcomes analyzed in the QRA can only be realized with ignition of the flammable mixture. The timing of the ignition affects the consequence outcome. For example, immediate ignition of the release may result in a pool fire or jet fire (or both); delayed ignition may result in a pool fire, flash fire, or explosion. For each scenario model, PHAST Risk calculates the outcome due to immediate ignition and delayed ignition. The immediate and delayed ignition probabilities used in this analysis are consistent with the guidelines published in the Dutch Purple Book [8].

The PHAST Risk default ignition probability data sets that were assigned for the release scenarios are described as the following:

- "Stationary" ignition probabilities were assigned for the pipeline incidents.
- "Road tanker" ignition probabilities were assigned for the truck incidents.
- "Tank wagon" ignition probabilities were assigned for the rail car incidents.

2.5 Flammable Effects Exposure for a Population

The flammable effects resulting from a flammable release include pool fires, jet fires, flash fires, and BLEVEs. The probability that an exposed population will suffer a fatality due to exposure to a flammable effect depends on the extent of exposure and protection of the population (indoor versus outdoor). For the IR calculations, it was assumed that the entire population is outdoors. For the SR calculations, the model assumes that 90% of the population is indoors and 10% is outdoors. All calculations assume that people are at ground level, so the ground level effect zones are used in calculating consequence outcomes.

The flammable effects and fatality consequences are calculated in PHAST Risk utilizing a grid cell system to calculate fatalities in effect zones, and the probability of fatality as a function of distance is calculated. The flammable effect in a grid is then compared to the populations in that grid to determine the probability and number of expected fatalities. For the IR calculations, the model only considers whether a person is located in a grid cell, which is always assumed to be yes. To obtain the SR outputs, the flammable effect consequences are integrated by the number of people present in the grid cell (defined as by the population density and size of the grid cell) to obtain the number of expected fatalities.

The flammable effect consequence methods used in this analysis are consistent with the guidelines published in the Dutch Purple Book. The Probit Method, which is dependent on radiation level and exposure time, is used to calculate the probability of fatality from flammable effects to exposed populations for BLEVE, pool fire, and jet fire effects. This method is applied to each grid cell independently and then the cumulative consequence outcome for a specific flammable effect is obtained by summing all the grid cells.

The consequence outcomes for the classes of flammable effects are summarized (the flame envelope is defined as the area between the lower flammable limit, LFL, and upper flammable limit, UFL):

- BLEVE, pool fire, jet fire—all persons, indoor and outdoor, within the flame envelope are considered fatalities. All persons, indoor and outdoor, exposed to radiation levels exceeding 11,000 BTU/hr/ft^2 (35 kW/m^2) are considered fatalities. For smaller radiation levels, the Probit method is utilized to calculate the probability of fatality.
- Flash fire—all persons, indoor and outdoor, within the flame envelope are considered fatalities. All persons, indoor and outdoor, outside of the flame envelope are not considered fatalities.
- Explosion—all persons, indoor and outdoor, exposed to overpressures exceeding 4.35 psig (0.3 barg) are considered fatalities. All indoors persons exposed to pressures exceeding 1.45 psig (0.1 barg) are considered have a probability of fatality 2.5% of the time. All other exposures are not considered fatalities. The TNT explosion method, with a 10% equivalency, is used to calculate the overpressure profile for explosion.

2.6 Release Conditions

As previously mentioned, the release events considered in the analysis include the following leak sizes/types:

- No release;
- Small spill, modelled as a 0.5-inch leak;
- Large spill, modelled as a 2-inch leak;
- Catastrophic leak, modelled as an instantaneous spill of the entire tank contents.

The rail tank car transport scenarios considered 30,000 gallons of LNG in each rail car. For the LNG, the methane was assumed to be stored at 90 psig and saturation temperature (-204°F). All releases were considered to occur at an elevation of 6 ft.

The road transport scenarios considered 10,000 gallons of LNG. For the LNG, the methane was assumed to be stored at 90 psig and saturation temperature. All releases were considered to occur at an elevation of 6 ft.

The pipeline considered assumed the LNG to be transferred at 60 psig and -260°F. Pump runout conditions were not considered for the releases, which may over-estimate the release flow for the large diameter releases. All releases were considered to occur at an elevation of 3 ft (i.e. it was assumed to be an above ground pipeline for the entire length).

2.7 Weather

The weather considered in the QRA analysis assumed an ambient temperature of 68°F and relative humidity of 50%. The wind was assumed to occur with equal probability in any direction at a speed of 2 m/s (no wind speed variations were examined) and a Pasquill Stability Factor F was used. The surface roughness factor used in the calculations was 1 m.

3 QRA Calculation Results

The quantitative risk for the transport of LNG, at a basis 430,000 gallons per day, via rail, truck, and pipeline was calculated using PHAST Risk. The Individual Risk (IR) was also calculated for each of the transport modes at each population density considered. These results are summarized in Table 14.

Table 14. Summary of the IR distances to various contours for the three transport modes considered here, as a function of population density.

	People/mi^2	**IR, yr^{-1}**		
		1×10^{-5}	**1×10^{-6}**	**3×10^{-7}**
		Distance to Contour, ft		
Truck	**100**	--	197	284
	1,000	--	206	307
	10,000	--	293	483
Pipeline	**100**	341	2,720	4,050
	1,000	1,820	4,210	4,660
	10,000	1,340	2,670	3,180
Rail	**100**	--	--	--
	1,000	--	--	135
	10,000	--	--	387

The results of the calculations for the Societal Risk (SR) are summarized in Figure 4 for the highway transport, Figure 5 for pipeline transport, and Figure 6 for rail transport. The SR is reported on a per-mile basis as a function of population density.

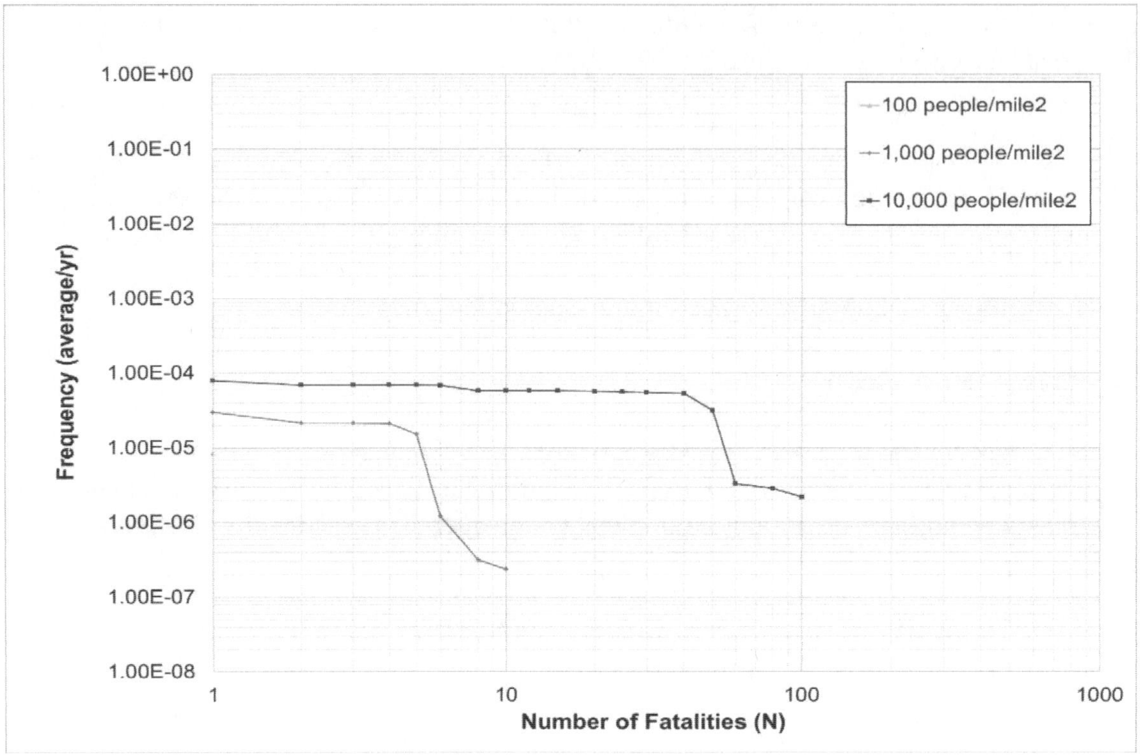

Figure 4. SR curves for the highway LNG transport as a function of population density, calculated on a per-mile basis.

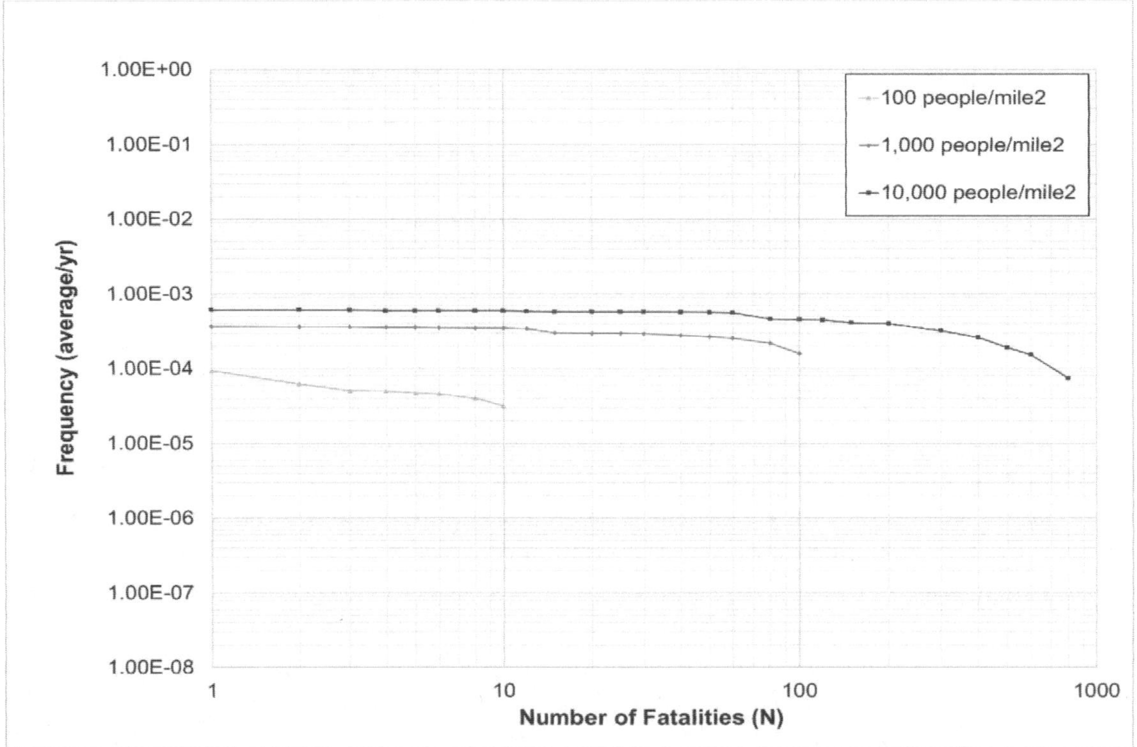

Figure 5. SR curves for the pipeline LNG transport as a function of population density, calculated on a per-mile basis.

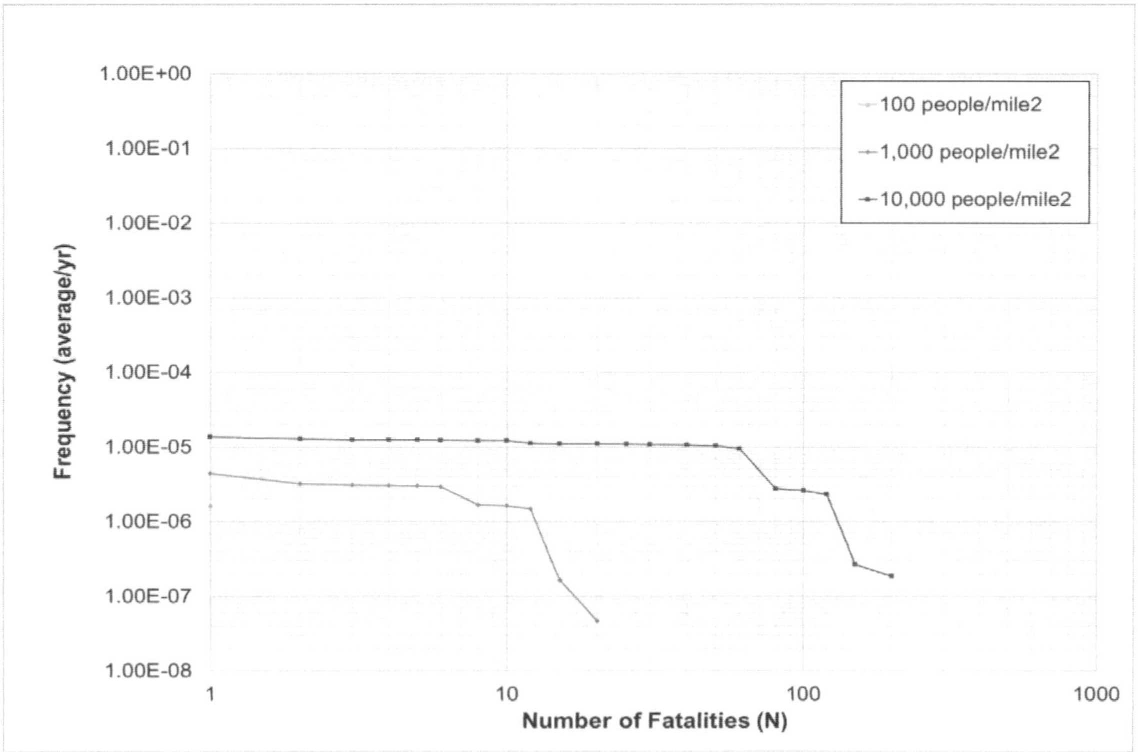

Figure 6. SR curves for the rail LNG transport as a function of population density, calculated on a per-mile basis.

For the 500-mile example route, the per-mile calculations were scaled (based on the composition of the route for rural, suburban, and urban populations) to provide the aggregate SR for the proposed transport operations along the entire route distance. These aggregate SR results are presented in Figure 7.

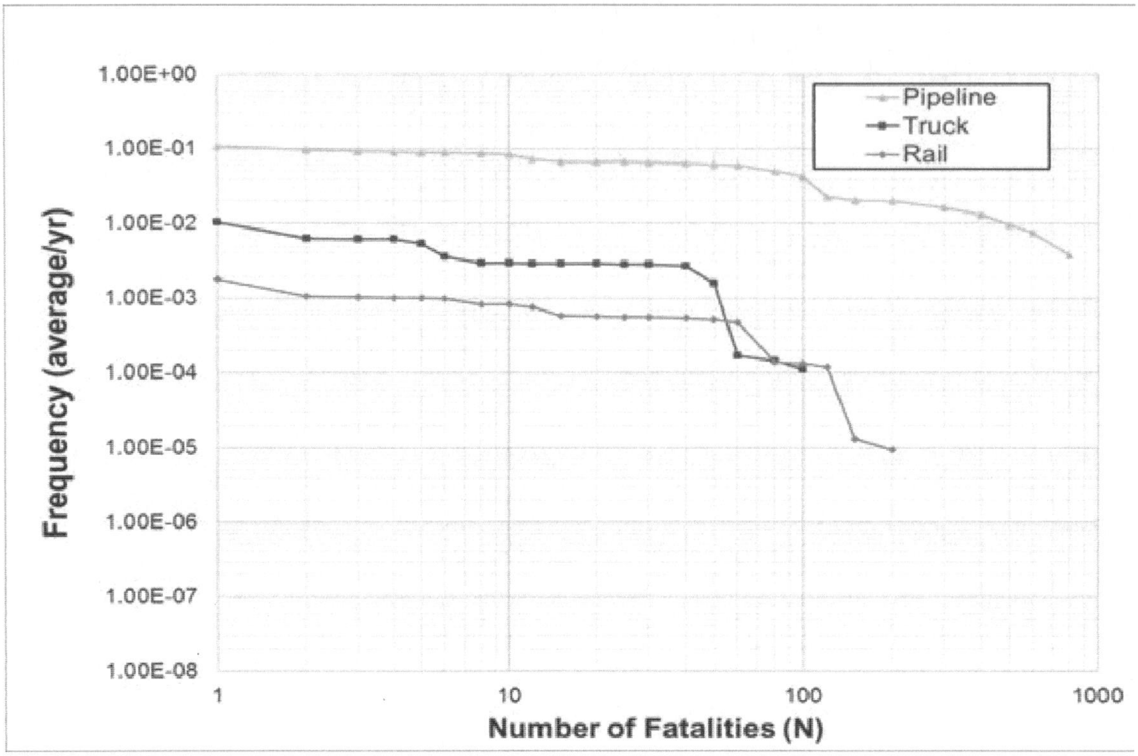

Figure 7. Aggregate risk for the transport of LNG along the example 500-mile route via each of the transport methods considered here.

4 Conclusions

The QRA of LNG transport modes was assessed for a basis of 430,000 gallons/day along an example route of 500 miles. The calculation results indicate that the transport of this amount of LNG along the route considered exhibits the highest risk for pipeline transport and lowest risk for rail transport. However, there are several influencing factors to consider that may have a significant impact on the calculation results. These factors are discussed below.

4.1 Reliability of Accident Model Data

Little to no data exists for the actual transport of LNG at the scales considered in this report. Engineering assumptions and simplifications were made in order to apply available data sets to the LNG transport risk analysis. It should be recognized that the surrogate data may not accurately reflect the performance of LNG-specific transport systems in real world accident situations. For example, LNG ISO containers are likely to be more robust in an accident relative to generic pressure tanker data owing to the unique construction and design requirements of the ISO containers. Similarly, onshore liquid hydrocarbon (i.e. oil) pipeline failure data was used for the LNG pipeline accident model. The basis document for the pipeline data makes reference to a significant portion of pipeline releases occurring later in the lifetime of the pipeline (i.e. 40-50 years), which may over-predict failures of new pipeline systems being design and constructed using modern methods. Additionally, the cryogenic liquid pipeline may be expected to experience releases at a different rate than oil pipelines, in part due to materials of construction.

4.2 Loading/Unloading and Local Delivery

The QRA presented here only considered the transport along an example 500-mile route. The analysis neglected the loading of LNG into the truck containers or rail cars at the production facility, the unloading of LNG at the receiving facility, and any potential local delivery that may be required.

For the case of pipeline transport, the LNG would likely be stored at the production facility in an onsite tank, which is a practice that is likely consistent with the other two transport modes. However, at the receiving facility, the LNG transported via pipeline would likely be received in a tank. There may be additional truck or rail transfer operations that could be required to deliver the LNG to the ultimate end-user. As these operations were not considered in the analysis, consideration of the entire LNG path from production to end-user may result in a higher risk than reported here. Comparing the transport modes with respect to this consideration, one would expect that the pipeline and rail transport modes may have higher risk values than reported here relative to truck transport which has the flexibility to deliver straight to any given end-user.

4.3 Selection of Release Properties

The LNG was modeled as a saturated liquid in the rail and truck scenarios, at what is expected to be the maximum allowable working pressure for the storage tanks. This was done in order to provide a large driving force for release. However, it shall be acknowledge that the selection of release temperature and pressure is likely to have a significant effect on the calculated consequences and, thus, the risk. Similarly, the pipeline releases did not account for pump limitation (i.e. runout) conditions that may result in reduced discharge rates.

4.4 Transport Routing

The analysis presented here assumed that all three transport modes were exposed to the same populations. Pipelines and rail corridors are likely to be less flexible relative to truck transport relative to the routes taken. For instance, it may be possible to route LNG transport routes in such a way to avoid high population centers which would result in a risk reduction (assuming the total route length is maintained). Additionally, the analysis assumed that the LNG pipeline was above ground for the entire length. Burying the pipe underground may significantly reduce the potential consequences of LNG releases from the pipeline and, thus, reduce the risk.

4.5 Transport Capacity

The analysis presented here considered a relatively large volume of LNG being transported on a daily basis via truck, which leads to the elevated risk profile when accounting for 43 truck trips per day, compared to an 8-inch pipeline operating continuously or one-unit train per week. Due to the abundance of trucks required to transport this volume, it inherently has an elevated risk profile due to the likelihood of having a truck accident. The normalized release frequencies are presented in Table 15 for all three transport modes. It should be noted that these are the release frequencies, which are different from the accident frequencies reported above, and consider only the frequency of observing a release. Recall that for the truck and rail accident rates, there is some probability that there is no release of LNG in the event of the accident. For example, only approximately 32% of train accidents result in a release and 23% of truck accidents result in a release.

Table 15. Summary of release frequencies for the transport modes considered here.

Transport Mode	Estimated Total Annual Route Length (miles/yr)	Release Frequency (accident/mile)	Calculated Annual Release Frequency (accident/yr)
Rail	26,000	4.08×10^{-7}	1.06×10^{-2}
Road	7,850,000	2.16×10^{-8}	1.70×10^{-1}
Pipeline	500	1.3×10^{-3}	6.5×10^{-1}

The accident rates indicate that when considering only a single trip, the truck transport mode is the least likely to result in a release of LNG. However, when scaled to account for the capacity considered as the basis in this analysis, it becomes more likely that truck transport results in an observed release relative to rail transport.

5 References

[1] Accessible via safetydata.fra.dot.gov.

[2] FRA Office of Safety Analysis, Report 3.16 – Summary of Train Accidents with Reportable Damage, Casualties, and Major Causes.

[3] FRA Office of Safety Analysis, Report 1.02 – Operational Data Tables.

[4] Accessible via hazmatonline.phmsa.dot.gov/IncidentReportsSearch/search.aspx.

[5] Section 3.3.3.3, Railways, page 3.13 in *Guideline for Quantitative Risk Assessment, Part Two: Transport* (Dutch Purple Book), Publication Series on Dangerous Substances, Ministerie van Verkeer en Waterstaat (2005).

[6] Hart, RJ and DR Morrison. "The Hazard We Know: Comparing Transport Risk of LPG and LNG," 11th Global Congress on Process Safety, Austin, TX (2015).

[7] "DNV Recommended Failure Rates for Pipelines," Report No. 2009-1115 (Rev 1), November 16, 2010.

[8] Chapter 4.7, Ignition, in *Guideline for Quantitative Risk Assessment* (Dutch Purple Book), Publication Series on Dangerous Substances, Ministerie van Verkeer en Waterstaat (2005).

18 AIChE Spring Meeting

14TH GLOBAL CONGRESS ON PROCESS SAFETY

ORLANDO FLORIDA 2018

Beyond HAZOP

Analyzing Common Cause and System Scenarios

Eloise Roche
SIS-TECH Solutions, LP
Houston, TX
eroche@sis-tech.com

Watson Dupont; Angela Summers
SIS-TECH Solutions, LP
Houston, TX
wdupont@sis-tech.com; asummers@sis-tech.com

Prepared for Presentation at
American Institute of Chemical Engineers
2018 Spring Meeting and 14th Global Congress on Process Safety
Orlando, Florida
April 22 – 25, 2018

Beyond HAZOP

Analyzing Common Cause and System Scenarios

Eloise Roche
SIS-TECH Solutions, LP
Houston, TX
eroche@sis-tech.coms

Watson Dupont and Angela Summers
SIS-TECH Solutions, LP

Keywords: hazards and risk analysis, HAZOP, LOPA, functional safety

Abstract

A production operation often consists of multiple unit operations, supported by plant-wide services such as instrument air, power, electronic communications, and cooling water. The production operation can be covered by multiple risk studies, which divide up the production operation based on non-technical reasons, such as unit management, study length and convenience. The use of multiple studies is particularly common with large production units or with integrated packaged equipment. The "divide and conquer" approach to the studies is good for planning and execution, but the resulting "pieces and parts" analysis can allow significant common cause scenarios to get lost in the clutter. Traditional HAZOP approaches generally focus on pairing individual causes with deviations and consequences. This traditional approach can result in inadequate consideration of more complex scenarios, such as those involving:

- Multiple automated system actions
- Global loss of service
- Transference of risk

These complex scenarios are often interrelated and thus are not well suited for analysis by HAZOP procedures that mandate a single cause-consequence relationship. This paper will propose extensions to the typical HAZOP process to address these challenges in a more holistic manner.

1 Introduction

Traditional HAZOP approaches focus on pairing individual causes with deviations and consequences [1, 2]. Unfortunately, more complex scenarios can and do occur in the process sector [3]. A single failure in a site-wide utility can be a common cause of simultaneous hazard scenarios in multiple unit operations. Likewise, due to the automation system design, there can be multiple equipment responses to a single failure that could create new hazard scenarios. These scenarios are easily overlooked when the facilitator and team inadvertently misapply the "double-jeopardy" exclusion or get excessively focused on single cause-consequence pairs [4].

The HAZOP can be broken up into multiple studies, potentially performed by different teams or even different organizations. This is a common practice for large projects or for projects utilizing packaged equipment. While it is possible that the multiple HAZOP teams might share some common membership, such as operations personnel, this is not always the case. This division of the process operation into multiple studies makes it more difficult to analyze scenarios that cross the study boundaries.

The team needs to collect information in a way that increases the likelihood of spotting more complex scenarios, regardless of how many interconnected process units there are or how divided up the process is from a study standpoint.

2 Complex scenarios

HAZOP teams using traditional techniques can insufficiently analyze or completely overlook complex scenarios [5], such as:

- Scenarios triggering multiple automated system actions
- Common cause scenarios triggered by global loss of service
- Scenarios that transfer risk upstream or downstream of node under study

2.1 Scenarios triggering multiple automated system actions

A single cause can trigger multiple safety actions by system design. These actions occur in an order related to how the process deviation propagates through the interconnected equipment. The process dynamics determine the sequence, concurrency, and timing of the process response and automated system actions. Deviations do not remain in the node and the process response to planned safety actions does not either.

Multiple automated systems actions are often quite intentional, such as when multiple instrumented safeguards are used to achieve the necessary risk reduction for a given scenario. There are two potential consequences of concern that the HAZOP team should be aware of. The first concern is a combination of otherwise individually safe actions that create a new hazard scenario. The second is a cascade of safe actions that result in conflicting states for the safeguard equipment. The HAZOP team should take particular care in addressing unintended side effects of the operator's response to alarms, since these actions are often under-documented during HAZOP.

2.1.1 Creating a new hazard scenario

Many instrumented safeguards address the hazard by stopping flow in a line through closure of a valve or turning off a motor. This final element action could be automated or the result of an operator responding to a safety alarm according to procedure. While this action can be perfectly correct for addressing the hazard scenario under review, there can be additional actions taken by the control or safety systems that could inadvertently isolate piping or equipment.

In the example shown in Figure 1A, the hazard being prevented is an overfill and possible overpressure of tank V2, initiated by loss of flow out of V2 for a variety of reasons. The HAZOP team has identified an interlock to be executed by the safety system that will close block valve B when there is a high level in V2. Closing block valve B completes the safety action; however, the overall automation strategy can, and often does, do more. If control valve A is allowed to simply respond to its measured variable, the process controller will drive control valve A open in response to the level in V1 rising. If control valve A is open when the process restarts, excess flow could be sent to V2, potentially initiating a hazard scenario in V2. Safe restart requires that the process be placed in an operational condition consistent with the startup procedure. It would be typical automation practice to configure the control system to automatically close control valve A when block valve B is closed.

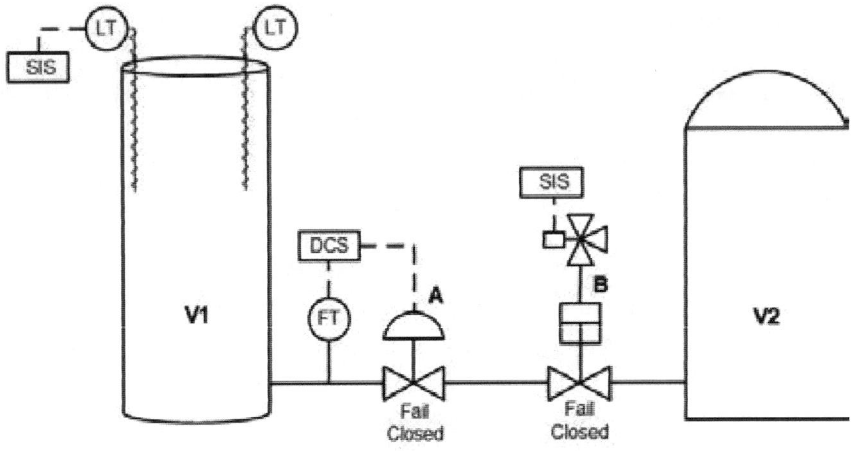

Figure 1A: Equipment under study for V2 overfill hazard

The logic created to ensure a safe restart could be a cause of a new hazard scenario. Pipe section C depicted in Figure 1B is now isolated. A new overpressure scenario is created when the piping is not designed for the thermal expansion of the liquid in the pipe at the maximum temperature.

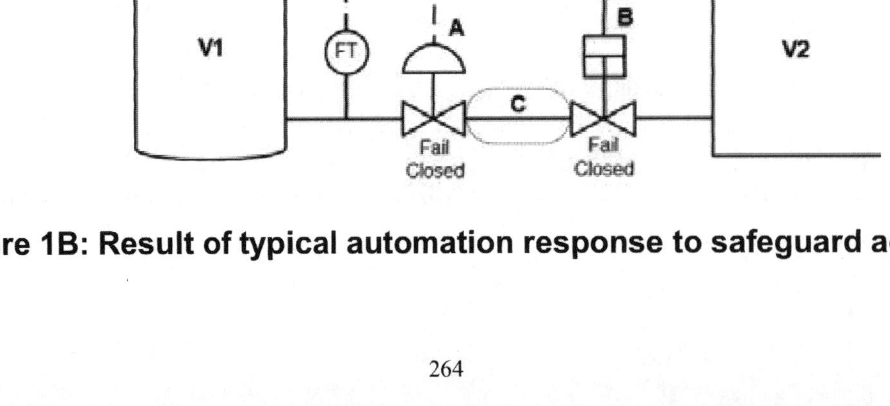

Figure 1B: Result of typical automation response to safeguard action

When the associated equipment is within a single HAZOP node or process unit operation, the interaction can be relatively easy to spot. The interactions become harder to spot when they are between the node/unit operation under study and immediately adjacent upstream or downstream equipment. If the potential interaction is with equipment that is not immediately adjacent to the node/unit operation under study, the resulting consequences become very difficult to spot unless the HAZOP team takes specific actions to support such an analysis.

2.1.2 Conflicting safeguard actions

For some process deviations, a cascade of consequences can propagate across multiple pieces of equipment. The HAZOP team should consider the possible interactions of the protection strategies for the different unit operations. Multiple instrumented safeguards can use the same field hardware but have conflicting safety actions for the final elements. Operator actions in response to a propagating hazard scenario can create additional conflicts.

In the example shown in Figure 2, the hazard scenario initially being prevented is the overfill and possible overpressure of tank V2, initiated by loss of flow out of V2 for a variety of reasons. A high-level interlock closes block valve B, preventing the overflow of V2 by stopping all flow from the upstream equipment. V1 is being covered in a separate HAZOP, during which the team identifies a potential reactive chemical event that could occur in V1 during a shutdown. This separate HAZOP team identifies an administrative procedure to de-inventory V1 through block valve B if there is an unplanned shutdown of the process. There is now a conflicting state for block valve B. One team wants block valve B to fail-open in response to a V1 challenge, but another team needs the same block valve to fail-closed to protect V2 from the overfill scenario.

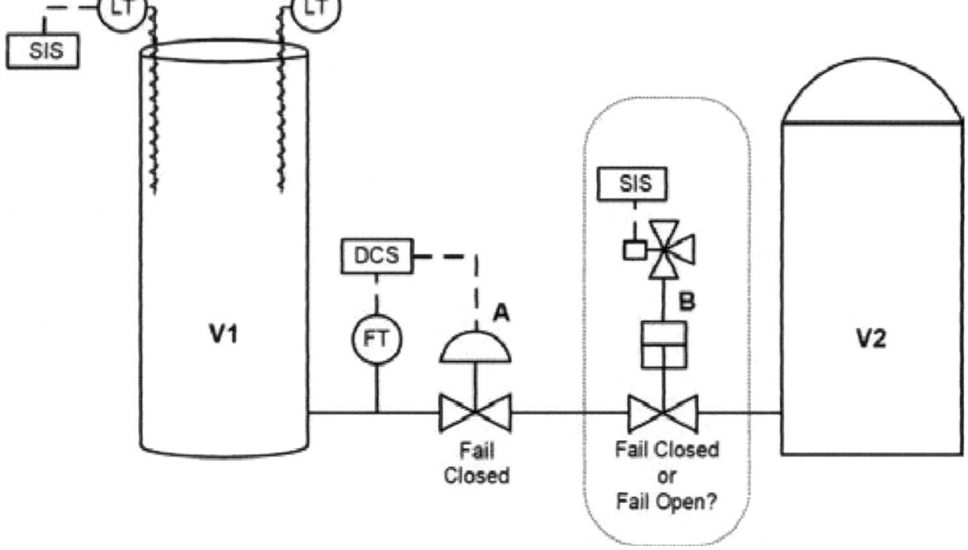

Figure 2: Final element with conflicting safe states

The protection strategy for one or both unit operations involved in this hazard scenario cascade should be rethought. Alternative safeguard strategies need to be identified so that both the initial overflow hazard and the subsequent reactive event hazard can be safely managed.

2.2 *Common cause scenarios triggered by global loss of service*

Traditional HAZOP practices address global loss of service scenarios at a very high level. Often these are evaluated in a global issues portion of a HAZOP using only generic prompts, e.g., global loss of steam. Understanding the impact of a global loss of service events requires understanding how the loss affects all the associated process control and safety systems. What appears to be a safe action for a single control loop to a loss of a utility might not be safe when all the control loops in the facility are reacting to the same loss of utility. HAZOP teams adhering to the minimum staffing requirements per OSHA PSM might not have the in-depth knowledge of the automation specification that is necessary to accurately predict the impact across an entire facility for such a vaguely stated cause.

In addition, the analysis of global loss scenarios is often limited to traditional utilities, such as power, instrument air, nitrogen and steam. There are other systems that are shared by multiple unit operations that can affect the operation of the overall facility in a similar way. Global loss of service analyses should consider:

- Loss of facility control system, safety system, or digital networking systems
- Loss of operations communication systems
- Unavailability of shared abatement systems (e.g., flares, thermal oxidizers, scrubbers, absorbers, or waste collection vessels)

2.3 *Scenarios that transfer risk upstream or downstream of node under study*

Transference of risk occurs when the process deviation, the safeguards against that deviation, and any changes in the process operating state as a result of the safeguard action are not part of the HAZOP node/unit operation under study.

One category of transference of risk includes situations where the safeguard preventing the scenario under study triggers a different hazard scenario partially or wholly outside of the node. Common examples of this are:

- Pressure relief device releases process contents to the atmosphere, resulting in a new hazardous situation to evaluate
- Blocking in the outlet of a tank (e.g., to prevent an underflow hazard) robs all downstream equipment of feed from this tank
- Blocking in the inlet of a tank (e.g., to prevent an overflow hazard or a reactive event) potentially deadheads upstream equipment

These hazard scenarios are largely addressed through traditional HAZOP as long as the node identification results in clear process conduits at each node boundary and the typical parameter deviations are rigorously applied. However, the scenario can result in a cascade of safeguard actions that spans multiple nodes, which should be evaluated to determine whether the scenario triggers multiple automation system actions

Another category of transference of risk occurs when the safeguard preventing the scenario is partially or wholly outside of the node. A common example of this is when a safeguard against overpressure of process equipment or piping is a relief device routed to an abatement system. In other cases, an instrumented safeguard could be located in another node/unit operation. When the HAZOP is being performed by different teams, it is possible that one team can be unaware that risk is being transferred from outside the study to their unit. This can result in an insufficient design specification and a safeguard that might not be fast enough [6] or have sufficient capacity to address the hazard [7].

2.4 Complex deviations are often interrelated

As seen above, these three types of complex scenarios cannot be meaningfully addressed independently from each other. Global loss of service scenarios resulting from a loss of utility, control, or communication over a facility can be a significant cause of multiple actions by the automated systems. The protection strategy for these scenarios often transfers risk, particularly to shared abatement systems. Likewise, mitigating the risk from new scenarios created by localized actions of the automated systems can result in a transference of risk. The HAZOP procedures should be designed to assist teams in evaluating more than single cause-consequence pairs that result in hazard scenarios in the node under study.

3 Proposed approach

Evaluating complex scenarios begins with adding targeted information to the node-by-node HAZOP. The node level information provides the HAZOP team the necessary information to more thoroughly look for study-wide scenarios. Effective reviews of transferred risk and loss of global services are dependent upon information from the node analyses, so these reviews need to be performed after the node analysis is completed. Figure 3 indicates a proposed order of analysis, which should result in less analysis recycle while ensuring the complex scenarios are sufficiently addressed. Failure to plan the resources needed to perform the resulting iterative hazard analysis work could increase the chance of hazard scenarios being overlooked.

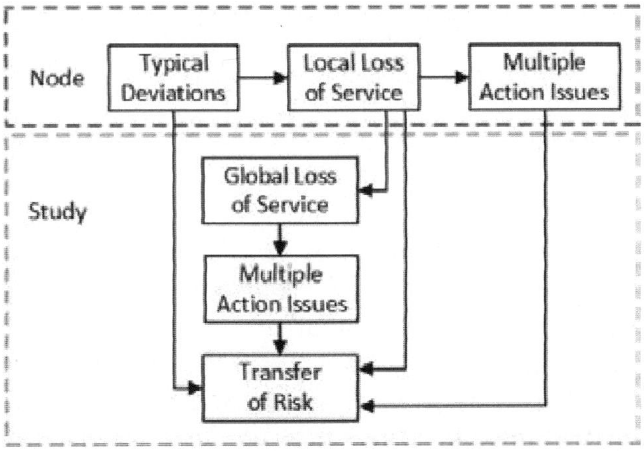

Figure 3: Modified HAZOP strategy to include complex deviations

3.1 Targeted node level techniques

Understanding the equipment response to global loss of service and whether there can be multiple automated system actions within the node under study requires initial evaluation at the node level. Prompts should be added as node-level deviations to ensure that these topics are not overlooked. Determining how the local equipment is expected to respond will involve using detailed design documents for the automated systems and the utility services. The typical HAZOP team membership might not include personnel with experience in using these documents. For example, working through the automated system actions requires knowledge of the logic narratives or the unit cause and effect matrix. As another example, understanding how global loss of power impacts the process equipment requires the ability to read power distribution drawings. When the HAZOP team identifies a complex scenario, it might be necessary to consult with additional personnel [2]. The facilitator should consider adding a notation to scenarios where risk transference is identified, so that scenarios can be filtered for the study-level analyses.

3.1.1 Transfer of risk identification

During the scenario development, the facilitator should note any scenario where one or more safeguards is in whole or in part within a different node or where the safeguard actions create a hazard scenario in a different node. For example, the facilitator could add a note in the comments for that scenario including unique text (e.g., <TRANSFER>) and a brief identification of the transfer of risk involved. If certain nodes are expected to frequently be the recipients of transferred risk (e.g., an abatement system), using a more specific label for these systems can help facilitate later analysis.

3.1.2 Local loss of service node deviations

The facilitator should create a unique deviation for each of the services necessary to control the equipment within the node and for safeguards to be effective. These deviations should cover utility and non-utility services, such as:

- Local deviation in a utility (i.e., power, instrument air, nitrogen, steam, etc.)
- Loss of facility control system (s), safety system(s), or associated communication system(s)
- Loss of shared abatement system(s)

At the node level, the additional deviation should assume the loss of each service is localized to the equipment in the node. The individual deviations are addressed using traditional techniques, while total plant loss of service is evaluated later in the study. If possible, the facilitator should describe the safety actions that occur when final elements (e.g., valves and motors) lose critical services. Special attention should be given to any identified safeguards within the node that would fail to perform under the defined loss of service [8]. If additional consequences are identified as a result of the loss of service, they should be evaluated before closing out the node.

3.1.3 Node-level multiple automated system actions

After the completion of other node-level deviations, a deviation for multiple automated system actions should be evaluated, for which the following would be asked:

1. Do any of the combination of safeguards responding to a deviation result in the creation of a new hazard scenario, such as process backflow, or the isolation of a vessel or segment of pipe?
2. Will the intended process control response to the deviation or related safeguard action result in a new hazard scenario?
3. Do any deviations share the same cause? If so, would the combination of safeguards responding to the deviations initiate a new hazard scenario?

"Yes" or "Unknown" to any of the above should result in an explicit evaluation of the new potential hazard scenario(s) to determine additional safeguard requirements.

3.2 Modified facilitation approach to global questions

Once the evaluation of all nodes had been completed, the facilitator should consider collecting and grouping the following information:

- Local loss of service scenarios, by service
- Identified transfer of risk scenarios not related to local loss of service

Doing so will provide a consolidated view that can assist the HAZOP team in spotting problematic safety actions that occur between nodes. It can also help in aggregating any scenarios that result in transferring risk to shared systems that might affect the design of the receiving system(s). For example, the protection strategy for some nodes might expect abatement equipment to be running that would itself be shut down by the same loss of service. Likewise, the node-level evaluation of the consequence for local loss of instrument power would typically assume that equipment in upstream or downstream nodes is still functioning.

3.2.1 Global Loss of Service Scenarios

Once the consolidated material is ready, the team should review the combined local loss of service scenarios from all nodes/unit-operations for each service. Instead of the traditional generic prompts, the following activities can be used to help the team identify potential hazard scenarios:

- Review the combined local loss of service scenarios from all nodes/unit-operations for each utility, control, or associated communications service. Are any of the node-level consequences for a global loss of service different than was identified for the local loss of service deviation? If yes, create a new summary for that node to use in the global loss of service analysis.

- Assuming the identified safeguards work as specified, review the aggregated local loss of service summaries to identify any scenarios triggering multiple automated system actions not assessed at the node level. Have any resulting hazard scenarios or conflicting safety actions already been resolved?

- Have the mitigation/abatement systems been sized to address the combination of scenarios that will result from this global loss of service?

"No" or "Unknown" should result in an explicit evaluation of the identified or potential unsafe state to define additional safeguard requirements. Particular attention should be given to the piping and isolation equipment between nodes.

3.2.2 Other Transfer of Risk

After completing the analysis for global loss of service deviations, the facilitator should have the team review the collection(s) of any transfer of risk scenarios not already covered. As above, a few targeted queries might focus the team on the purpose of this specialized task:

- Has the design of the safeguard(s) considered potential transfer of risk?

- Is any of the transferred risk the result of a common deviation? If so, has the safeguard design considered the combination of scenarios that will result from that cause?

"No" or "unknown" should result in a recommendation to address the potential design gap.

4 Conclusion

The outcome of a HAZOP is always dependent upon the makeup of the team conducting the analysis and the quality of the input information. Even with up-to-date documentation and an appropriately expert analysis team, there is no "silver bullet" to ensure that every possible scenario is identified. Complex scenarios involving multiple safeguard activations that are triggered by a single cause or where the consequence might be in equipment remote from the initiating cause are particularly prone to being overlooked. Planning the HAZOP work in a way that enables the team to target their analysis on the interrelated topics of multiple automated system actions, collective response to global loss of service, and transference of risk between nodes/unit operations can make these complex scenarios easier to recognize.

5 References

[1] CCPS/AIChE, Layer of Protection Analysis: Simplified Process Risk Assessment, Concept Series, Wiley-Interscience, New York (2001).

[2] CCPS/AIChE, Guidelines for Hazard Evaluation Procedures, Third Edition, Wiley-Interscience, New York (2008).

[3] Summers, Angela. "Risk Assessment Challenges to 20:20 Vision," Presented at AIChE 2014 Spring Meeting, 10th Global Congress on Process Safety poster session, New Orleans (2014).

[4] CCPS/AIChE, Guidelines for Enabling Conditions and Conditional Modifiers in Layer of Protection Analysis, Wiley-Interscience, New York (2013).

[5] CCPS/AIChE, Guidelines for Safe Automation of Chemical Processes, Wiley-Interscience, New York (2016).

[6] Roche, E. and Summers, A. "Are Your Safeguards Fast Enough to be Effective?" 13th Global Congress on Process Safety, AIChE. March 2017.

[7] Kaszniak, Mark, 2010. "Oversights and Omissions in Process Hazard Analysis: Lessons Learned from CSB Investigations," Process Safety Progress, Volume 29, No.3, pp. 264-269.

[8] CCPS/AIChE, Guidelines for Chemical Process Quantitative Risk Analysis, Second Edition, Wiley-Interscience, New York (1999).

Risk Awareness: The Key to Operational Discipline

Alfonsius Ariawan
DuPont Sustainable Solutions
1919 Minnesota Court, Mississauga, Ontario
Alfonsius.B.Ariawan@dupont.com

Lawrence Short
DuPont Sustainable Solutions
Greer, South Carolina 29650
Lawrence.S.Short@dupont.com

Prepared for Presentation at
American Institute of Chemical Engineers
2018 Spring Meeting and 14th Global Congress on Process Safety
Orlando, Florida
April 22 – 25, 2018

Risk Awareness: The Key to Operational Discipline

Alfonsius Ariawan
DuPont Sustainable Solutions
1919 Minnesota Court, Mississauga, Ontario
Alfonsius.B.Ariawan@dupont.com

Lawrence Short
DuPont Sustainable Solutions
Greer, South Carolina 29650
Lawrence.S.Short@dupont.com

Keywords: Operational Discipline, Risk Awareness, Engagement, Risk Sensitivity

Abstract

Operational Discipline is defined as conducting each task the right way, every time. The effectiveness of a management system requires Operational Discipline and rigor in execution. Typically, the 'right way' is interpreted as adhering to procedures. While adhering to procedures is certainly an important characteristic of Operational Discipline, it is not the only one. In our work with organizations in diverse industries, we have encountered many organizations that conduct tasks according to procedures. However, incidents still happen. Why is that? Based upon our experience, we find that a second critical characteristic of Operational Discipline is neglected - risk awareness and sensitivity. In order to reduce incidents, organizations must ensure that employees are aware of and sensitive to risks that may occur in the work places **and** have the procedures and processes in place to mitigate those risks. In this paper, we share our experience in building cultures that exhibit Operational Discipline through employee engagement strategies and the use of simple tools.

1. Introduction

Many of us have heard of the story dubbed the "Miracle on the Hudson", later popularized by Tom Hanks in the movie "Sully". Captain Chesley 'Sully' Sullenberger landed US Airways Flight 1549 safely on the Hudson river after his plane was struck by a flock of geese that disabled both engines shortly after take-off. All 155 people on board survived and Captain Sully was hailed a hero. The ensuing incident investigation found he made the correct decision to deviate from the available 'normal' engine loss procedure, which was designed for cruising altitude and not for immediately after take-off [1]. Captain Sully understood the intent and limitation of the available procedure and was aware of the risks associated with engine loss immediately after take-off. This allowed him to make an informed and ultimately, correct decision.

Having assisted numerous organizations with the management of their operational risks, we find that too many times, management believes the primary cause for their poor safety performance is operators not following available procedures or processes. Thus, we are often asked how we would improve existing culture in order to strengthen Operational Discipline (OD). Indeed, we do find that the lack of adherence to procedures is generally common in organizations that struggle with safety performance. The articulated need to strengthen operational discipline is, therefore, not an unreasonable one. However, just as in Captain Sully's case, if the desired outcome is an improvement in safety performance, then having operators following procedures is insufficient. The underlying cause of the lack of risk awareness needs to be addressed, as it is typically not practical to develop procedures for all situations that an operator may encounter. When a procedure does exist, operators need to know its limitations and must not forget to assess its effectiveness relative to the existing risks. This will help them make the correct decision, even if it means deviating from the procedure. Just because procedures are followed, it does not necessarily mean that the desired outcome of incident free operations will be achieved. In fact, we often find operators following processes and procedures just to 'check the box' with little ownership of the risks.

In this paper, we present our point of view that risk awareness is key to improving operational discipline. We will look at learnings gleaned from our work in raising the risk awareness level and embedding it into the culture of several of our client organizations.

2. Operational Discipline

Effective implementation of process safety management (PSM) does not happen just by designing good PSM systems. Effective implementation requires a continuous, consistent, and persistent application of these good systems in the day-to-day operations (Figure 1). The principle of "Operational Discipline" is recognized as a key success factor in driving effective PSM implementation. Organizations are increasingly realizing that operational discipline can result in rapid, substantial, and sustainable PSM improvements.

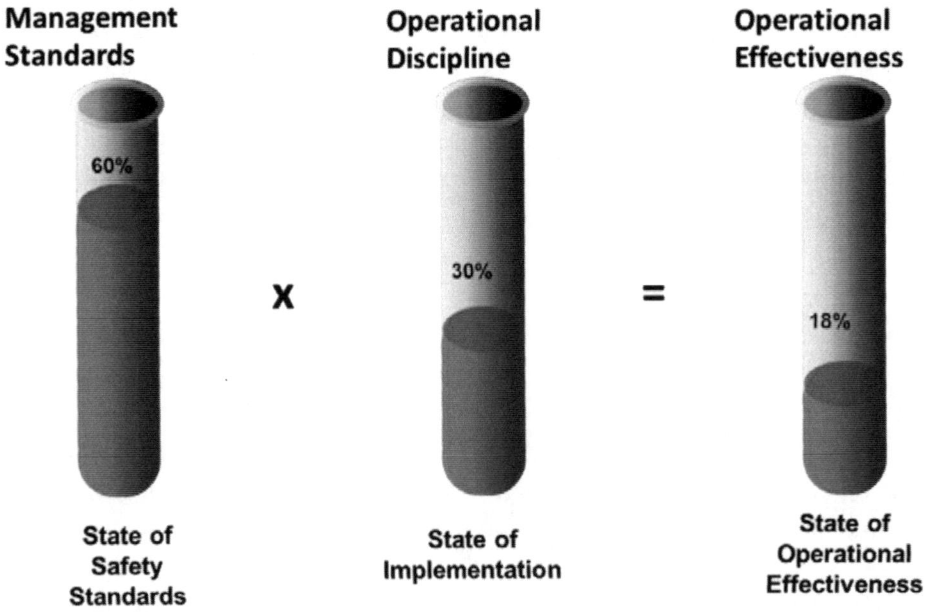

Figure 1. Performance is a function of the management system and level of execution© 2018 DuPont.

At DuPont, we define Operational Discipline as "the deeply rooted dedication and commitment by every member of an organization to carry out each task the right way, every time." Based on many years of cumulative work experience within DuPont and with other organizations, we have identified success characteristics that are the foundation of world-class Operational Discipline. These characteristics are:

- Leadership by Example
- Shared Values
- Strong Teamwork
- Organizational Pride
- Sufficient and Capable Resources
- Engaged Employees
- Active Communication
- Up-to-date Documentation
- Practice Consistent with Procedures
- Absence of Shortcut
- Excellent Housekeeping
- Innovative, Agile and Continuously Improving
- Just and Fair
- Risk Aware and Sensitive

Note that Operational Discipline is not defined as 'adherence to procedure.' Instead, its definition is framed from the perspective of the intended outcome – i.e. for tasks to be conducted 'the right way, every time.' One of the characteristics of OD, however, does include that of 'practice being consistent with procedures', notably complemented by one on being 'risk aware and sensitive.'

Absence of risk awareness, procedures and processes become items to comply to blindly. In situations where judgment is required, consequences may be fatal, with luck as our primary layer of protection. Culturally, when the lack of risk awareness is not addressed, it promotes a compliance mindset that results in complacency. We find that while some organizations struggle with adhering to processes and procedures, others wonder why despite the adherence, they are not achieving the desired level of safety performance, i.e. "if everyone follows procedure, we must be safe, right?"

So, how do we increase the level of risk awareness? And what will we learn when we do?

3. Safety vs. Risks

The first step in increasing the level of risk awareness is to shift the organizational mindset away from safety towards risks. The differences are subtle, but important (see Table 1). Safety is binary – safe or unsafe. Personnel often interpret being safe as the absence of incident. As such, organizations often fall under the illusion that they have everything under control when they proudly show their safety performance of zero incidents over the past impressive number of years. When a safety incident does happen, however, it is often a surprise, prompting the organization to react and look at deficiencies in current controls. It is reactive. Culturally, despite our constant preaching that everyone should own safety and despite the assurance of personnel's understanding of this concept, there is still a nagging feeling that personnel still see the safety group (or the maintenance, or technical group) as the real owner(s) of safety. Everyone is just there to do the job and follow procedures.

On the other hand, the idea of risks is more continuous – higher risk vs. lower risk. It is also proactive and discourages complacency. That is, regardless of the current safety performance, risks are still present and they need to be continually reduced and managed. In order to achieve the desired safety performance, everyone needs to be aware of the risks in their day to day operations and be sensitive to them. Risk is not another program, but a mindset that everyone must own individually. It is a concept that unifies the different layers of the organization – top management, middle management and operators. Incidentally, once the concept of risk has been internalized, personnel will start to see that it is no longer about safety vs. production, but rather, it is about safe production. The workforce begins to integrate other aspects of operational risks, such as productivity risks and quality risks, into their mindset.

Table 1. Differences between safety and risk thinking @ 2018 DuPont

SAFETY	RISK THINKING
All incidents can be prevented	You cannot totally eliminate risk
Management is responsible for Safety performance	Management is responsible for risk reduction. Everyone is responsible to recognize and communicate risks
Working safely is a condition of employment	Worker risk awareness is critical to working safely
Employee involvement is critical	You cannot realistically estimate risks without involving the people doing the work
Safety affects me, my colleagues and my customers	Risks affect people, quality, productivity and ultimately business performance

4. Raising the Level of Risk Awareness

We find that a bottom-up approach is more effective in raising the level of risk awareness and in transforming the culture of an organization. Recognizing that it is ultimately the hearts and minds of the frontline personnel that we are trying to influence, transforming mindset of frontline personnel from safety to risk, enables them to see risks more readily. Simultaneously, management personnel should be coached to provide the necessary support to the now engaged frontline workforce.

Our point of view is that frontline personnel have more knowledge about their operational risks than we typically give them credit for. They know which procedures work and which ones are typically not followed. At one point, they were new to a specific operation and had a heightened sense of risk awareness. Over time, they may become complacent and while they are aware of the risks, they no longer see them. Like a pattern on wallpaper, if you walk past the wallpaper often enough, you no longer see the pattern. Therefore, we spark engagement within them so that those risks are articulated and re-internalized.

In a high hazard environment, a process hazard analysis (PHA) is updated and revalidated at some regular basis with the involvement of operators as best practice. However, a PHA revalidation is an event. Effective risk management requires an on-going discussion of hazards and risks in both formal and non-formal setting. The intent of the continuous engagement of workforce on hazards and risks is not to replace the rigor of PHA, but to complement and normalize risk discussion. We simplify tools to enable better understanding and adoption, and create a trusting environment where frontline personnel are empowered [2].

After the transformation stabilizes, frontline personnel see themselves as layers of protection. Though not perfect, each one of them *owns* pieces of the operational risks and *internalizes* the roles they play in ensuring the integrity of controls, which makes for a more vigilant organization. Several characteristics of Operational Discipline are clearly observed. The frontline personnel no longer follow procedures blindly. In fact, it is common for them to identify hidden hazards and see the need to modify existing procedures.

That is, they do not only highlight risks, but also take the ownership to fix them. Because the engagement starts with frontline personnel and is driven by them, it is also more sustainable. In other words, organizations are less likely to exhibit ‘follow the leader’ syndrome, where changes in safety performance correlate to the movement of leaders.

As for the leaders, the now more risk-aware frontline personnel are holding them accountable to provide the necessary support to address the risks that have been surfaced. Leaders must become more transparent, communicating their plans more effectively to manage the identified risks. Failing to do so will damage the trust that has been created among the frontline personnel. The demand for accountability by the frontline personnel provides a meaningful reason for leaders to enhance their capability in interacting with them.

5. Conclusion and Key Learnings

Operational Discipline requires more than adhering to processes and procedures. Operational Discipline necessitates a number of other characteristics, one of which is risk awareness and sensitivity. To strengthen Operational Discipline, we propose an engagement that is focused on raising the level of risk awareness of the organization. The resulting transformation will reduce complacency and check-the-box mentality. Several key learnings from such engagements are highlighted below:

- A focus on risks (vs. safety) helps shift organizational mindset and individual decision making to be more proactive and reduce complacency.
- Demand from frontline personnel increases the effectiveness of leadership capability and accountability.
- Recognize that it is ultimately the heart and mind of frontline personnel that we want to influence. Start with the desired change at frontline organization and then design the required supporting structure and leadership capability accordingly (bottom up approach) to yield success.
- Engagement of frontline personnel in risk awareness can bring about a cultural transformation that is organic, characterized by high momentum and sustainable risk ownership – i.e. not only do they own the risk journey, but are a key part of reducing those risks as well.

6. References

[1] “Loss of Thrust in Both Engines After Encountering a Flock of Birds and Subsequent Ditching on the Hudson River”, Jan 15, 2009, https://www.ntsb.gov/investigations/AccidentReports/Reports/AAR1003.pdf (accessed March 8, 2018)

[2] D. Byrne and A. Ariawan. “Simplistic Yet Insightful Way to Present Operational Risks in Order to Facilitate Risk Discussions,” 13th Global Congress on Process Safety, San Antonio, 2017

Know Your Limits

Jerry Forest
Sr. Director, Process Safety
Celanese
225 E. John Carpenter Freeway Suite 1200
Dallas, TX 75234
(214) 417-1920
Jerry.forest@celanese.com

Prepared for Presentation at
American Institute of Chemical Engineers
2018 Spring Meeting and 14th Global Congress on Process Safety
Orlando, Florida
April 22 – 25, 2018

Know Your Limits

Jerry Forest
Sr. Director, Process Safety
Celanese
225 E. John Carpenter Freeway Suite 1200
Dallas, TX 75234
(214) 417-1920
Jerry.forest@celanese.com

Keywords: Safe Operating Limits, IOW, SOL

Abstract

Process safety professionals continue the elusive search for effective leading indicators. The search is for the one indicator to prevent process safety incidents. API RP 754 has already defined Tier 3 leading indicators as challenges to safety systems. Challenges include safe operating limit excursions as well as demands on safety systems. If chosen correctly, safe operating limits can be successful leading indicators for incident prevention. In fact, safe operating limits can be a leading indicator to other safety system challenges, including pressure relief devices and safety instrumented system activations.

The process Celanese used to revamp our safe operating limits (SOL) is called *Know Your Limits*. It involves a rigorous review of SOL set points; complete documentation of the technical basis for selection; documentation in standard operating procedures and operator certification/recertification materials; operator training; incident reporting of challenges; and integration with other process safety elements including alarm rationalization and management. A method to measure program effectiveness is presented. *Know Your Limits* has contributed to our success in reducing loss of primary containment incidents.

1 Introduction

Manufacturing facilities regulated by OSHA's process safety management (PSM) rule should be familiar with safe operating limits (SOLs) as "safe upper and lower limits for such items as temperatures, pressures, flows or compositions and an evaluation of the consequences of deviations" (COD). SOLs are required as part of the process safety information (PSI) for a PSM-covered process [1]. In addition, operating limits with COD and steps required to correct or avoid deviation are required for operating procedures. Tier 3 performance indicators listed in API RP 754 include safe operating limit excursions and demands on safety systems [2]. Careful selection of SOLs can provide operators an opportunity to act before a pressure relief (PRD) device or safety instrumented system (SIS) is activated, and to prevent loss of primary containment (LOPC) incidents. Since SOLs have the potential to prevent process safety incidents they are true leading indicators.

Recognizing the interconnectivity of SOLs with other elements of a process safety management system will contribute to the overall success of the program. SOLs are PSI and therefore they contribute to understanding hazards and risks during the process hazard analyses (PHAs). Operators should be trained in SOLs, and SOLs should be included in the standard operating procedures (SOPs). Instrumentation associated with SOL measurement should be part of an inspection, testing, and preventative maintenance (ITPM) program, and alarms associated with SOLs should be part of an alarm management program. *Know Your Limits* is a structured approach to identify and document SOLs and fully integrate them into a process safety management system.

2 Establishing the Need for Know Your Limits

Celanese has required all sites, regardless of being subject to OSHA rules, the establishment of SOLs, COD, and operator steps to correct deviation since the US process safety management regulation was promulgated in 1992. The requirements include development of PSI and inclusion of the limits in SOPs. In 2010, the Celanese definition of a process safety incident was changed to align with the newly-published API RP 754 standard. Concurrent with the change we categorized our incidents consistent with the American Fuel and Petrochemical Manufacturers (AFPM) advancing process safety 754 metrics collection criteria including but not limited to: point of release, consequence, mode of operation and type of material.

The point of release category includes atmospheric tank overflow. A Pareto analysis showed that between 2010 and 2012, loss of primary containment incidents due to atmospheric tank overflow accounted for 13% to 15% of the total Tier 1 and Tier 2 incidents within Celanese. This raised an interesting question. How could so many overflow incidents occur with a working SOL and alarm management program?

Subsequently, our process safety audits surfaced disconnects among the process safety elements and differences in various aspects of the SOL programs across our sites. For example, one site could have a process with many SOLs identified while a sister site utilizing the same technology may have few. There was no consistent definition of an SOL and each site had a different understanding of this type of PSI. Observations and data led us to develop the *Know Your Limits* initiative for implementation across our global sites.

3 What is a Safe Operating Limit?

It is a challenge to identify a definition of SOL and it is even more difficult to find methodology selecting SOLs. For example, *Guidelines for Risk Based Process Safety* mentions SOLs in several chapters without defining it or establishing guidelines on how to select them [3]. Perhaps the most succinct definition is found in the CCPS process safety glossary which defines SOLs as: "Limits established for critical process parameters, such as temperature, pressure, level, flow, or concentration, based on a combination of equipment design limits and the dynamics of the process" [4].

A second reference from *Inherently Safer Chemical Processes* provides a useful working definition of SOLs and describes them as: "Inherently safer designs ensure that the normal or quality operating limits are well within the safe operating limits, that the limits are within the instrumentation range, and that the instrumentation range is within the equipment containment limits" [5]. Figure 1 is adapted from Figure 5.5 of this reference.

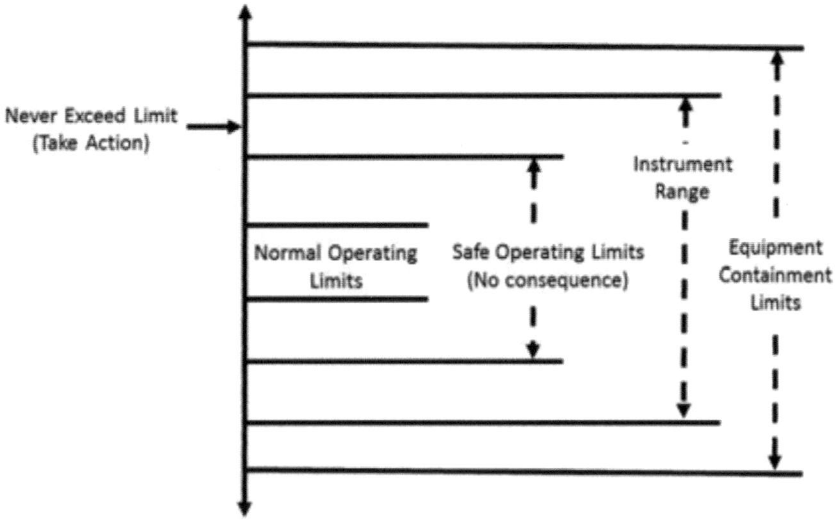

Figure 1: Operating ranges and limits

When the above definitions of SOLs are considered, the following characteristics emerge: 1) unique to the equipment, 2) unique to the dynamics of design, and 3) are hierarchical in nature and response times are considered.

In defining an SOL, Celanese chose the following characteristics:

- If challenged and not corrected, could result in significant consequences commensurate with a Tier 1 process safety incident
- Limited to a critical few and can be committed to memory – typically less than 20 for an operating unit
- Determined by consideration of equipment design limits
- Can be measured with instrumentation
- Includes an alarm and troubleshooting zone based on process dynamics which allows adequate operator response time and a specific pre-defined and mandated action to correct

Based on the characteristics, a guidance document and training package were assembled, clearly defining the various zones of the operating envelope including normal operation, troubleshooting, SOL, and unacceptable operating zones compared to the equipment and various alarm limits.

4 Operating Zones

An important aspect of the training package is defining a consistent set of operating zones. The definitions provide global consistency to the SOLs chosen. CCPS *Guidelines for Engineering Design for Process Safety* was chosen to serve as a guideline to define various operating zones. Figure 2 is adapted and simplified to emphasis the more important areas of operation [6].

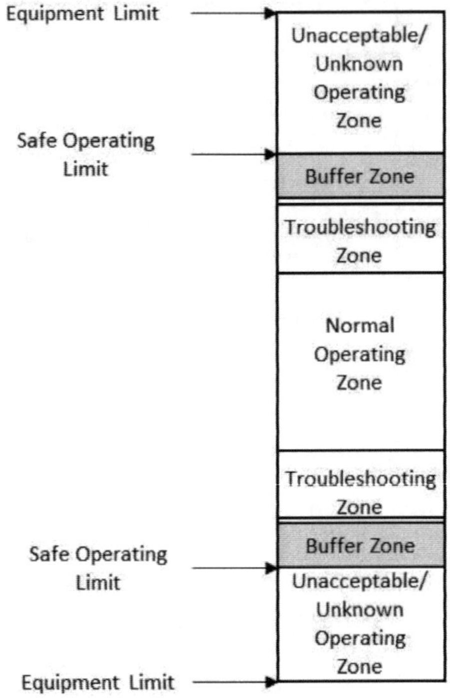

Figure 2: Operating Zones

Normal and Troubleshooting Zones
The zones are defined in such a way as to give operators time to respond to process deviations with pre-determined actions in order to return operation to normal. The zones help to define normal and critical alarms allowing time for operator response.

Safe Operating Limit
At this limit, operators respond to a challenge with pre-determined actions in order to return the process to a safe mode. The process should not be intentionally operated outside of the SOL limit. A result could be loss of primary containment.

Unacceptable/ Unknown Zones
Challenge of an SOL could lead to a process safety incident. By setting the SOL with an unacceptable/unknown zone, we provide an uncertain margin of error before the event occurs. Thus, exceeding an SOL is not guarantee of an incident, but certainly could lead to one.

5 Know Your Limits

The *Know Your Limits* initiative is a process to define and respond to SOLs consistently at our global sites. The initiative includes each affected element of our process safety management system. SOLs are chosen from high risk PHA scenarios. The PSI helps identify equipment design limits and process dynamics for response. The instrumentation used to measure the variable is included in an ITPM program. Output of the PHA includes an appropriate selection of the alarm set points allowing an operator adequate response time to correct challenges. Information of the consequence, operator steps to correct, and the instrumentation involved all become part of SOP which is used in operator certification and recertification. When SOLs are challenged, they should be investigated to understand and eliminate causes. The entire system should be audited. Each of the elements described above are documented as PSI which becomes input for future PHAs and management of change (MOC).

Pilot

Celanese piloted *Know Your Limits* at six sites representing our core technologies. Key deliverables were set to track progress and are discussed below. At the end of the pilot, the data was normalized to represent the time commitment in terms of one person working a full 40 hour work week to implement the program. The average time on the pilot phase was between 26 – 35 weeks, representing between 1,040 to 1,400 manhours. When the initiative was expanded to all facilities, the average time for complete implementation of the program was a little less than 3 years.

Initiative Deliverables

Table 1 lists the key deliverables of *Know Your Limits* and are discussed below.

Table 1: Key *Know Your Limits* Deliverables

Step	Deliverable
1.	Identification of SOLs/COD/operator actions
2.	SOL integration with alarm management
3.	SOP integration
4.	Verify and validate
5.	Operator training
6.	Incident reporting & investigation
7.	Evaluate effectiveness

<u>Identification of SOL, COD, and predefined operator action</u>

Pertinent PSI such as the unit PHA, equipment files, and P&IDs were used by a cross functional team to identify each SOL. To maintain consistency across the company, an SOL determination template was created and used to provide consistent documentation of the results of the exercise.

The template documents: P&ID, description, instrument tag, process variable units, safe limit low, normal operating range, safe limit high, alarm ranges, technical basis of selection, consequence of a challenge along with pre-defined operator actions to correct, monitoring requirements, and the reporting requirements.

Having all of the information in one place makes the document valuable PSI used to integrate the SOL into other process safety elements such as SOPs and training. Documentation of the technical basis of selection is of particular importance for maintaining the corporate memory and for future PHAs and MOCs.

SOL integration with alarm management

Alarms associated with SOLs are set to allow adequate operator response time to perform effective pre-defined corrective actions prior to a limit excursion. The set point for the alarm is established by considering the process dynamics, required operator response time to ensure effectiveness and the instrumentation detection and response lag. Low and high alarms are set at the boundaries of the normal operating zone shown in Figure 2. Low-low and high-high alarms might be set in the troubleshooting zone depending upon operator response requirements and process dynamics. Integration with alarm management includes defining appropriate location of SOLs in the operator alarm attention list and unique sounds and colors for SOL alarms on the distributed control system display.

A safety instrumented system (SIS) trip might be set beyond the SOL to prevent the consequence. The alarm characteristics show that operator response is possible prior to an SIS trip making SOL-related operator actions a leading indicator for SIS activations.

SOP integration & operator training

The SOLs, COD and predefined actions are included in operator certification and recertification materials and SOPs. Depending upon specific unit requirements, this information can either be located in a single SOP or dispersed throughout the unit procedures.

Incident reporting and investigation

The *Know Your Limits* program includes operator training on reporting SOL challenges for incident investigation initiation. All challenges to safety systems should be reported and investigated. SOLs are characterized in the incident database by: SOL adequate, SOL inadequate and SOL incident. The "adequate" tag is used when an SOL is challenged and the pre-defined operator actions return the process to the normal operating zone without incident. Similarly, the "inadequate" tag is used when the pre-defined operator actions do not return the process to the normal operating zone, but no incident occurs. An "Incident" is a characterization used when a process safety incident occurs and either the pre-defined operator action did not prevent, or in cases where no SOL exists and the intended response to an SOL could have prevented an incident. The incident database is used to collect data and perform analysis on the types of variables challenged (e.g., temperature, level, flow, etc.)

6 Evaluate Effectiveness

Effectiveness of *Know Your Limits* is evaluated using the number and type of challenges (leading indicators), and reduction of actual incidents with *Know Your Limits* causes (lagging indicators). An evaluation using a protocol designed to test the key deliverables (Table 1) is used to evaluate initiative effectiveness.

Metrics

In the first year of implementation, incident reporting of SOL challenges increased 80% and by the third year of implementation, the ratio of SOL challenges to Tier 1 plus Tier 2 events leveled out at 6:1.

Since the program began, the most frequently reported challenge is level. The next most frequent challenge is pressure. Figure 3 shows a four year compilation of SOL challenge by type. Of the challenges categorized as adequate/inadequate, 96% are adequate – meaning, the operator actions to return the parameter back into control were effective. Continued reporting of the level challenges has driven the actual atmospheric vessel overflow incidents to zero in the first year of full implementation of the program as shown in Figure 4. Figure 5 shows the reduction of incidents that could have been prevented had an effective SOL program been in place otherwise known as "*Know Your Limits* Causes."

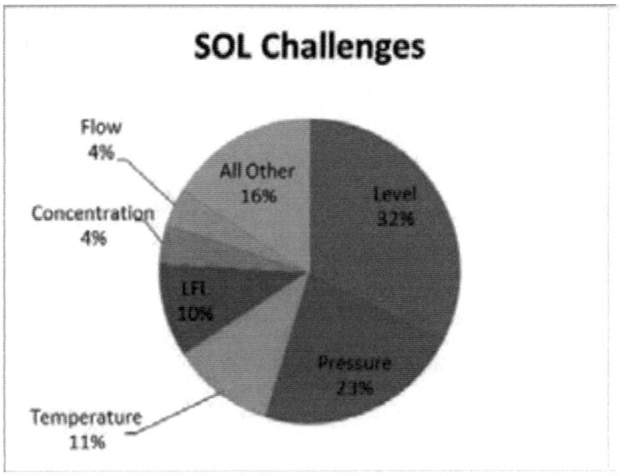

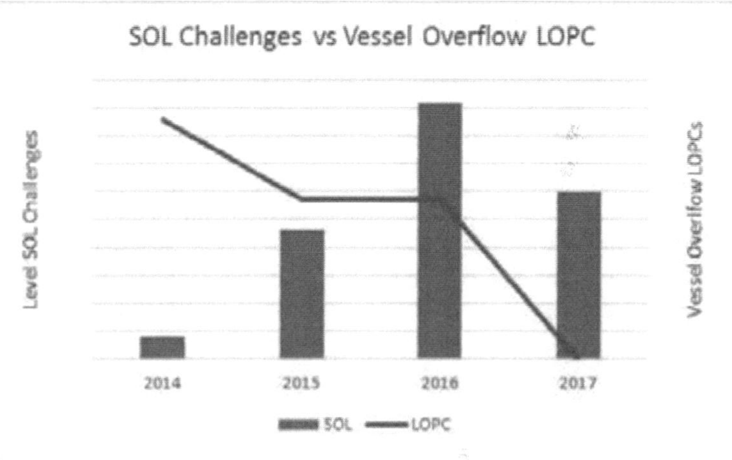

Figure 3: Four Year SOL Challenge History **Figure 4: Level Challenges vs Level Incidents**

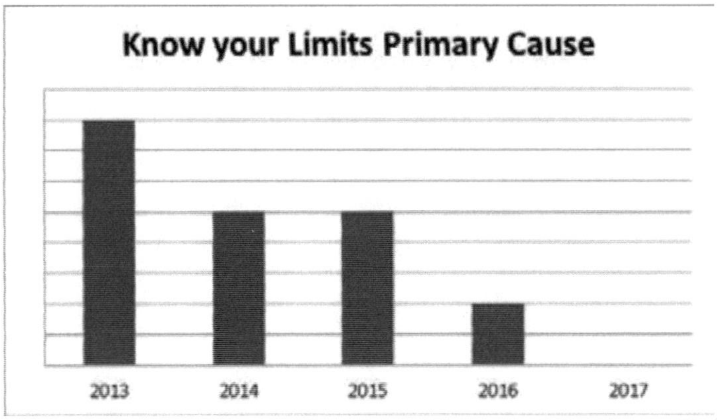

Figure 5: Tier 1 and Tier 2 incidents with *Know Your Limits* Causes

Audit evaluation

A thirteen question audit protocol was developed to validate the effectiveness of implementation of the key deliverables defined in Table 1. The protocol groups the questions in categories: identification of *Know Your Limits*, protective measures, documentation, reporting and training.

Figure 6 illustrates the trend of findings following a review of the completed protocols, which clearly illustrates a future area for improvement in how we complete operator training for *Know Your Limits*. In the figure, "Tier 3" refers to site procedures, and not API RP 754 incident Tiers.

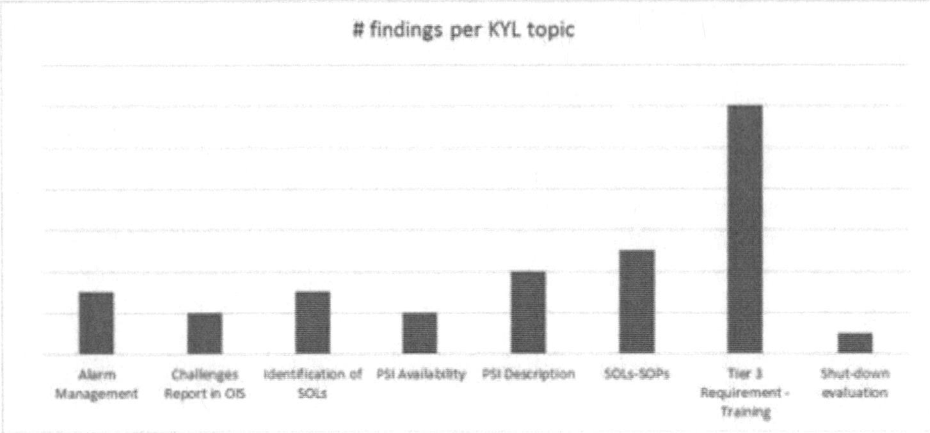

Figure 6: Effectiveness audit, findings

7 Conclusion

Know Your Limits illustrates two important aspects of any process safety initiative. First, data should drive actions. Understanding true causes of lagging indicators and developing solutions to eliminate the causes creates results. Analysis of LOPCs incidents showed a relatively high percentage of vessel overflow events could be prevented. By establishing effective safe operating limits, *Know Your Limits* incidents have been reduced to zero since this initiative was launched.

The second lesson from the initiative is defining how various process safety elements work together. SOLs do not exist as a standalone entity. They are first PSI. As such, they are used in PHA and MOC. They are included in SOPs and should be included in operator certification and recertification. SOLs are used as a basis for alarm management and the instruments measuring the limit are part of a sites mechanical integrity program. Challenges are reported and investigated and the causes are analyzed for continuous improvement. *Know Your Limits* and help reduce SOL excursions LOPC incidents.

8 References

[1] Process Safety Management, OSHA 3132, 2000 (Reprinted), https://www.osha.gov/Publications/osha3132.html (accessed January 8, 2018)

[2] API. API RP 754 Process Safety Performance Indicators for the Refining and Petrochemical Industries. ANSI/API Recommended Practice 754, second edition, April, 2016

[3] CCPS. *Guidelines for Risk Based Process Safety*. Center for Chemical Process Safety, American Institute of Chemical Engineers, Newark, NJ, 2007

[4] CCPS Process Safety Glossary, https://www.aiche.org/ccps/resources/glossary/process-safety-glossary/safe-operating-limits (accessed January 8, 2018)

[5] CCPS. *Inherently Safer Chemical Processes A Life Cycle Approach* . Center for Chemical Process Safety, American Institute of Chemical Engineers, Hoboken, NJ, 2009.

[6] CCPS. *Guidelines for engineering Design for Process Safety*. Center for Chemical Process Safety, American Institute of Chemical Engineers, Hoboken, NJ, 2012

Acknowledgments

The author wishes to thank Dawn Wurst, Vice President Process and Asset Safety, Georgia Pacific, for the use of the creative name *Know Your Limits*.

510908 How Human Factors and Operational Discipline Impact Barrier Management

Abstract:

Building an effective barrier management strategy and safety culture within any organization can often be challenging, especially since a significant aspect to each involves people. We've all heard the fundamental elements of a successful safety (and operational) program includes people, process, and plant (or equipment); however, it's the 'people' aspect that always seems to give us problems. After all, people are unique, complex, unpredictable, and come with a wide array of behaviors and risk tolerance. All of these characteristics create a challenge for the organization throughout the risk management process as human error and the potentially negative consequences that may follow have to be a core consideration along the way.

Human factors and operational discipline are not synonymous terms, but have very similar human elements that can strengthen the performance (both safety and operational) of an organization when properly applied in a management system. This presentation will review the fundamental concepts of human (and organizational) factors and show how they relate to operational discipline in an effective barrier management plan or strategy.

18 AIChE Spring Meeting

14TH GLOBAL CONGRESS ON PROCESS SAFETY

ORLANDO FLORIDA 2018

Inherent Safety and the Elements of PSM Programs

Michael J. Hazzan, P.E., CCPSC, CPSA
AcuTech Consulting Group Inc.
1919 Gallows Rd, Suite 900, Vienna, VA 22182
mhazzan@acutech-consulting.com

Prepared for Presentation at
American Institute of Chemical Engineers
2018 Spring Meeting and 14th Global Congress on Process Safety
Orlando, FL
April 22 – 25, 2018

Inherent Safety and the Elements of PSM Programs

Michael J. Hazzan, P.E., CCPSC, CPSA
AcuTech Consulting Group Inc.
1919 Gallows Rd, Suite 900, Vienna, VA 22182
mhazzan@acutech-consulting.com

Keywords: Inherent Safety, IST, process safety management, RBPS, PSM

Abstract

Inherent safety (IS) is implicitly part of PSM/RBPS programs (henceforth simply referred to as PSM programs). It is not a discrete element, like Management of Change, but it should be part of the thought process for the design of these programs and how they are applied and implemented. Traditionally, IS has been regarded as an intrinsic part of the design of the process, i.e., part of the "hardware" aspect of the process or part of the properties of the process with respect to the hazards presented by the hazardous materials used in the process. However, they can also be applied to the thought process of each element of a PSM program when the elements are being conceived and planned, or later as they are revised and refined. That is, the same principles of IS can be applied to aspects of the process that are largely programmatic, i.e., the design and execution of the PSM program policies, practices, and procedures. This is a relatively new application of inherent safety concepts that were first proposed by Trevor Kletz in 1977.

Introduction

The four IS strategies, i.e., *Substitution, Minimization, Moderation,* and *Simplification* have usually been applied in the context of the chemicals and the hazards/risks that they present, e.g., substituting one chemical with another that presents less of a process safety hazard, or minimizing the inventories of the chemicals that can create process safety hazards. However, the same application can also be made to the various elements of a process safety program. For example, *Minimization* has traditionally been associated with reducing the inventory of hazardous materials, which is a physical aspect of the process. However, we can pose the following question: Should shift schedules and their working times be *minimized* to avoid fatigue? This is a possible application of *Minimization* to a non-physical or programmatic/procedural aspect of the process. Another example: Can a different type of non-destructive testing be *substituted* to increase the likelihood of detecting a particular type of corrosion. This is an example of how the inherent safety strategies can be used in the planning of a mechanical integrity program to reduce hazards and improve performance. Another possible use of the four inherent safety strategies is their use in guidewords, deviations, checklists, and other process safety analytical and assessment tools. In particular, the HIRA/PHA, MOC, and Incident Investigation elements are examples of such possible use.

For some PSM elements the relationship with the IS strategies is direct and relatively strong, e.g., HIRA/PHA, whereas for other elements the relationship is more indirect, e.g., Measurement and Metrics. While the use of the four main inherent safety strategies in this non-traditional manner represents a novel use of the concepts, it is not intended to supplant their possible use in traditional, physical aspects of the process, but to supplement them.

The interrelationship between inherent safety and PSM programs has been previously described by Kletz and Amyotte (Ref 10. Kletz 2010), as well as Hendershot, et al (Ref 3. Amyotte). These authors argue that inherent safety is implicitly part of PSM programs, that it not be included as a discrete element, but as part of the thought process for the design of these programs and how they are applied and implemented. The goal of their research was to make inherent safety explicit in the management systems that form the framework of a process safety program.

Examining each element of a PSM program for the use or impact of inherent safety would be beyond the scope of this paper. Therefore, the following elements have been selected for examination herein:

- Process Hazard Analysis/Hazard Identification and Risk Analysis
- Management of Change/Operational Readiness
- Asset Integrity/Mechanical Integrity
- Conduct of Operations/Operating Procedures

Inherent Safety & Process Hazard Analysis/Hazard Identification and Risk Analysis

Process Hazard Analysis/Hazard Identification and Risk Analysis (PHA/HIRA) is the PSM element with the most direct relation to the four IS strategies and the element where IS has been used the most in the PSM community (other than in the security of chemical/processing facilities). The routine use of IS checklists in PHAs, the use of the four IS strategies in the formulation of guidewords/deviations or What-If questions during PHAs/HIRAs should be considered. An example of such a checklist is shown in Table 1. (Ref 10. Kletz 2010). Similar tools have been published by CCPS (Ref 6. CCPS). However, the application of IS in PHAs/HIRAs does not address the programmatic aspects of PHAs/HIRAs or how they fit into the facility PSM program. Rather, it is an application of IS concepts to examine the state-of-the-art of the process technology, including the equipment design, and the chemicals used in the process. This is the traditional application of IS. At the end of this section the possible use of the four strategies of IS to the programmatic aspects of PHA/HIRA is addressed.

In addition to the general use of IS checklists in general PHAs/HIRAs, reviews devoted only to IS are sometimes performed. In these specialized studies the four IS strategies are examined explicitly to determine if there are feasible ways to use them. Such IS-only reviews are currently required by regulation in three jurisdictions (New Jersey, California - refineries only, and Contra Costa County, CA). Often the format and study approach for these IS-only reviews is PHA-like, i.e., worksheets, guidewords/deviations, and similar techniques and formats are often used for conducting them.

Contra Costa Health Services (CCHS), in their IS guidance document, includes analytical activities on how IS should be addressed for new vs. existing processes. For new processes, the guidance identifies the following different project phases when reviews should be performed to confirm if IS is applicable: 1) during the chemistry forming (synthesis) phase for product/process research and development to focus on the chemistry and process; 2) during the facilities design scoping and development prior to completion of the design basis to focus on equipment and configuration; and 3) during the basic design phase of the project. The CCHS guidance contains more detailed questions and review items to apply during each of these project phases. The guidance does recognize that IS may not be applicable for all facilities and project phases in the same way or in the same sequence. IS should be examined as early as possible to minimize the impact on the project.

For existing processes, the CCHS guidance notes that the consideration of IS is limited to scenarios where a Major Chemical Accident or Release (MCAR) could reasonably occur. This can be accomplished and documented by denoting which scenarios identified in PHAs/HIRAs have consequences (without crediting any active safeguards) that rise to that level. In most chemical/processing industry PHAs that are part of formal PSM programs, this is documented in PHA worksheets by describing the consequences so that offsite and onsite impacts that result from major releases are clearly identified. This is also usually supplemented by a qualitative or quantitative consequence or severity ranking that clearly indicates severe consequences that are the hallmark of a MCAR. Facilities can perform one of the following methods to ensure that IS is considered and documented for the covered processes:

- An independent IS analysis that is done in addition to a PHA. Checklists or guideword analysis contained in the guidance document that incorporates ISS can be used to accomplish this analysis.
- An IS analysis that is incorporated into the existing PHA review process. This would require that the PHA for each process be re-done in its entirety to include an initial IS analysis. Revalidated PHAs that examine only portions of the process may not be sufficient to satisfy the initial IS review if the whole process is not evaluated. Checklists or guideword analysis contained in the guidance document that incorporates IS can be used to accomplish this analysis.

Whichever type of IS analysis is implemented by the facility the CCHS guidance specifies that the analysis be conducted and documented to 1) document the qualifications of the team facilitator/leader and team makeup, 2) document the ISS's considered as well as those implemented, document the method used for the analysis if independent IS analysis was performed, 3) if a checklist for the ISS was used, for items that were not considered, document why those items were not considered (i.e., not applicable or were already considered in previous consideration), 4) document the grounds that were used to make the feasibility determination, 5) document the IS considered, the IS implemented, and the IS not implemented, 6) revalidate the IS at least once every five years. (Ref 8. CCHS)

Amyotte and Kletz also state that PHAs for capital project reviews, as well as other PHAs that are performed should include the following purposes:

- to identify hazards, hazardous situations and specific events that could produce undesirable consequences,
- to examine the currently available safety measures to deal with the identified hazards and events, and
- to suggest alternatives for risk reduction based on inherent, as well as engineered (add-on) and procedural, safety. (Ref 10. Kletz 2010)

Facility siting is usually examined qualitatively during PHAs/HIRAs using checklists. It is also addressed quantitatively in most facilities as a separate analytical activity. The vulnerability to the effects of explosions of the administration and control buildings at Flixborough and the casualties suffered in contractor trailers at the BP Texas City refinery are often cited as examples of the importance of this topic. The use of segregation in the hierarchy of controls and the use of distance in facility siting are examples of *Moderation*.

Amyotte and Kletz offer another important consideration regarding the use of IS strategies on engineered projects. It is a common practice for operating companies in the chemical/process industry to outsource or contract engineering services, especially for capital projects. This is often the case on large projects where partnerships and joint ventures are formed, but may also apply to smaller projects through the awarding of subcontracts. The degree of commonality in risk management practices among the different parties may vary widely and there may not be a common set of expectations for safety performance and risk-awareness for different proejcts. By extension, differences in approach to inherent safety and the corporate value placed on this concept can also vary widely, resulting in an uneven application of IS strategies on any given project. This problem can be partially solved by including detailed requirements for IS reviews and the use of the four main strategies in the project technical specifications. However, engineering contractors may have a less developed culture and experience of applying IS in their work than the operating companies they work for. Of course, the opposite can also be true and the engineering company may propose IS solutions to hazard/risk reduction that were not specified by the operating company. (Ref 10. Kletz 2010)

Table 1. Inherent Safety Checklist for PHAs

Guideword	Checklist Question
Minimize	• Is the storage of all hazardous gases, liquids and solids minimized? • Are just in time deliveries used when dealing with hazardous materials? • Are all hazardous materials removed or properly disposed of when they are no longer needed or not needed in the next X days? • Are all tasks involving working at heights done in as few trips as possible? • Are all tripping hazards minimized and all walking surfaces tractional during all weather conditions? • Are low noise equipment and machinery taken into consideration when making new purchases? • Is shift rotation optimized to avoid fatigue? • Are awkward positions and repetitive motions minimized? • Are attempts made to completely eliminate raw materials, process intermediates or by-products? • Are elbows, bends, and joints in piping minimized?
Substitute	• Can a less toxic, flammable or reactive material be substituted for use? • Is there an alternate way of moving product or equipment as to eliminate human strain? • Can a water based product be used in place of a solvent or oil based product? • Are all allergenic materials, products and equipment replaced with non-allergenic materials, products and equipment when possible?
Moderate	• Can potential releases be reduced via lower temperatures or pressures, or elimination of equipment? • Are all hazardous gases, liquids and solids stored as far away as possible to eliminate disruption to people, property, production and environment in the event of an incident? • When purchasing new equipment, are acceptable models available that operate at lower speeds, pressures, temperatures or volumes? • Are workplaces designed such that employee seclusion is minimized? • Are food and beverages offered at the workplace appropriate to minimize negative impact and maximize positive impact? • Are all power tools de-energized when not in use for extended periods? • Are all lights, sprinklers, accesses/exits, ventilation ducts, electrical installations and machines clear of piled materials?
Simplify	• Are all manuals, guides and instructional material clear and easy to understand, especially those that are used in an emergency situation? • Are equipment and procedures designed such that they cannot be operated incorrectly or carried out incorrectly? • Are machine controls located to prevent unintentional activation while allowing easy access for stopping the machine? • Are all machines, equipment and electrical installations easily isolated of all sources of power? • Is the workplace designed for consideration of human factors (i.e. an ergonomically designed workplace)?

In most PSM programs human factors is addressed in the PHA/HIRA element by analyzing both human error potential in the formulation of PHA causes and by addressing human factors engineering issues in the equipment and operations. Human factors are also an important aspect of the Operating procedures and Conduct of Operations elements. The use of human factors checklists, particularly to analyze human factors engineering issues, should include the four IS strategies, particularly *Minimization* and *Simplification* in the design of the man-machine interface. An example of the importance of human factors is demonstrated by the following incident that occurred at a refinery:

> A plug valve is used primarily for on/off, and some throttling services. The valve is opened or closed with the use of a valve wrench. A quarter turn of the valve blocks the flow path. The typical arrangement for such a valve is to have the valve operator be in-line with the flow path when the valve is open, and to have the valve operator be perpendicular with the flow path when the valve is in the closed position. In preparing the pump for maintenance, an operator relied on the visual position of valve wrench to determine that the suction valve was open. He moved the wrench to what he believed was the closed position with the wrench perpendicular to the flow of product. The valve was actually open. (Ref 11. CSB, 2005)

Figure 14.1 Plug Valve Operator

This incident demonstrated not only an important human factors issue, but also a weakness in the MOC program at the refinery where it occurred. Allowing a valve operator to be installed in the opposite manner that convention would require for this type of valve almost guaranteed that an incident would occur. When the valve was re-configured the MOC should have revealed this serious (and obvious) error, but did not. Additionally, in this situation facilities should only procure and install quarter-turn ball/plug valves (a common valve application) where the valve is designed so that the valve operator can only be installed in the normal convention. Internal engineering specifications for valves should reflect this convention.

Finally, the application of IS in PHAs/HIRAs, whether they are the regular studies performed by the facility, or they are separate studies will require that the facility or its parent company be well versed in the concept of "feasibility" and have a definition of the term and how the definition should be applied in each study performed. This term and how it affects the application of IS is addressed in more detail in another paper presented at this conference (Ref 9. Hazzan).

Beyond the traditional use of IS in PHAs/HIRAs, there are opportunities to apply the four strategies of IS in how these studies are planned and executed. For example, using a different PHA method, e.g., HAZOP instead of What-If, based on the complexity of the process or the uniqueness of the hazards/risks is an application of *Substitution*. Using different PHA/HIRA team members with specific expertise in the hazards/risks to be studied is another application of *Substitution*, e.g., substituting existing team members with those who more thoroughly understand the emergency response plan and communications systems during the global/general node of a HAZOP study. Another example is substituting team members who are knowledgeable in the state-of-the-art of the process technology being addressed while addressing IS during a PHA/HIRA. Adjusting the node list in a PHA/HIRA to subdivide the process in more detail to reduce the effects of a complex process design during the PHA/HIRA is an example of *Simplification*.

Inherent Safety & Management of Change/Operational Readiness Programs

Management of Change and Operational Readiness (or Pre-startup Safety Review as it is often referred to) are often combined elements by organizations in the same overall MOC procedure in PSM programs. The OR/PSSR is typically one of the last steps in the MOC process. These elements offer several ways to incorporate the four IS strategies. First and foremost, MOC is required to be used whenever changes to existing equipment are being proposed, including IS changes. However, the reverse should also be true, that is, whenever a change is being contemplated to existing equipment, the MOC process should include IS considerations. This is just as important as looking for IS opportunities while processing a MOC for other reasons. The use and consideration of IS strategies should be incorporated into the identification of prospective changes and their technical basis, the MOC safety and hazards reviews or PHAs conducted to examine the potential process safety impacts of prospective changes, and the modification of operating and other procedures pursuant to an equipment change. Specifically, the safety review of the proposed change can employ checklists with IS questions similar to what has been described for PHAs/HIRAs above. The four IS strategies should also be included in MOC/PSSR procedure checklists. The overall MOC process and where IS guidewords and checklists can be used in it is shown in Figure 1 (Ref 10. Kletz 2010). Table 2 shows a list of MOC questions that can be used when reviewing proposed changes. The use of IS strategies and concepts in the design of a process should be clearly documented because they are inherent and thus not obvious to those who did not participate in the weighing of design options. Therefore, the design could be compromised and the process safety risk increased by later changes where the rationale for the original design is unknown. This should be reflected in the Process Knowledge Management (or Process Safety information) element of the PSM program.

Figure 1. Inherent Safety-Based Management of Change Process

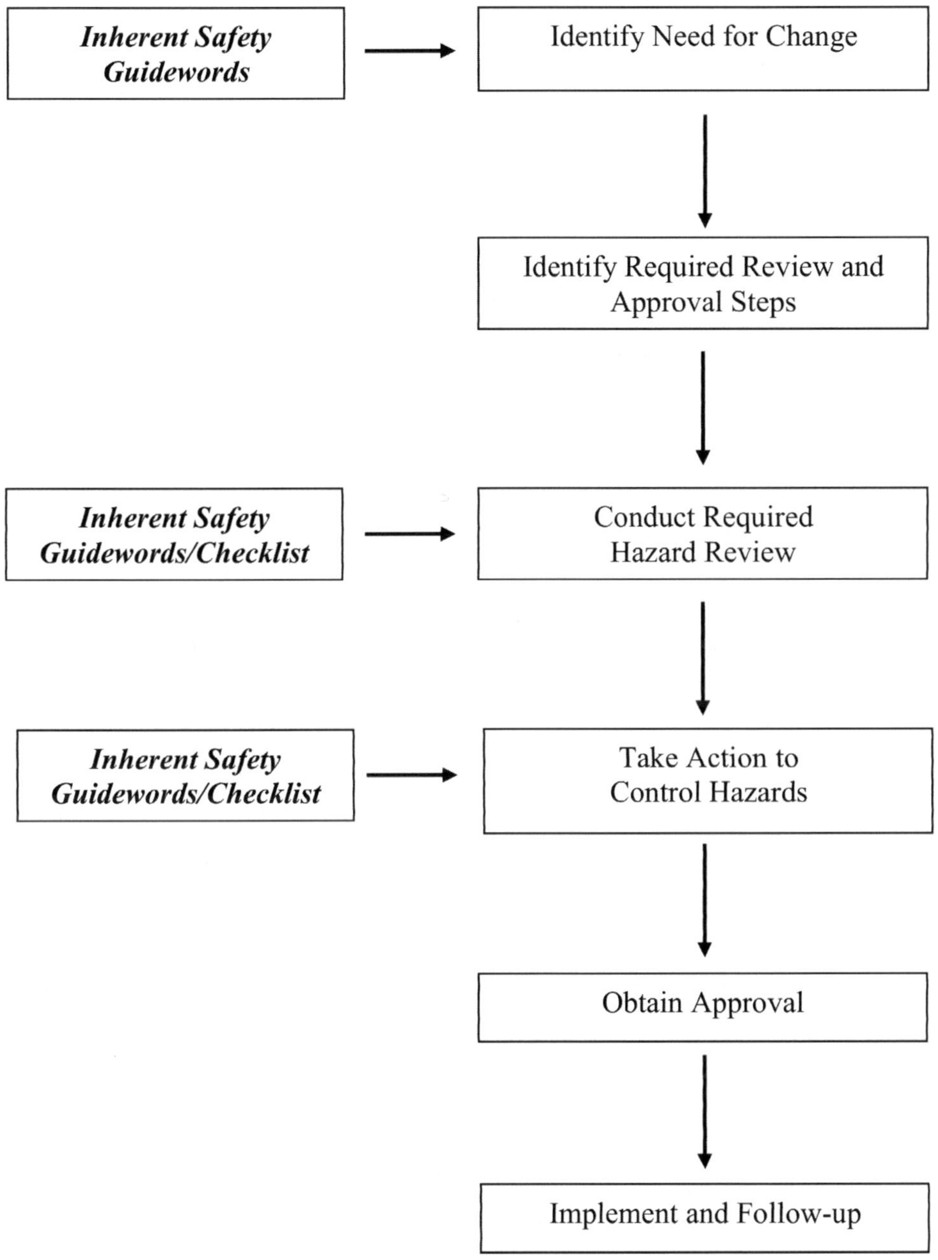

Table 2. Inherent Safety Checklist for MOCs

IS Strategy	MOC IS Checklist Question
Minimize	• Have the inventories of hazardous materials and materials included within the scope of the PSM program been minimized in the design of the proposed change? • Have piping dead legs and low flow portions of piping been minimized as the design of the proposed change? Can additional piping dead legs and low flow portions of piping be eliminated while implementing the change? • Can just-in-time deliveries be used when dealing with hazardous materials in the modified process? If so, has the possible risk transfer to transportation sectors been evaluated? • Have ignition sources been eliminated or minimized to the extent possible in the modified process? • Are the human interfaces with the altered process been evaluated so reduce fatigue? Will the existing shift rotation require modification as a result of the change? • Has the opportunity to completely eliminate hazardous raw materials, process intermediates or by-products been evaluated as part of the change even if the change was proposed for a different reason? • Has the opportunity to eliminate or minimize ignition sources been evaluated as part of the change even if the change was proposed for a different reason? • Has the opportunity to reduce inventories of hazardous materials and materials included within the scope of the PSM program been evaluated in the equipment that is appurtenant to or connected to the equipment being changed (even if the change being proposed was for a different reason)?
Substitute	• In the review of the proposed change can a less toxic, flammable or reactive material be substituted for use? • Can alternative materials of construction be substituted that will offer higher resistance to (or even immunity from) known corrosion and damage mechanisms in the equipment being changed? If so, is it possible to extend the alternative materials of construction be extended to additional portions of the process at the same time? • Can a water based materials be used in place of a solvent or oil based product?
Moderate	• Can lower temperatures, pressures, speeds or other process parameters be used to achieve the desired reactions in the modified process? • In reviewing the proposed change are all hazardous gases, liquids and solids stored as far away as possible to eliminate disruption to people, property, production and environment in the event of a release?
Simplify	• Have all procedures and other written guidance for operating, maintaining, or performing other human interface with the modified process been evaluated to make these documents as simple and easy-to-use as possible, particularly upset/emergency procedures, including emergency shutdown procedures. • Are equipment designed such that they cannot be operated incorrectly or carried out incorrectly, e.g., are valve operators on quarter-turn manual valves configured so that when the valve operator is in-line with the flow direction the valve is open and when the valve operator is perpendicular to the flow direction the valve is closed? • Have DCS and other control panels been designed following weel-known conventions for displays, colors, and other characteristics that will make operations intuitive and easy-to-understand and therefore reduce human error? • Have DCS and other control panels been evaluated with respect to alarm management principles and conventions? • Can rotating equipment and other machinery be stopped locally and from central control rooms or locations?

IS Strategy	MOC IS Checklist Question
	• Have other design and operational conventions been followed so as to obviate human errors? • Has the opportunity to simplify procedures associated with the process, particularly emergency procedures, been evaluated to simplify the procedures, even if the change was proposed for a different reason?

Inherent Safety & Asset Integrity/Mechanical Integrity Programs

Asset integrity (or sometimes referred to as Mechanical Integrity) includes a broad range of technical activities in a PSM program, including:

- Identification of equipment to be included in the asset integrity program.
- Written maintenance procedures.
- Inspection, testing, and preventive maintenance (ITPM) programs.
- Maintenance personnel training and qualification.
- Management of asset integrity deficiencies.
- Asset integrity quality assurance (QA), which involves the design, fabrication, and installation of the equipment, as well as the replacement of materials.

Each of these sub-areas of asset integrity can involve consideration of the four IS strategies. For example, *Substitution, Moderation, and Simplification* can be applied in the selection of the materials of construction when designing and fabricating equipment. *Minimization* can be applied in developing and implementing risk based inspection programs where ITPM frequencies and schedules are optimized. *Minimization* can also be employed in the development of ITPM programs, e.g., the *minimization* of dead/ low flow legs in piping systems, which will reduce the number of locations that are susceptible to galvanic corrosion in no flow/low flow areas of the processes. Also, *Substitution* can be used by employing new, or different, and more effective ITPM methods, such as the use of improved non-destructive testing (NDT) techniques.

The QA portion of an asset integrity program is often where the traditional uses of IS strategies are employed because it is when equipment is designed and specified. The preliminary or conceptual design stage of a project is when the basic chemistry is determined and this offers the best opportunity to use the *Substitution* and *Moderation* strategies if possible. It is also during this phase of a project when the basic equipment sizing is determined based on the desired throughput of the process and *Minimization* can be used in this phase.

The opportunity also exists during the preliminary or conceptual design stage to make decisions about the capacity of the equipment to inherently withstand certain upset conditions, particularly the robustness of the pressure retaining boundary, i.e., the primary containment of the process. Examples include:

- If the maximum pressure achievable in the process given abnormal events in with respect to the chemistry, and other abnormal events such as external fires is 200 psig. If the primary containment is designed to mechanically withstand 300 psig without relying on any active safeguard such as a relief device, this can be regarded as a use of *Moderation*. Some would regard this as a design choice rather than a IS option. However, those choices, when they are intentionally made to avoid a certain hazard(s) without relying on any other active or passive safeguard can be regarded as inherent in nature.

- The intentional choice of certain materials of construction during preliminary design. If the materials of construction are selected to specifically make the process immune to certain types of corrosion or damage mechanisms that would affect other materials of construction, that can also be regarded as an inherent solution using *Substitution*.
- Adjusting the operating temperature of a process (either higher or lower) to avoid the temperature where corrosion under insulation (CUI) can occur, given the materials of construction. For some materials of construction CUI is possible at or around ambient temperatures, so the only option is to increase the operating temperature. This is a novel, and perhaps counterintuitive use of the IS strategy of *Moderation*, which usually involves lowering temperatures.

The QA portion of AI/MI programs is also driven by RAGAGEPs that govern the design and construction of processes. The IS strategies are indirectly built-into many codes, standards, and other RAGAGEPs by establishing design limits for process parameters, establishing the criteria for dimensional characteristics (e.g., the minimum thickness of pressure retaining equipment) and setting the design margins for the process/equipment. By specifying known and higher-than-required design margins for pressure, temperature, and other process conditions, the process will have a known level of robustness in its design. Therefore, the application of the design principles contained in the governing RAGAGEPs are examples of *Moderation*. Internal company design standards often reflect this concept and provide even more inherent protection in the design of processes/equipment. An common example is the specification of schedule 80 piping (i.e., extra thickness) where schedule 40 piping (normal thickness) will suffice.

Inherent Safety & Conduct of Operations/Operating Procedures

Conduct of Operations (COO) is a PSM element that covers a broad range of process safety program activities. This is the element of the program where operating discipline and adherence to written and approved procedures as well as adherence to strong operating principles is established. It is also a process safety program element that is very sensitive to the prevailing culture in an organization. For the purposes of this paper, Conduct of Operations has been combined with Operating Procedures, although in most PSM programs these are treated as separate elements. The four IS strategies are or can be part of the COO/SOP element and its related activities as follows (Ref 5. CCPS COO):

Minimization:

Alarm management. In recent years modern control systems, particularly integrated distributed control systems (DCS) have come under increasing review for the number, ranges, and human factors aspects of their process alarms. Problems in this area have been contributors in several process safety incidents. The elimination of nuisance alarms, the widening of alarm setpoints so that multiple simultaneous alarms during upset or abnormal conditions are minimized or avoided, the selection of alarm displays, how alarms are silenced, how alarms are recorded, and other human factors considerations have received increased attention. The modification of the control systems to make improvements have resulted. This entire set of activities is often referred to as alarm management. An industry standard devoted to this topic has been issued (Ref 1. ANSI). Alarm management is an important application of the IS strategies of *Minimization* and *Simplification*.

<u>Fatigue management.</u> Overtime can be a sensitive subject in many facilities. Sometimes there is pressure to work excessive overtime from facility or company management to help solve more basic staffing problems. Sometimes there is peer pressure in favor of excessive overtime for the financial benefits that result. Regardless of where the pressure originates, when it exists a culture issue also exists. The CCPS RBPS Guidelines and Conduct of Operations and Operational Discipline book includes fatigue as a COO issue (Ref 7. RBPS) (Ref 5. CCPS COO). Overtime can also be an organization work rule or a collective bargaining agreement issue, in which case the solution to the problem will require changing policies and procedures, or re-negotiation of the labor contract (if the workforce is represented by a union). The Baker Panel (Ref 4. Baker) noted that fatigue was a contributor to the BP Texas City incident. Also, this problem was not limited to hourly employees and contractors, as salaried employees, including supervisors had also worked excessive amounts of hours. Industry has recognized that fatigue is an important issue, and API RP-755 (Ref 2. API 755) has been published to provide guidance for facilities to detect and counter fatigue in those parts of the workforce working night shifts, rotating shifts, extended shifts/days, or call-outs and where their job requires process safety actions. This reference establishes a fatigue risk management system so that the process of managing fatigue among those positions may be sensitive to it is a continuous and sustainable process. However, if there is underlying pressure being brought to bear on personnel to work excessive overtime, then there is a safety/process safety culture issue that must be solved first. Fatigue management is an important application of the IS strategy of *Minimization*.

<u>Fitness for duty.</u> Fitness-for-duty refers to a range of possible influences on worker performance, such as impairment due to drug or alcohol, fatigue (addressed above), illness, distraction due to personal issues, and mental condition (Ref 5. CCPS COO). No external influences on performance should be allowed for facility personnel. This includes the use of alcohol or drugs while on duty, horseplay, harassment, and other aberrant behaviors. No tolerance for these behaviors goes beyond simply forbidding them in written policies and stating these prohibitions during training/orientation sessions. It means living them at each level of the organization, including self-policing within peer groups and during off shifts. Minimizing the external negative influences on personnel behavior and performance and establishing the minimum requirements for fitness for duty expectations is an important activity. Therefore, fitness for duty is also an important application of the IS strategy of *Minimization*.

<u>Shift differences and shift turnover.</u> The culture of different operating shifts within a given facility sometimes varies widely. What would never be tolerated on one shift might be routine on another. This often represents the difference between staffing levels during normal working hours (i.e., day/weekday and off-hours (i.e., night/weekend/holiday shifts), and the presence and attention of technical staff and management during the day shift of normal weekdays. Most of this difference can be attributed to the attitudes and beliefs of supervisors, foremen, and other mid-level

management assigned to off shifts. Having to accomplish the same level of production with fewer resources sometimes fosters the attitude that shortcuts and other deviations from approved procedures are acceptable in order to achieve those goals. Sometimes this situation fosters a abnormal "can-do" attitude where independently finding any solution to a problem and getting around a production delay/issue quickly is seen as a positive attribute, regardless of what rules, procedures, and norms are disregarded in the process. Yet the process conditions and the potential hazards they present are exactly the same regardless of the time of day or day of the week. One of the definitions of the process safety culture is doing the right thing even when nobody else is watching. This definition captures the conditions that exist in many facilities during off shifts, but not in all facilities.

The root causes of some major process safety incidents have included inadequate shift turnover (Ref 4. Baker). Some parts of the shift turnover issue are procedural and some are cultural. Personnel need to understand both the importance of this activity and that it must be formally transacted. In recent years there has been a recognition of the importance of shift turnover and widespread improvement in this activity. Many facilities now have written procedures that specify the minimum requirements for shift turnover, i.e., how and where is to be accomplished, including its documentation. Minimizing the differences between the culture and operating practices between shifts and any differences in shift turnover will help reduce incidents. For these reasons, differences between shifts and shift turnover represent important situations where the IS strategy of *Minimization* can be applied.

<u>Emergency shutdown & stop work authority.</u> The emergency shutdown of a process or unit (a rare but critical event) is an extremely important aspect of operational discipline. This action can be a controversial activity, or at least regarded as controversial, but should not be. At its core, this is mostly a process safety culture problem, rather than an operations problem. Some operators so not truly believe that there will be no negative ramifications if they activate an emergency shutdown, or they simply believe that the decision to initiate such crucial operations should only be made by a supervisor. A similar culture issue exists with respect to stop work authority. A contributing factor in many serious incidents is facility personnel not realizing that they could have or should have stopped themselves or others from continuing with an operation or task that was becoming dangerous or had the potential to become dangerous. In most companies, high level statements of safety or process safety policy emphasize the importance of emergency shutdown and clearly state that operators have the sole authority to activate when conditions warrant. Similar statements encourage personnel to exercise their good judgement and be proactive about stopping an operation or task from proceeding when they recognize an error or other conditions that could make the situation worse. However, many employees are hesitant to actually exercise their judgement and stop work, especially when the work of others is involved. Minimizing the hesitance to activate an emergency shutdown when warranted, and minimizing the hesitance to exercise stop work authority are important process safety cultural issues. Therefore, emergency shutdown and stop work authority are critical applications of the IS strategy of *Minimization*.

<u>Other process safety cultural issues.</u> The normalization of deviation and loss of a sense of vulnerability are important core principles of process safety culture. Deviations from approved rules, procedures, codes/standards, or other formal guidance that are tolerated so that they do not

arouse curiosity, are not questioned or challenged, and become part of the everyday way of operating the facility then become accepted or "normalized." These deviations are likely to eventually become latent failures in a process safety incident scenario. Also, sustaining a healthy level of respect for the hazards and risk of the facility and its operations, particularly over extended periods of time maintains a sense of vulnerability. A reduced sense of vulnerability results from complacency over time in the absence of incidents or near misses. This is particularly true when high consequence – low frequency events occur that are in the risk profile of the facility or company. The low frequency of such events, which includes serious process safety events results in few, or likely no personnel working in such facilities ever having experienced such an event. Therefore, over a lengthy period of time, they tend to "forget" that such events are possible. This tendency is also characterized by an erosion of the respect that this type of risk should constantly command. Minimizing the occurrence of these cultural situations is an important aspect of operational discipline. Therefore, minimizing the normalization of deviation and the loss of a sense of vulnerability are important applications of the IS strategy of *Minimization*.

Simplification:

Operating Procedures. Operating procedures should be subject to periodic review to 1) ensure that they are current and accurate, and 2) to the extent possible be as simple as possible (*Simplification)*. Operating procedures that are written in as simple a style and format as possible will be easier to use and hence will be used more often by the operators. Operating procedures that are unnecessarily lengthy, complex, and written in a language that is difficult for the intended users to follow will not be used as often. Another consideration for writing operating procedures is the reading comprehension level of the intended users. Vocabulary and syntax can be selected to match a desired reading comprehension level. This is an important consideration as engineers tend to write the initial drafts of many operating procedures, yet it is not the engineers but the operators who are the primary audience for the operating procedures. However, simplicity should not be achieved at the sacrifice of procedure clarity. For example, "open valve slowly" can mean something different to different users because "slowly" is a general term. If the process is sensitive to degree of valve position, then perhaps more detailed directions would be appropriate, e.g., "open valve 10% initially and then 10% every 30 seconds until you reach 70%, then fully open." Operating procedures prepared with IS principles in mind tend to be easier to use and preclude the possibility of operators taking short cuts. This section has been written in the context of operating procedures. However, the same concepts of *Simplification* apply to any PSM-related procedure, e.g., maintenance procedures, emergency response plans, safe work procedures/permits, etc.

Operating Limits. The operating limits and limiting conditions for operation that are included in the operating procedures should be subject to *Moderation,* however, this will occur not because the procedures are modified but because the operating conditions in the process are modified (i.e., moderated) and the published values in the procedures will follow suit.

Human Factors. The design of control systems displays, and other ergonomic issues that allow operators to accomplish what the operating procedures require easily. Sometimes attention has to be paid to these types of design issues in order to make it possible to accomplish the what the procedures require. For example, if an emergency operating procedure requires that four steps in widely separated locations be performed rapidly and an operator would have only a few minutes to accomplish them, then the process design must be simplified to reduce the time necessary for the operator to get from one location to another quickly, else the function in question should be automated. *Simplification* is an important design consideration for the human-machine interface.

Bypass/removal of safety or protective features. As with operating procedures and control of work/SWP procedures, a procedure to manage and control the bypass or removal of safety or protective features should be as simple and straightforward as possible, including any permits associated with them. This is another example of the importance of the IS strategy of *Simplification* to procedures and permit systems.

Conclusion

There are ample opportunities to apply the four strategies of IS in the planning, development, and implementation of each of element of a PSM program. This requires that the application of the strategies to programmatic aspects of facility operation, rather than to the physical aspects of the facility and its chemicals, which is the traditional use of IS. A different thought process about how each element works and how it can be improved will be required. The management system procedures for each PSM element should reflect how the IS strategies will be employed, as this is not a one-time application but an ongoing process as the facility/company process safety program evolves over time.

References

1. American National Standards institute (ANSI), ANSI/ISA–18.2–2009, *Management of Alarm Systems for the Process Industries*, 2009.
2. American Petroleum Institute (API), *Fatigue Risk Management Systems for Personnel in the Refining and Petrochemical Industries*, API RP-755, 2010.
3. Amyotte, P, Goraya, A, Hendershot, D, Khan, F., *Incorporation of Inherent Safety Principles in Process Safety Management*, Proceedings of 21st Annual International Conference – Process Safety Challenges in a Global Economy, Orlando, FL, April 2006.
4. Baker, J. et al., *The Report of BP U.S. Refineries Independent Safety Review Panel*, January 2007 (Baker Panel Report).
5. Center for Chemical Process Safety, *Conduct of Operations and Operational Discipline*, American Institute of Chemical Engineers, 2011.
6. Center for Chemical Process Safety, *Inherently Safer Chemical Processes: A Life Cycle Approach, American Institute of Chemical Engineers, 2nd Ed.*, 2009.
7. Center for Chemical Process Safety (CCPS), *Guidelines for Risk Based Process Safety*, American Institute of Chemical Engineers, 2007.
8. Contra Costa Health Services (CCHS), California, *Industrial Safety Ordinance Guidance Document,* Section D, June 2011.
9. Hazzan, M., *The Definition of Feasibility as it Applies to Inherent Safety*, 18th Global Congress on Process Safety, Center for Chemical Process Safety, Orlando, FL, 2018.
10. Kletz, T., Amyotte, P., *Process Plants – A Handbook for Inherently Safer Design*, 2nd Ed., CRC Press, 2010.

11. U.S. Chemical Safety Board (CSB), *Oil Refinery Fire and Explosion*, Case Study 2004-08-I-NM, October 2005.

Encouraging the Use of Inherent Safety

David A. Moore, PE, CSP
President & CEO
AcuTech Group, Inc.
1919 Gallows Road, Suite 900
Vienna, VA 22182
dmoore@acutech-consulting.com
www.acutech-consulting.com

Prepared for Presentation at
American Institute of Chemical Engineers
2018 Spring Meeting and 14th Global Congress on Process Safety
Orlando, Florida
April 22 – 25, 2018

Encouraging the Use of Inherent Safety

David A. Moore, PE, CSP
President & CEO
AcuTech Group, Inc.
1919 Gallows Road, Suite 900
Vienna, VA 22182
dmoore@acutech-consulting.com
www.acutech-consulting.com

Keywords: Inherent Safety, IST, process safety management, CCPS, PSM, PSM Excellence, Process Safety Regulations

Abstract

Many organizations are still not taking full advantage of the concept of inherent safety and using it as a fundamental and comprehensive tool to reduce process safety hazards. As a result, the industry may be leaving some "risk on the table" that could be reduced, potentially for "free", by using the methods of inherent safety. As companies are trying to set performance targets to achieve world class "operational excellence", inherent safety should be a critical element of their process safety strategy as it has the potential to contribute to beneficial hazard reduction.

The priority for the adoption of inherent safety is the recognition of the meaning of ISD, how it can be helpful, how it is applied, and when and where it can be used. Then the organization must decide how to take full advantage of the opportunities of implementing inherent safety.

There may be a perception that inherent safety is not a suitable strategy since the core technology or chemical cannot be changed and that is seen as all is it used for. A common perception is that it may be threatening to the very existence of some facilities or companies in chemical manufacturing, storage, or use if forced to implement it and so it is not attempted. This may be true if there is a regulatory-driven situation where it is required despite the company believing it is infeasible. However, it may be a misconception that it is only about the full substitution of chemicals because it may not be appreciated that inherent safety philosophy credits any move in an inherently safer direction using the four strategies of inherent safety (minimize, simplify, substitute, and moderate) as a positive step in hazard reduction. Inherent safety can be applied on a macro or micro level, and most opportunities with existing and modified plants are not being realized but could be if more incremental inherent safety applications were identified.

What may have been lacking is additional guidance clearly defining a process for practically applying inherent safety principles at all stages of a process lifecycle. In addition, the concept that inherent safety is ingrained in the mindset of all employees as a "way of doing business" should be promoted.

CCPS recognized this and formed a committee to update the leading "Guidelines for Inherently Safer Chemical Processes: A Life Cycle Approach, 2nd Edition. This culminated in 2017 in the start of a new CCPS guideline project to produce an updated 3rd Edition which is being led by AcuTech. The new guidelines will make the case that there are many ways to apply the principles of inherent safety and they should be encouraged in day to day activities as a matter of principle for all employees. This paper will explain the key concepts of getting inherent safety to be actively and more completely applied to the advantage of the corporation.

1 Introduction to the Opportunities of Inherent Safety

Many organizations are still not taking full advantage of the opportunities for inherent safety (referred to as Inherently Safer Design or ISD) as a tool to eliminate or reduce hazards in their Process Safety Management (PSM) initiatives. Technologists have long recognized the value of eliminating or reducing hazards, applying ISD without calling it by that name; they simply considered it to be good design (1). However, a concerted effort to apply modern concepts of ISD must be understood, defined, documented, and nurtured to be successful. As a result, the industry may be leaving "risk on the table" that could be reduced, potentially for "free".

Creative ideas from the employees of the company may be keys to step changes in process safety performance and hazard reduction and may be readily implemented. Understanding the real potential of ISD and then making it an integral part of the safety philosophy of the organization is the key to taking maximum advantage of the tool.

2 Background on ISD

2.1 The Concept of Inherent Safety

Inherently Safer Design presents a holistic approach to making the development, manufacture, and use of chemicals safer. It involves such practical applications as substituting more benign chemicals at the development stage, minimizing risk in the transportation of chemicals, using safer processing methods at the manufacturing stage, and decommissioning a manufacturing plant. The concept of ISD is not new, nor is it unique to the process industries. It is part of the overall process safety management system of an enterprise.

The modern approach to chemical process safety is to apply risk management systems theory. This includes recognition of the hazards posed by the process and a continual effort to analyze the risks and to reduce or control them to the lowest levels practical while considering the balance of other objectives of the business. A hazard is classically defined as a 'situation with the potential for harm'. Risk is defined as 'the likelihood that a defined consequence will occur'.

Process risk management is the term given to the collective efforts to manage process risks through a wide variety of strategies, techniques, procedures, policies, and systems. Chemical process hazards are defined to come from two sources:

- hazards that are characteristic of the materials and chemistry used, and

- hazards that are characteristic of the process variables -- the way the chemistry works in the process.

In general, the strategy for reducing process risk, whether directed toward reducing the frequency or the consequences of potential accidents, can be classified into five categories:

- Inherent - Eliminating the hazard by using materials and process conditions that are non-hazardous; i.e., substituting water for a flammable solvent;
- Segregate – The use of distance and barriers to reduce the effects of hazards.
- Passive - Minimizing the hazard through process and equipment design features that reduce either the frequency or consequence of the hazard without the active functioning of any device or requiring human input; i.e., providing a dike wall around a storage tank of flammable liquids;
- Active - Using controls, alarms, safety instrumented systems, and mitigation systems to detect and respond to process deviation from normal operation; e.g., a pump which is shut off by a high-level switch in the downstream tank when the tank is 90% full. These systems are commonly referred to as engineering controls or safeguards, although human intervention can also function as an active layer;
- Procedural - Using policies, operating procedures, training, administrative checks, emergency response, and other management approaches to prevent incidents, or to minimize the effects of an incident; i.e., hot work procedures and permits. These approaches are commonly referred to as administrative controls and require humans that interface with the process (primarily operators) to recognize and react to abnormal or adverse events in a prompt and correct manner as directed by the approved procedures and the training on those procedures.
- All five categories can contribute to the overall safety of a process. Ideally, the steps of analyzing, reducing, and managing risk will be considered in a hierarchical manner as shown in Figure 1 (2).

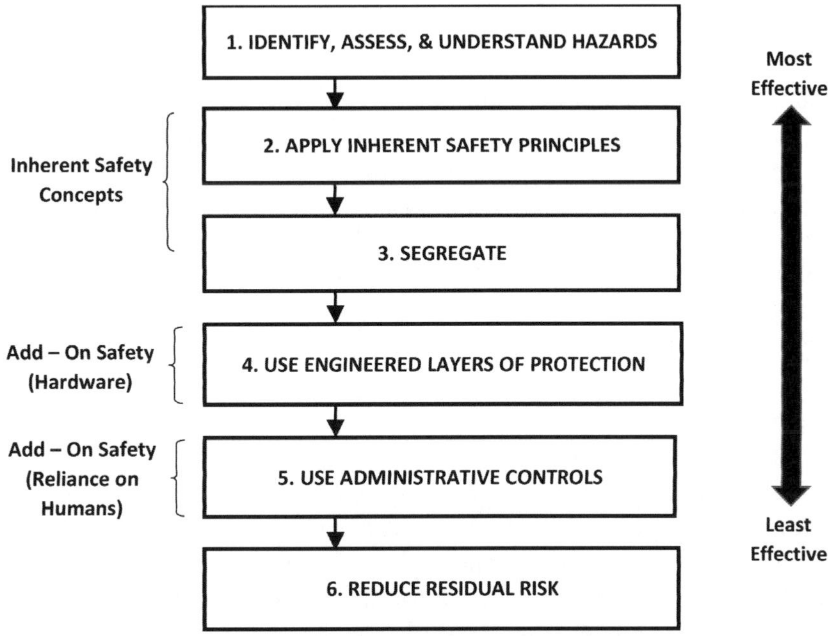

Figure 1 Hierarchy of Controls

2.2 *The Strategies of Inherent Safety*

Approaches to applying inherently safer processes and plants have been grouped into four major inherent safety strategies. The names or titles of these strategies were established in the 1st and 2nd editions of the CCPS guidelines on inherent safety (2). These groupings follow the original treatment of the topic of inherent safety as published by IChemE and IPSG (3) and Kletz (4), (5). However, in this earlier, seminal work different names/titles were used. In general, these may be considered subsets of the four core strategies. These different names/titles and how some strategies have been incorporated into the four used herein are presented in Table 1for clarity and as a reference in interpreting older work on inherent safety:

Table 1. Inherent Safety Strategies

INHERENT SAFETY STRATEGIES	
ISD STRATEGY	**DESCRIPTION**
Minimize	Use smaller quantities of hazardous substances (also called intensification).
Substitute	Replace a material with a less hazardous substance.
Moderate	Use less hazardous or energetic processing or storage conditions, a less hazardous form of a material, or facilities that minimize the impact of a release of hazardous material or energy (also called Attenuation and Limitation of Effects).
Simplify	Design facilities which eliminate unnecessary complexity and make operating errors less likely, and which are forgiving of errors that are made (also called Error Tolerance).

Inherently safer design is a fundamentally different way of thinking about the design of chemical processes and plants. It focuses on the elimination or reduction of the hazards, rather than on management and control. This approach should result in safer and more robust processes, and it is likely that these inherently safer processes will also be more economical in the long run (Kletz, (3), (4). It must be recognized, however, that applying an inherently safer concept to a plant does not necessarily mean that plant is the most safe, secure, efficient, or economical.

The essential issue with the concept of inherent safety is that the focus should be on reducing or eliminating hazards by changing the materials, chemistry, and process variables such that the reduced hazard is characteristic of the new conditions. This compares with adding layers of safety to a process to reduce the risk but not reducing the nature of the hazard directly.

The cost or feasibility of changing an existing design to a more inherently safer technology may be unjustified or difficult to justify from an investment standpoint. For this reason, the options must be holistically weighed, and the total lifecycle costs and risks be analyzed for completeness.

2.3 *Why Not Widely adopted?*

The priority to the adoption of inherent safety is the recognition of the meaning of and defining a full set of opportunities for applying inherent safety. There may be a perception that inherent safety is not a suitable strategy since the core technology or chemical cannot be changed and that it may be threatening to the very existence of some facilities or companies in chemical manufacturing, storage, or use if forced to do so. This is a misconception because it may not be appreciated that inherent safety philosophy is that any move in an inherently safer direction using the four strategies of inherent safety (minimize, simplify, substitute, and moderate) is a positive step. Inherent safety can be applied on a macro or micro level, and most opportunities with existing and modified plants are not being realized.

What may have been lacking is additional guidance clearly defining a process for practically applying inherent safety principles at all stages of a process lifecycle. CCPS recognized this and formed a committee to update the leading "Guidelines for Inherently Safer Chemical Processes: A Life Cycle Approach, 2nd Edition. This culminated in 2014 in the start of a new CCPS guideline project "Essential Practices for Developing, Strengthening and Implementing Process Safety Culture – Project # 249" led by AcuTech. The new guidelines will be make the case that there are many ways to apply the principles of inherent safety. This paper will explain the use of the guideline and the key findings for an application of the approach. Since the publication of the original concept book in 1996, there have been many developments on the concept of inherent safety. This new edition provides the latest knowledge so that engineers can derive maximum benefit from inherent safety.

3 The Dilemma of the Regulation of Inherent Safety

Inherent safety as a regulatory requirement has known complicating issues, such as the lack of metrics to judge the adequacy of the ISD employed. More so, experience has shown that the public and regulators may not be satisfied with incremental safety improvements even if they are substantial reductions in risk. The key expectation with most persons exposed to a potential release is that the reduction of risk results in zero exposure to them. The final expectation is the often the reduction of consequences versus the reduction of risk in total. Inherent safety becomes the preferred mechanism in the eyes of the public for achieving that reduction. The aura of impractical regulations requiring inherent safety causes most companies to fear the methods rather than embrace them.

While this is a natural expectation on the part of the public to reduce consequences to zero through ISD, it is hardly achievable in any practical way for most of complex industrial facilities without substantial sacrifice on the part of the site, particularly if it is an existing installation. From the public's opinion, these sacrifices are expected in their interest of public safety. Incidents also cause an impetus for emboldening the public to demand greater safety and security, and create step changes in the evolution of process safety. For these reasons it is preferred to adopt ISD and encourage its use voluntarily to make actual hazard reductions and to show responsible behavior in the management of process hazards. It is also essential to make the point that ISD has a role in the holistic reduction of risk along with the other strategies of the hierarchy of controls.

These challenges are mitigated by the adoption of 1) a broad definition of ISD across the entire lifecycle versus a one-time challenge to the chemistry of the process, 2) voluntary efforts by industry to encourage the thorough application of ISD, 3) reasonable regulatory requirements that recognize a broad definition of ISD and degrees of achievement possible, and 4) proof that ISD is making an effective contribution to risk reduction as is truly feasible.

4 ISD During the Process Lifecycle

The application of ISD is proven to be possible along the entire lifecycle of a process. A common misconception is that ISD is only about the substitution of the hazardous chemical in use with a less hazardous one, but that is not always feasible. If the perception is that is the only opportunity for applying ISD, and it is highly infeasible to even consider it (i.e., substituting the production of a basic chemical that has toxic or reactive properties is the objective of the operation and there is no good substitute available), then the conclusion is ISD is not possible. This is shortchanging the many applications of ISD even if on more of a smaller scale given the inherent nature of the hazards of the primary process chemistry.

Figure 2 illustrates the many opportunities for applying ISD. For new, modified or continual improvement projects the best opportunities to make substantial, cost-effective, and lasting changes may be during the concept, R&D, and design development stages. However, it requires a concerted effort and focused direction to prefer using creative research and engineering on alternatives or moderate applications of chemical processes when traditional methods may be simply repeated. “Proven” or traditional applications of technology are not necessarily inherently safer than alternatives and may be repeating inherent hazards needlessly. During the rapid development of a concept to the engineering and operational phases there may not be any effort spent on ISD unless there is education on how and why to do this, encouragement on finding process designs with less hazards, and a holistic view taken of the application of ISD.

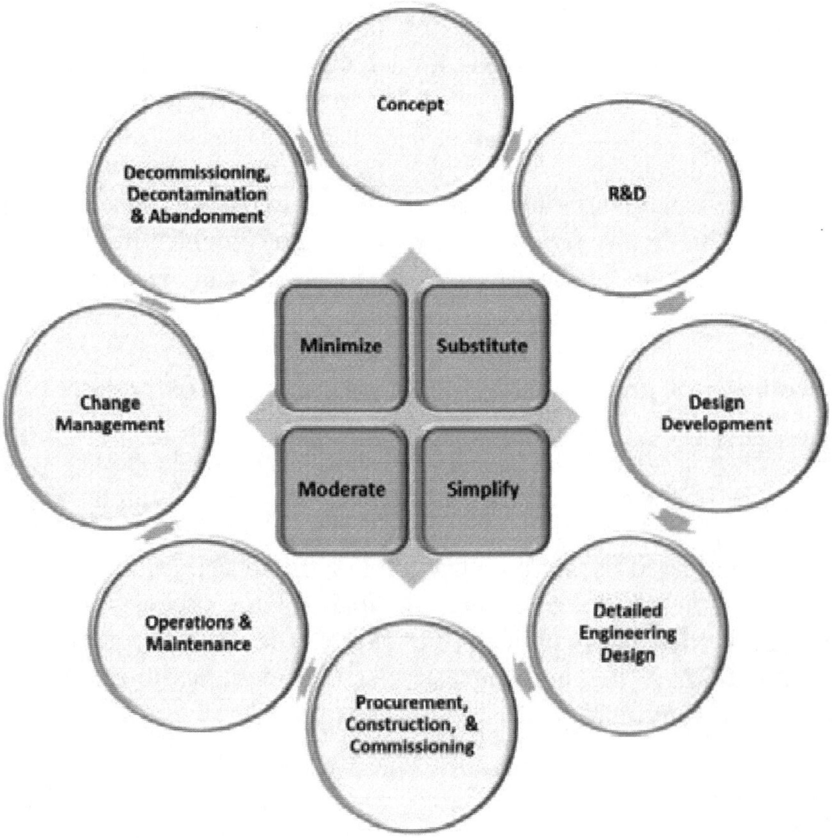

Figure 2. Inherently Safer Lifecycle

During the detailed engineering phase and procurement, construction & commissioning there are many chances to find ways to reduce hazards through ISD. For example, the selection of a sub-technology for the chosen process design is an opportunity for hazard reduction (the technology, vendor design, chemistry, etc., are all variables). Reducing the complexity of the design through the ISD strategy of simplification should be examined and may greatly reduce the likelihood of a process safety incident.

5 Applications of ISD

5.1 Process Hazards Analysis

Perhaps the most practical and effective way to ensure ISD is being considered and applied throughout the lifecycle is by requiring it during the conduct of process hazards analysis (PHA), and to ensure through auditing that that process is being effectively done. Inherently Safer reviews can be conducted as separate studies, using a HAZOP-like technique, or incorporated into HAZOP studies or revalidations. This involves the application of inherent safety guidewords as "deviations," using, for example, the Guideword Matrix provided in Table 2.

As with the use of HAZOP methods for conducting regular PHAs, it can be used at any level of subdivision of the process, including node-by-node. While thorough, this can also increase the amount of time required to conduct the study. It also requires a degree of brainstorming by a team that understands IS concepts well and can apply them to a process based on the P&ID and knowledge of the operation. The CCPS ISD Guidelines provide examples of a combination HAZOP/checklist worksheet that includes review of potential IS opportunities applicable to the node and its hazardous materials.

Table 2: Guideword Matrix (6)

	Minimization	Substitute/ Eliminate	Moderate	Simplify
Raw Material				
In-process Storage				
Product Inventory				
Process Chemistry				
Process Controls				
Process Piping				
Process Equipment				
Process Conditions				
Maintenance				

6 Inherent Safety Culture

Failings in process safety culture has been a root cause in many significant incidents that have affected the chemical processing industry. Process safety culture forms the underlying tone for whether the process safety program is truly functional, whether its personnel believe in it and support it, and whether the management systems established to institutionalize and implement the program are successful. There is a strong link between the quality of leadership demonstrated by program participants and the quality of the process safety culture (or whether it is positive or negative), and hence the quality and health of the process safety program itself. Being a leader in a process safety program requires more than simply managing the program – it requires demonstrated leadership. (7)

Inherent safety as a necessary and desirable concept should be ingrained in the mindset of all employees to make it a "way of doing business ". An organization has truly adopted inherent safety to the fullest when it is evident it is an "inherent theme" of the day to day operations for all personnel. Everyone in the organization can understand, apply, and encourage the uses of ISD and so having that culture allows ISD to be a continual force-multiplier activity rather than only a one-off, special study that is only considered and dismissed as impractical for the core process technology.

7 How to Make ISD Part of the Culture

7.1 Embedding ISD into the Organizational Culture

AcuTech believes a new focus on inherent safety will prove to be an invaluable means of achieving excellence in PSM. What may have been lacking is several management system aspects to ensure it is effectively used:

1) a recognition by management of all the opportunities for ISD to benefit the organization
2) education to understand the principles
3) visible leadership commitment to the use of ISD including governance of the issue and reinforcing and rewarding for the use of inherent safety
4) additional guidance clearly defining a process for practically applying inherent safety principles at all stages of a process lifecycle
5) a systematic and sustained approach to evaluate and implement inherent safety
6) support and rewards for employees that make justified inherently safer suggestions
7) ensuring that the use of ISD is evaluated as part of the governance process for the PSM system, such as in metrics and as evaluated during audit.

The overall ISD effort should be embedded into the PSM program, not only as a special study. Guidance should be directed to immersing ISD in the operating philosophy of engineers, operations, maintenance and managers. The goal is to make inherent safety a 'way of doing business' that everyone understands, appreciates, and voluntarily and enthusiastically follows.

One of the important aspects of maintaining a vigilant process safety culture is to thoroughly understand the risks posed by a facility and to reduce those risks to an appropriate level (i.e., act upon them). Most process safety programs contain an element where the hazards/ risk is analyzed and then action items are developed to reduce the risk to some tolerable level. These elements go by different names, e.g., process hazard analysis (PHA), hazard analysis and risk identification (HIRA), among others. Management of change hazard reviews, incident investigations, and audits are other elements of PSM programs and are part of the process of understanding facility hazards/risks and acting upon them. Some of these activities are required by various PSM-related regulations. Some are voluntary or company required activities. In any case, ISD should be an integral part of those analyses. Companies should have a formal ISD process, whether it is a standalone study or integral to their PHA methodology.

8 Conclusion

The application of ISD may not be fully utilized by most companies in the process industries. A formal recognition of its value is essential to encourage management to implement directives requiring its use. But the real adoption of ISD will be with firms that engrain the concept into the culture of the organization throughout the entire lifecycle of their processes.

9 References

[1] Hendershot, Dennis, Inherently Safer Design, An Overview of Key Elements, Professional Safety, February 2011, ASSE.

[2] "Guidelines for Inherently Safer Chemical Processes: A Life Cycle Approach, 2nd Edition, American Institute of Chemical Engineers, Center for Chemical Process Safety, 2004.

[3] The Institution of Chemical Engineers & The International Process Safety Group, Inherently Safer Process Design, 1995.

[4] Kletz, T.A., Cheaper, Safer Plants, or Wealth and Safety at Work. Rugby, Warwickshire, England: The Institution of Chemical Engineers, 1984.

[5] Kletz, T.A., Plant Design for Safety, Rugby, Warwickshire, England: The Institution of Chemical Engineers, 1991.

[6] Contra Costa County, CA, Industrial Safety Ordinance Guidance Document, Attachment D, June 15, 2011

[7] Moore, David A., Essential Practices for Developing, Strengthening and Implementing Process Safety Culture, Global Congress on Process Safety, April 2015, American Institute of Chemical Engineers, Center for Chemical Process Safety.

APPLICATION OF A PROPERTY-BASED INHERENTLY SAFETY QUANTIFICATION FRAMEWORK FOR INTEGRATING RISK ASSESSMENT INTO PROCESS SAFETY LIFE CYCLE

Dr. Monzure-Khoda Kazi, Dr. Fadwa Eljack, Eng. Ahmed AlNouss*
Department of Chemical Engineering,
Qatar University, Doha, Qatar

Introduction

Process plant involves different types of risks which need to be managed appropriately to avoid dangerous incidents. Some risks are associated with design decisions such as choice of unit, chemicals or design layout etc. Some are related to the management of proper operating conditions within safely operational limits. It is advantageous to consider the safety aspects and to assess the risks associated with potential hazardous sources from the early stage of process safety life cycle to avoid unfavorable conditions and complexity on the later phase [1]. Process safety metrics provide a good balance between sophisticated quantitative evaluation and simple application. The metrics can be easily adopted, enable comparison between different alternatives and be integrated in optimization for the supply chain. There have been multiple efforts to develop process safety metrics, which are made by both industrial and academic agencies [2-8].

The concept of inherent safer design is very helpful to reduce the hazardous conditions by safer design principles instead of controlling them by add-on protective systems and procedures. Degree of freedom for the decision for inherently safer process design continuously decreases as life cycle proceeds. Therefore, changing plant design in a later stage will cost more than in the early stage. But the implementing of inherent safer design principles from early stage of design remains challenging due to the lack of proper tool/methodology which can assess the safety performance continuously with the change of mixture properties, operating conditions and even for the unit types. The significant challenges of the problem related to the safety parameters are the possible means to represent the associated results for different scenarios and processes. One safety parameter which is very important to a process may not be the critical issue for others. The scope to identify the key root safety parameters from the historical accidental data base can overcome this limitation. Again, most of the time it is very hard to do the techno-economic analysis simultaneously due to the lack of continuous equations to comply with the whole system's model.

In this work, an **I**nherently **S**afer **D**esign **T**ool (*i*-SDT) is presented for early stage process synthesis to characterize and track the risk associated with different life-cycle phases of industrial processes and products. It also helps to develop characteristic equations for different safety parameters (i.e., flammability, explosiveness, toxicity, etc.) and provides cluster safety parameter score for doing inherent safer design during early stage of design using very limited amount of process information. This property-based inherent safety quantification framework is a tailor made semi-quantitative safety analysis tool which will provide safety assessment in continuous manner to overcome the subjective nature of the existing available safety metric. The proposed safety metric has the flexibility to operate by identifying the major accident-prone units of a process, as well as the major safety and operating parameters. Therefore, in the future it can be embedded to any techno-economic framework to do the cost and safety analysis simultaneously using available materials, design and accidents information. The developed i-SDT tool was used to compare different technologies and variable capacity of ammonia processes to identify the safer alternative in terms of risks associated with the accident-prone unit/section and to highlight the areas of improvement in any existing process using the inherent safer design principles.

Inherently Safer Design Tool (*i*-SDT)

The key deliverable of this work is a property based inherently safer design tool (*i*-SDT). Ultimate target of the *i*-SDT tool is to work as a design tool to assess inherently safer designs during research and early stage of process plant design using very limited design and historical accident data. It will provide safety assessment in a continuous manner to overcome the subjective nature of the existing available safety metric. The tool has the potential to be embedded or integrated with any techno-economic framework to do the technical, cost and safety analysis simultaneously using available materials, design and accidents information.

Methodological Formulation

This *i*-SDT tool can be used to evaluate the safety performance of key unit or key section of the process. But, it is not limited to specific unit/section; it can be extended to the entire process or to the supply chain (e.g. storage and transportation). The tool estimates a score for each key safety parameters, C_{SP} that is a function of two key components: (1) The risk assessment of available constituents (such as chemical or unit). This is determined from accident data bases as a probability and (2) The safety issues associated with these constituents. In this formulation it is represented by a safety parameter index, I_{SP}. I_{SP} values are estimated using a property-based platform. Safety parameters such as flammability, reactivity, toxicity and explosiveness are all function of key thermos-physical properties. We have developed characteristic equation for each safety parameter that is function of constituent(s), lower flammability limit (LFLi), upper flammability limit (UFLi), threshold limit value (TLVi) and reactivity score (SR,i). Safety parameter index for mixture ($I_{SP,mix}$) that is a function of mixture property (P_{mix}). The properties of the mixture, $P_{mix,SP}$ is defined to have suitable mixing rules. The use of surrogate properties was introduced in Property Integration framework as a way to tracking properties, which are not conserved. Therefore, the final formulation of cluster safety parameter score (C_{SP}) is:

$$C_{SP,section/process} = \alpha_{max} \times \sum_{u=1}^{N_u} \gamma_u \times I_{SP,mix} \qquad \text{Eq. 1}$$

$$I_{SP,mix} = Ce^{\pm kP_{mix,SP}} \qquad \text{Eq. 2}$$

$$P_{mix,SP} \equiv \sum_{i=1}^{N_c} f(T, P, x_i) \qquad \text{Eq. 3}$$

The working flow of the i-SDT tool is given in Figure 1:

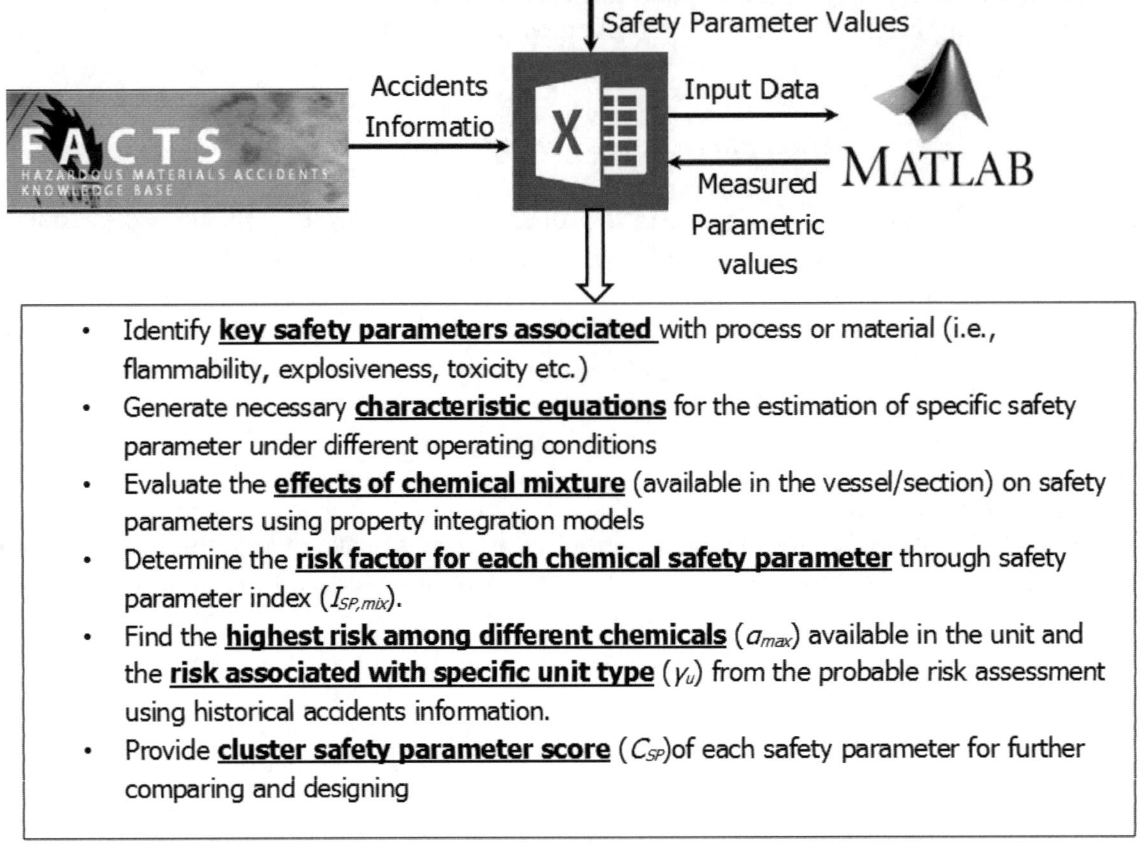

Figure 1. Working Flow Diagram of *i*-SDT Tool

Application of *i*-SDT Tool for Inherent Safer Design

An ammonia process synthesis example is used here to illustrate the application of the *i*-SDT tool for identifying the key units or sections of the process to facilitate the inherent safer design during early stage with limited process information. Process flow diagram of the base case Ammonia process and a reliable historical accidents database are available. Besides, operational information was obtained via simulation flowsheet and safety parameter information was gathered from chemical safety literatures, NFPA and MSDS sheets.

The newly developed *i*-SDT tool can identify the hazardous effect of each chemical and chemical mixture under different operating conditions in continuous manner. Therefore, designer can identify those materials and can think about their alternatives to achieve safer operation without hampering plant profitability. For example, designer wants to examine some key units of a stream reforming (SR) ammonia process to identify which is the dangerous material in that specific section, so that he can think about better materials for safer operation.

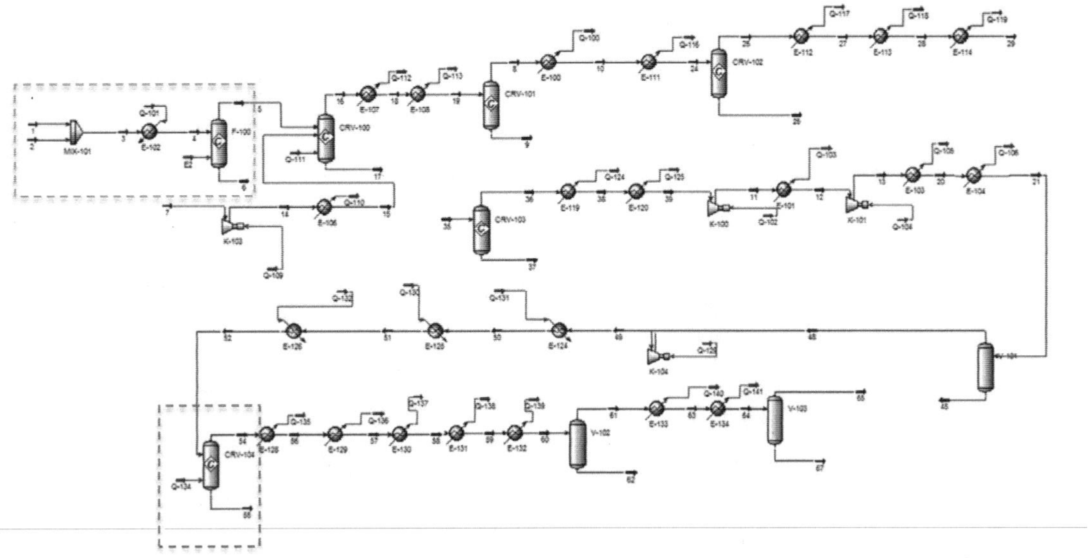

Figure 2. PDF of steam reforming (SR) ammonia process

Let's assume designer needs to know the state of materials safety in a single unit (high pressure reactor in steam reforming process, see red dotted box in Figure 2) and in a section of the process (multiple units consisting of conversion reactor, heater and mixer, see blue dotted section in Figure 2). After analyzing with the *i*-SDT tool, the following values of cluster safety parameters can be found for different process safety parameters (i.e., flammability, explosiveness, toxicity, reactivity) in the following figures (Figure 3-4) where final results are shown in cluster safety parameter score ($C_{SP,Unit}$ and $C_{SP,Section}$). Individual contributions of each unit to final C_{SP} values are also visible. By looking at those values the specific unit can be identified based on vulnerability to any safety incidents.

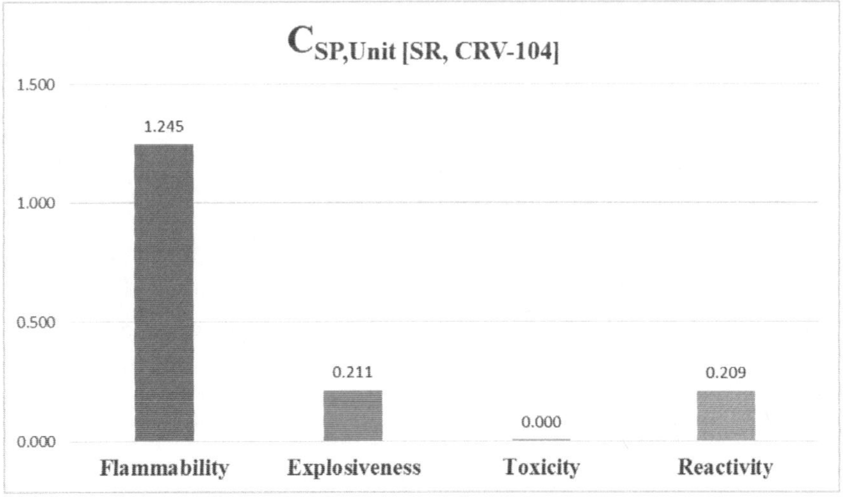

Figure 3. Cluster safety parameter score of conversion reactor in SR ammonia process

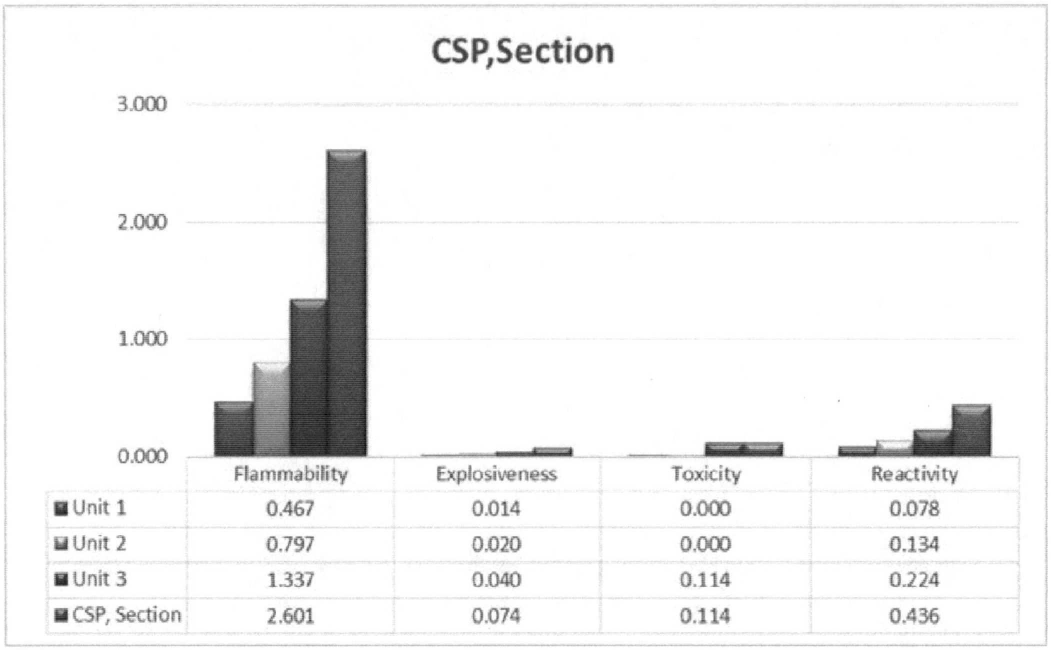

	Flammability	Explosiveness	Toxicity	Reactivity
Unit 1	0.467	0.014	0.000	0.078
Unit 2	0.797	0.020	0.000	0.134
Unit 3	1.337	0.040	0.114	0.224
CSP, Section	2.601	0.074	0.114	0.436

Figure 4. Cluster safety parameter score of a section in SR ammonia process

Finally, designer can look for alternative techniques or components which can be replace those units/components to make it safer without compromising the plant profitability. Usually, the profitability analysis is done after completing the detail designing when the scope of design modification reduces significantly due to the process complexity. This type of *i*-SDT tool can provide continuous assessment of the design with the help of very limited process information (only using PFD and probability using accident database). Moreover, this tool is dynamic in nature which make it very useful during process integration and optimization. Proposed *i*-SDT tool also has the capability to compare the units/section of the different technologies.

Conclusion

The developed inherent safer design tool (*i*-SDT) can characterize and track the risk associated with different life-cycle phases of industrial processes and products. It also generates characteristic equations for different safety parameters (i.e., flammability, explosiveness, toxicity, etc.) under various operating conditions. The final output of this *i*-SDT tool is the cluster safety parameter score which is useful for doing inherent safer design during early stage of design using very limited amount of process information. Moreover, the formulation of the *i*-SDT tool can be embedded and enhanced in the multi-objective optimization formulation to reconcile competing objectives. In future, this tool can be extended to develop cumulative supply chain models for accounting the safety aspects with the techno-economic, and environmental evaluation of each stage of the supply chain.

References

1. Roy, N., et al., A review of safety indices for process design. Current Opinion in Chemical Engineering, 2016. 14: p. 42-48.
2. AIChE, Introduction: The Fire and Explosion Index System, in Dow's Fire & Explosion Index Hazard Classification Guide. 1994, John Wiley & Sons, Inc. p. 2-2.

3. Lewis, D.j., The Mond fire, explosion and toxicity index – a development of the Dow Index. AIChE on Loss Prevention, New York, 1979.
4. Heikkila, A.-M., Inherent safety in process plant design, in D.Tech Thesis. 1999, Technical Research Centre of Finland: VTT Publications 384.
5. Khan, F.I., T. Husain, and S.A. Abbasi, Safety Weighted Hazard Index (SWeHI): A New, User-friendly Tool for Swift yet Comprehensive Hazard Identification and Safety Evaluation in Chemical Process Industrie. Process Safety and Environmental Protection, 2001. 79(2): p. 65-80.
6. Khan, F.I. and P.R. Amyotte, I2SI: A comprehensive quantitative tool for inherent safety and cost evaluation. Journal of Loss Prevention in the Process Industries, 2005. 18(4–6): p. 310-326.
7. Tugnoli, A., G. Landucci, and V. Cozzani, Key performance indicators for inherent safety: Application to the hydrogen supply chain. Process Safety Progress, 2009. 28(2): p. 156-170.
8. Gentile, M., W.J. Rogers, and M.S. Mannan, Development of a Fuzzy Logic-Based Inherent Safety Index. Process Safety and Environmental Protection, 2003. 81(6): p. 444-456.

18 AIChE Spring Meeting
14TH GLOBAL CONGRESS ON PROCESS SAFETY
ORLANDO FLORIDA 2018

Learning from Lubes Plant Fire - Importance of HAZARD Identification and Preparation

Presenter/Masaki Iguchi
JXTG Nippon Oil & Energy Corporation
Address: 1000, Hatsushima-cho-hama, Arida, Wakayama, 649-0393 Japan
Email address: iguchi.masaki@jxtg.com

Other Author/ Hisashi Shibuya
JXTG Nippon Oil & Energy Corporation
Address: 1-1-2 Ohtemachi, Chiyoda, Tokyo, 100-8162 Japan
Email address: shibuya.hisashi@jxtg.com.

Prepared for Presentation at
American Institute of Chemical Engineers
2018 Spring Meeting and 14th Global Congress on Process Safety
Orlando, Florida
April 22 – 25, 2018

Learning from Lubes Plant Fire - Importance of HAZARD Identification and Preparation

Presenter/Masaki Iguchi
JXTG Nippon Oil & Energy Corporation
Address: 1000, Hatsushima-cho-hama, Arida, Wakayama, 649-0393 Japan
Email address: iguchi.masaki@jxtg.com

Other Author/ Hisashi Shibuya
JXTG Nippon Oil & Energy Corporation

Keywords: **Lubes Plant Fire**

Abstract

In 2017 January, JXTG Wakayama refinery experienced a lubes plant fire. The fire broke out at 15:37 as gas fire and transitioned to oil fire/ pool fire. The initial opening was caused on high pressure cold separator overhead line of Lube Solvent Extract Unit Hydro-Treater Section (PT-2HD) per ammonium bisulfide corrosion.

Direct cause of this was gradual increase of heavy feed processing rate that resulted in increase of ammonium hydrogen sulfide (ammonium bisulfide) concentration at moisture condensation part. As a measure, equipment improvement that is composed of water injection facility installation, etc. was performed. Also, improvement of corrosion control that is composed of inspection change from spot thickness measurement to RT or UT scanning, etc. was conducted.

As for indirect cause, it was identified that HAZARD Identification for lubes hydro-treater ammonium bisulfide corrosion was inadequate because evaluation of the unit was omitted in past risk assessment as of its lower desulfization and denitrogenation rates than fuel hydrofiners'. As a measure for this, improvement of Hazard Identification and Risk Assessment (HIRA) related education was conducted. Also, increment of knowledge on ammonium bisulfide corrosion was attained.

As BLEVE potential vessels located around the fire area, municipal office issued resident evacuation instruction very quickly from the fire occurrence, though risk of causing BELVE was not so high throughout the incident. This quick decision making was enabled through previous communication with the refiner on higher consequence risk assessment result.

From all of above, it can be said that HAZARD Identification and preparation are important. For that purpose, Wakayama refinery will try further to enhance Process Safety Management System.

1. Introduction

Wakayama refinery started operation in 1941 and has current crude processing capacity of 127.5KBD with largest class lubes plants in Japan. The refinery experienced a fire incident on Integrated Lubes Plant (ILP) in Jan 22, 2017. Equipment that located in the area of approximately 900 square meters suffered significant damage, though no personnel injury occurred. Next is summary of the incident, measures, and lessons learned.

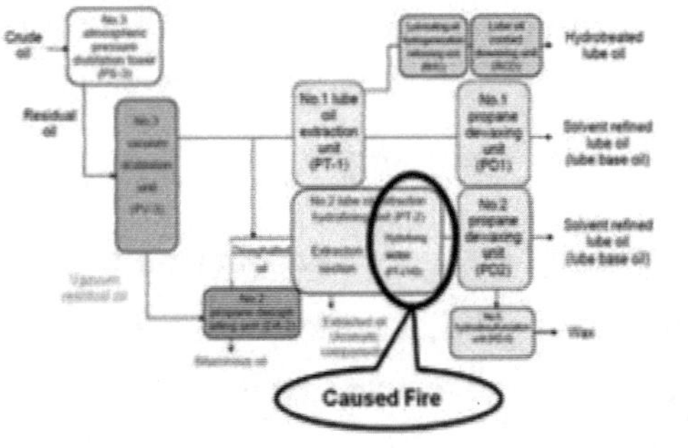

Figure1. Wakayama refinery's lube oil production process

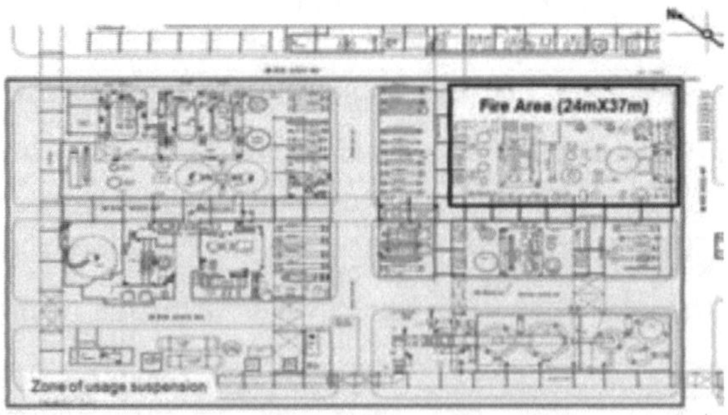

Figure2. Fire area

2. Incident Overview

2.1 Phase 1: Gas fire resulted from flammable gas leakage and its ignition (15:37-47)

PT-2 HD (Lube Solvent Extraction Unit Hydro-Treater Section) high pressure purge gas line at D-253 (high-pressure low-temperature separator drum) overhead failed and a large amount of flammable gas leaked and ignited. The incident investigation revealed that the failure initially took place on the inlet pipe of the control valve PICV-251A in the overhead line. The leaking gas was released at a critical velocity and immediately ignited. The gas likely ignited with some static spark. White smoke was observed as the fire develops.

2.2 Phase 2: Transition to oil fire (15:48-16:13)

PT-2 HD D-252 (high-pressure-high-temperature separator drum) /D-253 were filled with oil from upstream and the oil overflew into the high-pressure purge gas line. Approx. 5.3 KL oil leaked from the perforated pipe opening and caught fire.

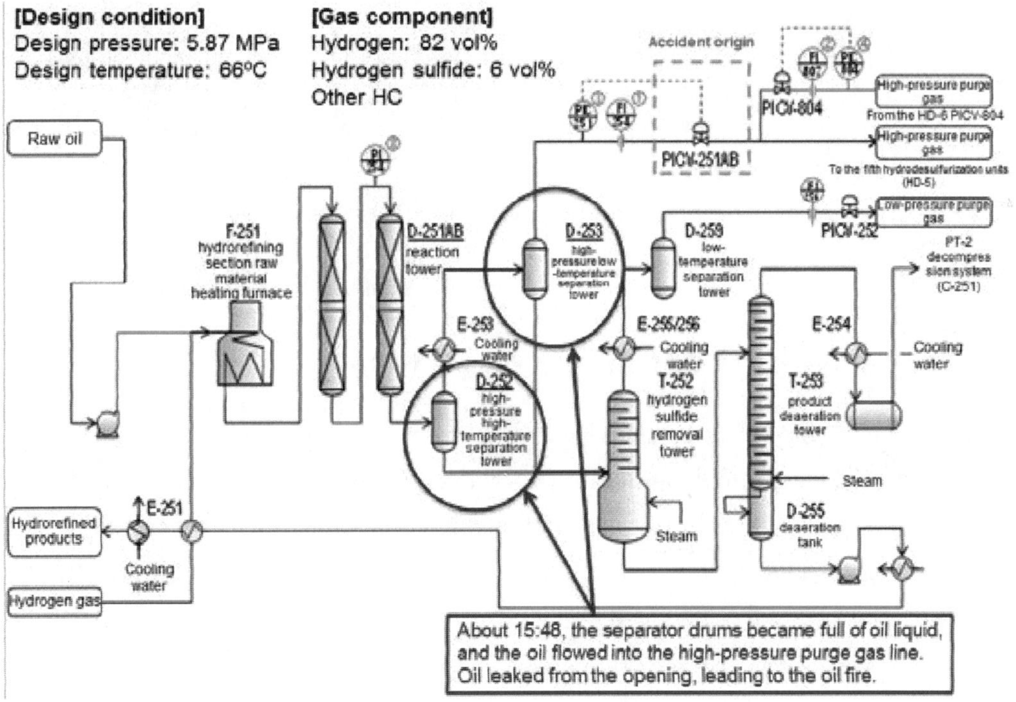

Figure3. Transition to oil fire

2.3. Phase 3: Transition to pool fire (16:14-)

As adjacent pipe lines were exposed to flame, PT-2 feed line pipe ruptured with severe bulging just upstream of flow control valve FICV-201. Leaking feed oil and the fire spread inside the spill wall and developed to pool fire with black smoke. Then approx.102 KL oil leaked with catching fire. The leaking feed oil and the fire spread inside the spill wall and developed to pool fire with black smoke observed.

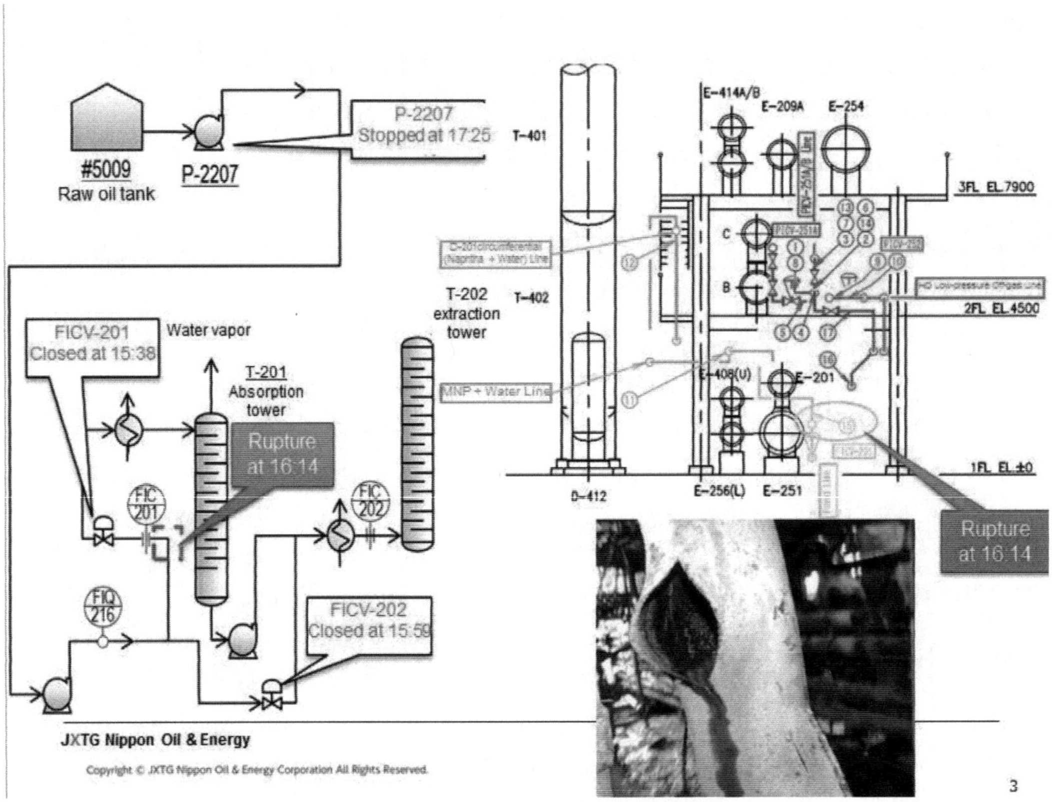

Figure 4. Transition to pool fire

2.4. Opening location of high-pressure purge gas line

Eighteen openings were found via field investigation. These were screened by internal pressure drop timing and trace of fire causing metal temperature rise per metal inspection result. Twelve openings were found in the high pressure purge gas line where pressure dropped first and only one opening was not appeared to be affected by heat of the fire. It was opening #7. Based on microscopic sectional view of opening #7, wide thinning of inner surface was observed. Also, groove like corrosion trace was observed. (Figure7)

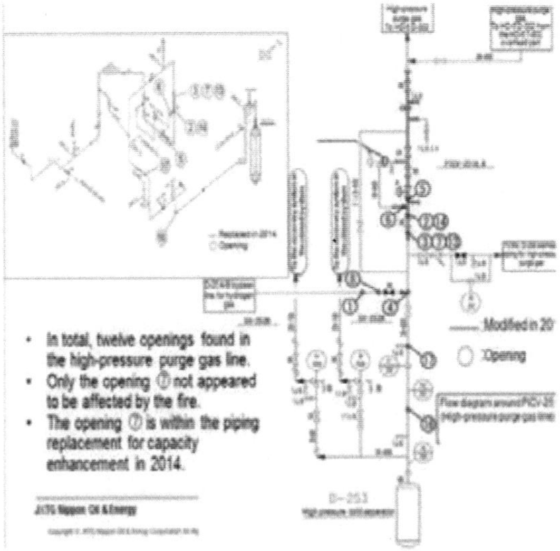

Figure5.Opening location of high pressure purge gas line

Figure5.Opening location of high pressure purge gas line

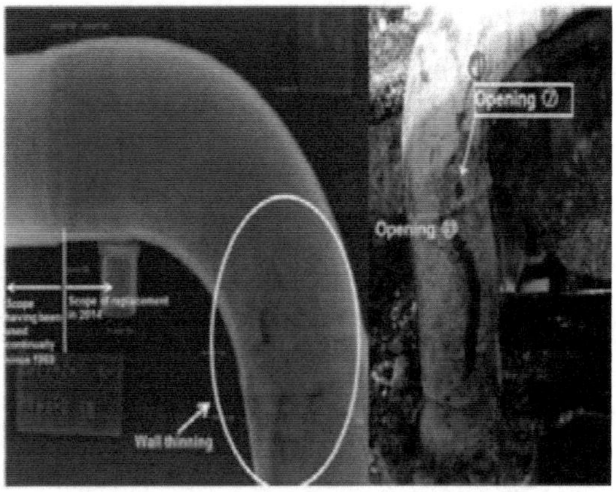

Figure6. Opening location of high pressure purge gas line

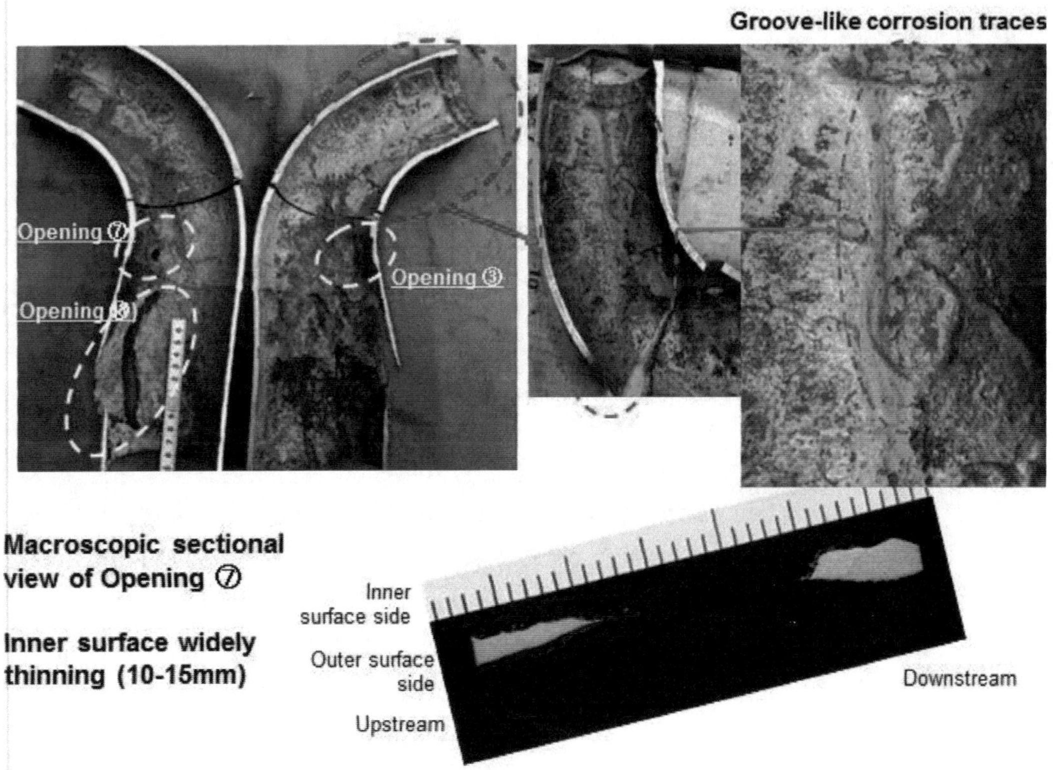

Figure7. Opening location of high pressure purge gas line

2.5.Evacuation of residents

Evacuation instruction was made by the municipal office just after 1hour and 34minutes later, approximately, from the fire occurrence based on the past communication on severer scenario incident risk assessment result. Taking into account that in reality residents' evacuation always takes some necessary time, it is thought that it will be difficult to safely complete evacuation if an evacuation instruction is issued after a BLEVE possibility has increased. Therefore, we think that it is important in future to issue a preventive evacuation instruction depending on a state of outbreak of a fire, when safety of peripheral residents should be considered first.

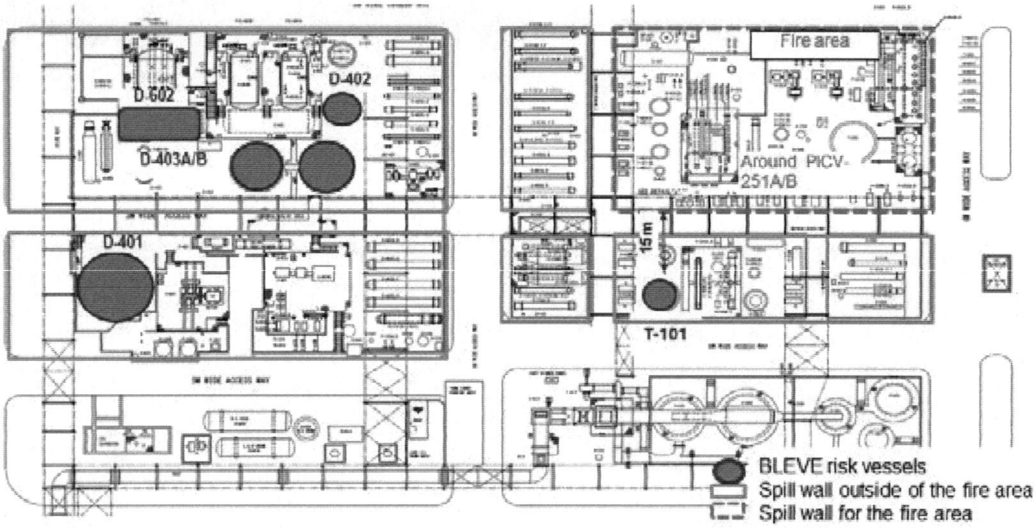

Figure 8.Fire area and location of BLEVE potential Vessels

Figure 8 indicates vessels that have scenario of significant consequence level of BLEVE that had been identified via. Higher Consequence Risk Assessment. As water release to cool BLEVE potential vessels had been performed at appropriate timing based on pre-developed firefighting strategy and tactics, etc. and no BLEVE sign was observed per the vessels inside pressure monitoring throughout the incident, the risk was low.

3. Summary of causes

3.1. Summary of the direct cause

Processing rate of heavy feedstock in the hydrogenation unit increased gradually

(Figure 9), and a corrosive environment with high concentration of ammonium hydrogen sulfide (ammonium bisulfide) was created at moisture condensation part, causing the severe alkaline sour water corrosion. As shown in Figure 10, corrosion rate at 2016 became more than 3times of it at 2002.

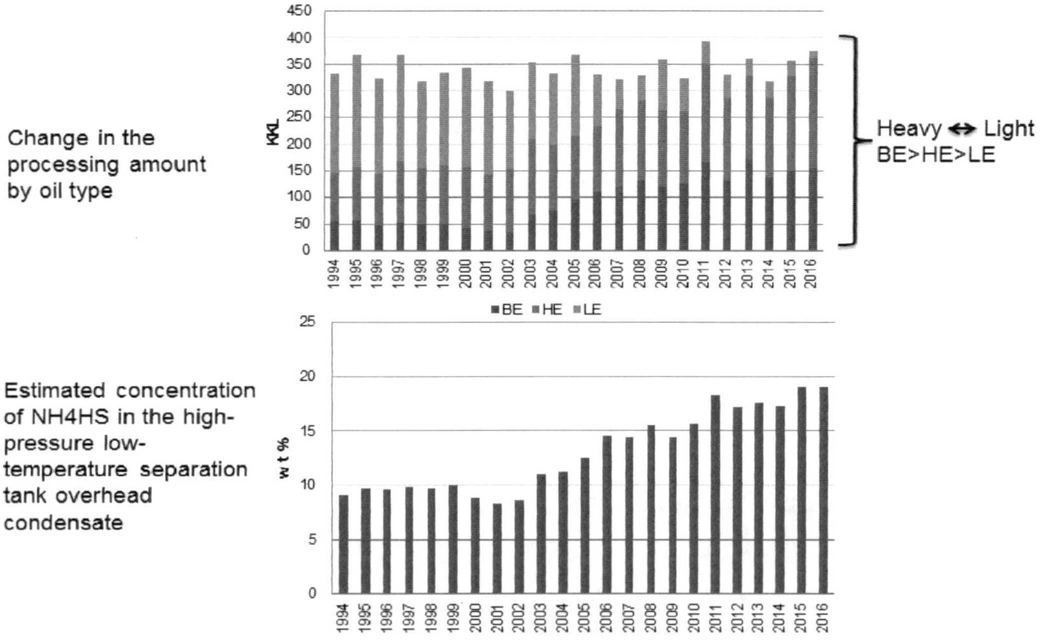

Figure9. Changes in the operation of the hydrogenation unit and the estimated concentration of NH4HS

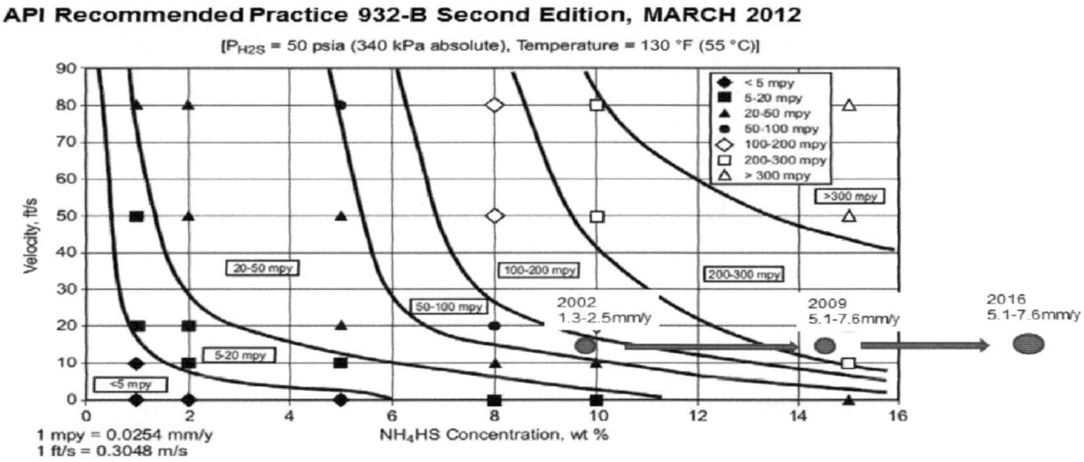

Figure10.Changes of estimated concentration of NH4HS and corrosion rate

3.2. Summary of the indirect cause

Evaluation of the equipment in question was omitted in past risk assessments due to lower desulfurization and denitrogenation rate than those of fuel oil desulfurizing units. Cause of this may be regarded as cognitive bias or silo effect. From this, it was concluded that quality of the HAZARD Identification was not adequate.
As the corrosion control was conducted depending on only thickness measurement of representative point measurement, it was regarded that knowledge about ammonium bisulfide corrosion in the lube oil manufacturing unit was also not adequate.

4. Measures against the causes

4.1 Measures against the direct cause

<u>4.1.1 Equipment improvement</u>

One of measure against the direct cause is installation of washing-water injecting and waste water sampling equipment to improve operating condition on the API RP 932-B chart. (Figure 11&12)

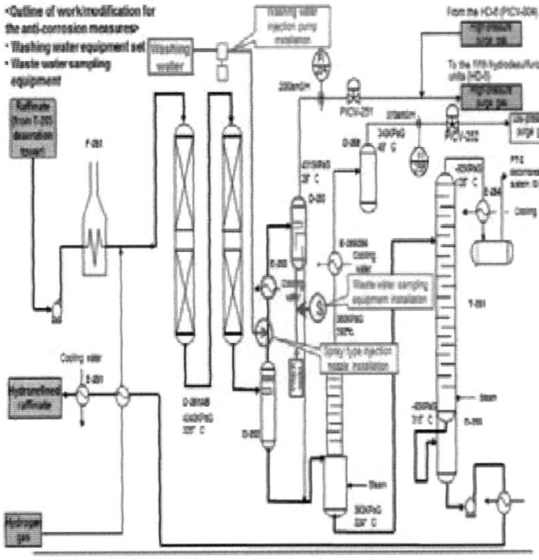

Figure11. Measures against the direct cause: Equipment Improvement

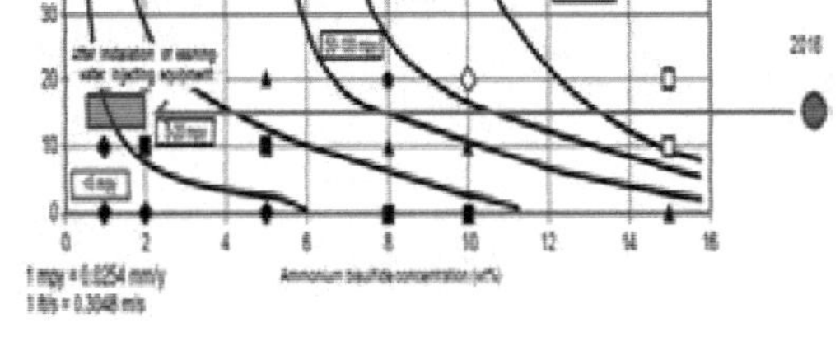

Figure12. Corrosion rate Improvement

<u>4.1.2 Improvement of corrosion control</u>

Another measure is corrosion control improvement from the thickness measurement-based to comprehensive corrosion control including operation management. Before the fire, ultrasonic spot thickness measurement was conducted. This will be changed to RT or UT scanning. Simulation based corrosivity evaluation is also introduced. As for operation management, ammonium bisulfide concentration measurement will be introduced.(Table 1)

Table 1. Measures against the direct cause: Improvement of corrosion control

	Inspection			Operation management	
	Ultrasonic/radiographic inspection	**Visual inspection of inside surface**	**Introduction of corrosion evaluation software**	**Washing water**	**Measurement of NH4HS concentration**
Before the fire	Spot (In-house inspection every 10 years, safety inspection every year)	At the time of overhaul inspection	X	X	X
After the operation restart	· Completely shifted to the radiographic inspection · After the operation restart, the radiographic inspection is conducted every month for half a year. · For the subsequent half a year, the radiographic inspection is conducted quarterly. · The frequency and scale of the inspection after one year from the operation restart shall be determined based on the operation management state.	At the time of overhaul inspection	Corrosivity evaluation by the simulation software	O	Every month

4.2 Measures against the indirect cause

4.2.1Improvement of Hazard Identification and Risk Assessment (HIRA) related education

Based on what have been observed, themes of the improvement are ”Elimination of cognitive bias among different level of people", "Enhancement of knowledge among different level of people on HIRA" and "Activation of sense of responsibility among different level of people on process safety, etc." To attain above, improvement of HIRA education system was performed.

Table 2. HIRA related education system/ Existing and New (Planed)

Target level	Education Contenet / Target	Risk Assessment & Management		Process Hazard Identification		Maintenanc Work Preparation Hazard Identification	
		Existing Education	Education(Black: Existing, Blue: New(Plan), Block: SOMS-6 Item))	Existing Education	Education(Black: Existing, Blue: New(Plan), Block: SOMS-6 Item))	Existing Education	Education(Black: Existing, Blue: New(Plan), Block: SOMS-6 Item))
Junoir	✓New engineer ✓TA Coordinator ✓In charge of work ,etc. Trainer: Day tme SSV, GH(Maint., Tech.)Class	None	①Risk Matrix Application Manual(Prob., Cons. and Approver portion) equivalent	Exists	③Hazard Recognition Training per each area ④TA training for 1st time people(Experimental learning) ⑤MTS repetitional educatiuon(H2S, S,etc.) ⑥Basis of Combustion and Extinguishment equivalent	Exists	⑨Incident case education per Head quarter SHE ⑩Pre-TA Safety Education(IH included) ⑪High Risk Work Review EducationEquivalent
Supervisor	✓SSV, Deputy SSV(PC, Staff) ✓Maint. Team Lead ✓Tech. Team Lead, etc. Trainer: Manager(GM) Class	Exists	②Risk Assessment Initial Training: Trainer SA (Reflesh training per 3 years)	None	⑦1st line Supervisor Training equivalent(Including teaching side of Basis of Combustion and Extinguishment (Teaching side)	N/A	⑪High Risk Work Review EducationEquivalent
Manager	✓GM ✓SSI ✓Vice Ref. Manager ✓Ref. Manager Trainer: SA/ H'dq'r SE	–	✓Seminars and Lectures	N/A	⑦' 1st line Supervisor Training equivalent	N/A	N/A
Contractor	✓Supervisor ✓Worker Trainer:BM(GM Class)	–	N/A	–	⑥' Basis of Combustion and Extinguishment equivalent ⑧TA training for 1st time people (Experimental Learning: Target scoping under going)	–	⑪' High Risk Work Review EducationEquivalent

Characteristics of the improvement are reinforcement of education program from view point of 3 aspects; "Risk assessment and management", "Process Hazard Identification" and "Maintenance Work Preparation Hazard Identification", to targeted people, grouped as "Junior", " Supervisor", "Manager" and "Contractor". To let people have active sense of responsibility, senior level people who have competency act as trainers to low experience level people.

4.2.2 Increase knowledge about ammonium bisulfide corrosion

Another measure is to increase knowledge about ammonium bisulfide corrosion. Corrosion mechanism about ammonium bisulfide shall be learnt with more research work and improve how lesson leaned be applied when new findings be obtained.

5. Conclusion

HAZARD identification and preparation are important because of prevention of a process safety incident. To attain this, "Cognitive bias elimination/ Silo penetration",

"Enhancement of knowledge" and "Repeated stimulation of every level of people" has to be performed.

It is also important because of minimization of damage if a process safety incident occurs. To attain this, "Conduct Higher Consequence Risk Assessments", "Preparation of strategy & tactics for firefighting", "Training of people and perform firefighting" and "Communication of result with local authority" has to be performed.

Wakayama Refinery established "Safety day" on Jan. 22 to call back and memorize the pain. We will try further to enhance process safety culture, people's competency and Process Safety Management System.

Tsunami-induced oil spill fire simulation in the vicinity of petroleum industries in Japan

Tomoaki NISHINO
Disaster Prevention Research Institute, Kyoto University
Gokasho, Uji 611-0011, Kyoto, Japan
nishino.tomoaki.3c@kyoto-u.ac.jp

Yohei TAKAGI
Yokohama National University
79-5 Tokiwadai, Hodogaya-ku, Yokohama 240-8501, Japan
takagi-yohei-hn@ynu.ac.jp

Prepared for Presentation at
American Institute of Chemical Engineers
2018 Spring Meeting and 14th Global Congress on Process Safety
Orlando, Florida
April 22 – 25, 2018

Tsunami-induced oil spill fire simulation in the vicinity of petroleum industries in Japan

Tomoaki NISHINO
Disaster Prevention Research Institute, Kyoto University
Gokasho, Uji 611-0011, Kyoto, Japan
nishino.tomoaki.3c@kyoto-u.ac.jp

Yohei TAKAGI
Yokohama National University

Keywords: tsunami, oil spill, fire spread, numerical modeling

Abstract

This paper summarizes the development and validation of a numerical simulation model for tsunami-induced oil spill fires in the vicinity of petroleum industries. The tsunami following the 2011 Tohoku Earthquake caused large-scale oil spill fires spreading over Kesennuma Bay, Japan. A total of 7,530 kL of marine diesel oil were spilled from 13 fuel tanks destroyed by the tsunami, and contributed to the fires by igniting on the sea. According to previous experiments, burning floating debris, such as wooden building components, were likely to ignite the oil floating on the sea by acting as a candle wick, which at ordinary temperatures is not flammable. The oil spill fires on the sea caused secondary fires, such as building fires, ship fires, and wildland fires, resulting in approximately 247 ha of burned-out area. In particular, some tsunami vertical evacuation buildings were damaged by the oil spill fires, and people who had escaped to the buildings from the tsunami were exposed to the fires. Nevertheless, safety measures against the tsunami-induced oil spill fires have not been sufficiently considered in Japan because of the lack of the methods for predicting the damage from the tsunami-induced oil spill fires. Therefore, we have been developing a numerical simulation model for tsunami-induced oil spill fires that is a coupled simulation of tsunami, oil spill, and fire spread. In order to validate the proposed model, we numerically reconstructed the tsunami-induced oil spill fires at Kesennuma Bay in the 2011 Tohoku Earthquake. As a result of comparison of the calculated results with the observed records, we concluded that the proposed model can reasonably predict the dynamic states of the tsunami-induced oil spill fires. Moreover, we applied the model to the petroleum industries at Osaka Bay, Japan in the tsunami following the future plate-boundary earthquake that will occur along the Nankai megathrust in the south of Japan for the model verification.

1 Introduction

A tsunami following an earthquake with the moment magnitude of 9.0 off the coast of Japan on March 11, 2011, which is often referred to in Japan as the 2011 Great East Japan Earthquake and Tsunami, caused large-scale oil spill fires spreading over Kesennuma Bay in Miyagi Prefecture as shown in Figure 1. A total of 7,530 kL of marine diesel oil were spilled from 13 fuel tanks destroyed by the tsunami, and contributed to the fires by igniting on the sea. The oil spill fires on the sea caused secondary fires, such as building fires, ship fires, and wildland fires, which destroyed approximately 247 ha of the areas [1]. In particular, some tsunami vertical evacuation buildings were damaged by the oil spill fires, and people who had escaped to the buildings from the tsunami were exposed to the fires. Nevertheless, safety measures against such tsunami-induced oil spill fires have not been sufficiently considered in Japan because of the lack of numerical simulation methods for the fires. For reasonable planning of the fire safety measures, it is important to develop a numerical simulation method to predict the big picture of tsunami-induced oil spill fires and assess the validity of the method from the standpoint of the reproducibility of the past tsunami-induced oil spill fires.

Figure 1. Oil spill fires at Kesennuma Bay in the 2011 Great East Japan Earthquake and Tsunami (provided by Ryosuke Onodera).

We have been developing a numerical simulation model for tsunami-induced oil spill fires that is a coupled simulation of tsunami, oil spill, and fire spread [2]. In order to validate the proposed model, we numerically reconstructed the tsunami-induced oil spill fires at Kesennuma Bay in the 2011 Great East Japan Earthquake. As a result of comparison of the calculated results with the observed records, we concluded that the proposed model can reasonably predict the dynamic states of the tsunami-induced oil spill fires [2]. Moreover, we applied the model to the petroleum industries at Osaka Bay, Japan in the tsunami following the future plate-boundary earthquake that will occur along the Nankai megathrust in the south of Japan for the model verification.

In this paper, we summarize the development and validation of the numerical simulation model for tsunami-induced oil spill fires [2] and describe the calculation examples of oil spill fires in the vicinity of petroleum industries at Osaka Bay, Japan in the tsunami following the future plate-boundary earthquake.

2 Numerical simulation model for tsunami-induced oil spill fires [2]

Figure 2 is a schematic of a computational model for large-scale oil spill fires on water in tsunamis. In this model, assuming that the oil spill fires on water in tsunamis are the assembly of burning oil particles floating on water, the location and burning behavior of the individual oil particle are predicted in time series under the tsunami that is simulated using a proven numerical code by Goto et al. [3] . Goto et al.'s model solve the nonlinear shallow water equations using a leapfrog staggered-grid finite difference scheme and a moving-boundary approach at the run-up front. Therefore, the model consists of the following sub-models: (1) an existing model for tsunami numerical simulation (tsunami flow model), (2) a model for the motion of the oil particle in tsunamis (oil spread model), and (3) a model for the burning behavior of the oil particle on water (oil burning model). Here, the following information are given as input conditions: initial water surface elevation, topography, oil spill conditions such as rate, amount and location, ignition conditions such as time and location.

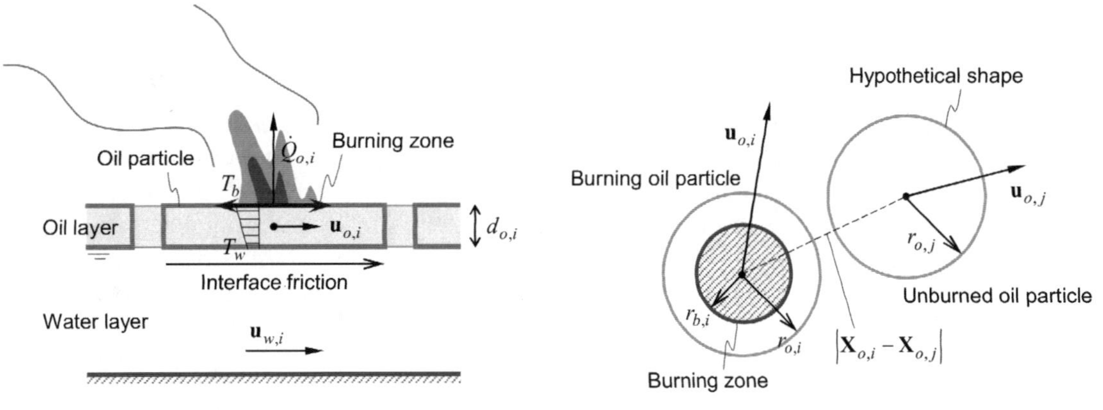

Figure 3. Schematic of a particle-based numerical simulation model for tsunami-induced oil spill fires [2].

Here, we make the following assumptions for the model formulation:

Assumption A: Tsunamis are two dimensional water flows of which velocity in the $x-$ and $y-$ directions are already known at an arbitrary point and time, i.e., water flow velocity in the $z-$ direction is neglected.

Assumption B: Oil floats on water, i.e., oil and water are clearly separated, and chemical and biological state changes of oil are neglected such as dissolution and emulsification.

Assumption C: Oil is the assembly of disc-shaped particles.

Assumption D: Thickness and radius of the oil particle depend on the number of oil particles per area in the mesh.

Assumption E: The oil particle travels horizontally due to the interface friction between oil and water, i.e., the spreading due to gravity and viscous force is neglected.

Assumption F: Locations and times of fire origins are already known.

Assumption G: The fire of the burning oil particle expands the burning zone concentrically at velocity depending on the thickness.

Assumption H: The fire of the burning oil particle spreads to the unburned oil particles when the burning zone contacts with the unburned oil particles.

Assumption I: Mass and volume of the burning oil particle decrease due to the gasification.

Assumption J: The fire of the burning oil particle is extinguished when the thickness is smaller than 1 mm.

2.1 Tsunami flow

Tsunamis are mainly caused by the deformation of the sea bottom due to earthquakes at several kilometers below the sea surface. The deformation covers a horizontal distance of tens of kilometers to hundreds of kilometers. Therefore, tsunamis are usually described as long waves. In a theory for long waves, the vertical acceleration of water particles is negligible compared with the gravitational acceleration. Consequently, the vertical motion of water particles has no effect on the pressure distribution. The following equations for conservation of mass and momentum with integrated form, where tsunamis are approximated by two-dimensional incompressible and inviscid fluids, are widely used as the governing equations for tsunami propagation and inundation [3]:

$$\frac{\partial \eta}{\partial t} + \frac{\partial M}{\partial x} + \frac{\partial N}{\partial y} = 0 \quad (1)$$

$$\frac{\partial M}{\partial t}+\frac{\partial}{\partial x}\left(\frac{M^2}{D}\right)+\frac{\partial}{\partial y}\left(\frac{MN}{D}\right)+gD\frac{\partial \eta}{\partial x}+\frac{gn^2M}{D^{7/3}}\sqrt{M^2+N^2}=0 \tag{2}$$

$$\frac{\partial N}{\partial t}+\frac{\partial}{\partial x}\left(\frac{MN}{D}\right)+\frac{\partial}{\partial y}\left(\frac{N^2}{D}\right)+gD\frac{\partial \eta}{\partial y}+\frac{gn^2N}{D^{7/3}}\sqrt{M^2+N^2}=0 \tag{3}$$

where η is the vertical displacement of water surface above still water level, D is the total water depth, M and N are the discharge fluxes in the $x-$ and $y-$ directions, respectively, g is the gravitational acceleration, and n is the Manning's roughness. The discharge fluxes M and N are defined as follows, letting the velocity of water in the $x-$ and $y-$ directions be $u_{w,x}$ and $u_{w,y}$, respectively:

$$M = u_{w,x}D, \qquad N = u_{w,y}D \tag{4}$$

2.2 *Motion of the oil particle in tsunamis*

The location of the oil particle $\mathbf{X}_{o,i}$ is calculated as follows in the same manner as the conventional Lagrangian models, assuming the sum of the deterministic displacement due to the equilibrium of forces and the probabilistic displacement due to the turbulence of water flow:

$$\mathbf{X}_{o,i} = \mathbf{X}_{o,i}^{I} + \int_{t_{s,i}}^{t} \mathbf{u}_{o,i}dt + \sum_{l=m_{s,i}}^{m} \sqrt{24\boldsymbol{\kappa}_i\Delta t}\left(\boldsymbol{\xi}_i-\frac{1}{2}\right) \tag{5}$$

where i is the identification mark of the oil particle, $\mathbf{X}_{o,i}^{I}$ is the initial location of the oil particle, $\mathbf{u}_{o,i}$ is the velocity of the oil particle, $\boldsymbol{\kappa}_i$ is the diffusion coefficient, t is the time, $t_{s,i}$ is the start time of the spill of the oil particle, Δt is the time increment, $\boldsymbol{\xi}_i$ is the uniform random number from 0 to 1, $m_{s,i}$ is the number of steps at the beginning of the spill of the oil particle, and m is the number of current steps.

Assuming that the oil particle performs translational motion due to the interface friction between oil and water, which is proportional to the square of the relative velocity according to two-layer flow models [4, 5], the equation of motion is established for the individual oil particle as follows:

$$m_{o,i}\frac{\partial \mathbf{u}_{o,i}}{\partial t} = C_f \rho_{o,i} A_{o,i}(\mathbf{u}_{w,i} - \mathbf{u}_{o,i})|\mathbf{u}_{w,i} - \mathbf{u}_{o,i}| \tag{6}$$

where $m_{o,i}$ is the mass of the oil particle, $\rho_{o,i}$ is the density of the oil particle, $A_{o,i}$ is the contact area of the oil particle with water, $\mathbf{u}_{w,i}$ is the velocity of water, and C_f is the friction coefficient.

Letting the thickness of the oil particle be $d_{o,i}$, the mass of the oil particle $m_{o,i}$ is expressed as follows:

$$m_{o,i} = \rho_{o,i} A_{o,i} d_{o,i} \tag{7}$$

Thus, the prediction formula of the velocity of the oil particle $\mathbf{u}_{o,i}$ is expressed as follows in inverse proportion to the thickness of the oil particle $d_{o,i}$:

$$\frac{\partial \mathbf{u}_{o,i}}{\partial t} = C_f \frac{1}{d_{o,i}}(\mathbf{u}_{w,i} - \mathbf{u}_{o,i})|\mathbf{u}_{w,i} - \mathbf{u}_{o,i}| \tag{8}$$

The thickness of the oil particle changes depending on the number of oil particles per area in the mesh. Here, the average thickness in the present mesh of the oil particle $\overline{d_k}$, which is calculated as follows, is given to the thickness of the oil particle $d_{o,i}$:

$$d_{o,i} \equiv \overline{d_k}\,, \qquad \overline{d_k} = \sum_{i=1}^{n_o} \delta_{ik} V_{o,i} \Big/ A_k \tag{9}$$

where k is the identification mark of the present mesh of the oil particle, $V_{o,i}$ is the volume of the oil particle, δ_{ik} is the dummy variable that is 1 if the oil particle i exists in the mesh k, otherwise 0, A_k is the effective area of the mesh excluding the occupancy area of floating building components, and n_o is the number of the oil particles.

The radius of the oil particle also changes depending on the number density of surrounding oil particles in the present mesh. Noting that the oil particle is regarded as the disc-shaped particle, the radius of the oil particle $r_{o,i}$ is calculated as follows:

$$r_{o,i} = \sqrt{\frac{A_{o,i}}{\pi}} = \sqrt{\frac{V_{o,i}/d_{o,i}}{\pi}} \tag{10}$$

2.3 *Burning behavior of the oil particle on water*

According to previous experiments using kerosene and crude oil [6-9], the flame spread rate over oil floating on water hardly changes when oil is thick, but becomes smaller when oil is thinner because the mechanism of convective flow changes. Although it is unclear whether these experimental results can be applied to different types of oil, we use these results temporarily regardless of oil type, noting that the flame spread rate is small compared with the velocity of oil carried by tsunamis. Here, the radius of the burning zone of the oil particle $r_{b,i}$ is calculated by using the following approximation equation:

$$\frac{\partial r_{b,i}}{\partial t} = \begin{cases} 1.2 d_{o,i} + 0.0016 & (0.001 \leq d_{o,i} < 0.007) \\ 0.01 & (0.007 \leq d_{o,i}) \end{cases} \tag{11}$$

When the thickness of the oil particle is smaller than 1 mm, the fire of the oil particle is extinguished and, thus, the flame spread does not occur.

The fire of the burning oil particle spreads to the unburned oil particles when the burning zone contacts with the unburned oil particles. As the plane shapes of the oil particle and the burning zone are treated as circles, the fire spread between the oil particles is predicted by investigating the positional relationship between two circles. Thus, the condition of the fire spread between the oil particles is expressed as follows, letting the burning oil particle and the unburned oil particle be i and j, respectively:

$$\left|\mathbf{X}_{o,i} - \mathbf{X}_{o,j}\right| \leq r_{b,i} + r_{o,j} \quad and \quad d_{o,j} \geq 0.001 \tag{12}$$

Mass loss rate per unit area of oil in pool experiments becomes constant when a pool diameter is large, because the emissivity of a flame is closer to 1 with increase in the diameter [10, 11]. However, such a result cannot be applied to oil floating on water in tsunamis without change, because heat loss to water decreases the mass loss rate. Here, assuming that the heat loss rate from a burning surface of oil to water is proportional to the temperature difference and is inversely proportional to the thickness of the oil particle, the mass loss rate per unit area of the oil particle $\dot{m}''_{b,i}$ is calculated as follows by adding the heat loss term to the mass loss rate obtained from pool experiments:

$$\dot{m}''_{b,i} = \dot{m}''_{b,pool} - \frac{1}{L_v}\left\{\frac{k_o(T_b - T_w)}{d_{o,i}}\right\} \qquad (d_{o,i} \geq 0.001) \tag{13}$$

where $\dot{m}''_{b,pool}$ is the mass loss rate per unit area of oil obtained from pool experiments with a sufficiently large diameter, T_b is the boiling point of oil, T_w is the temperature of water, k_o is the effective heat transfer coefficient, and L_v is the heat of vaporization. Here, the water in the vicinity of the burning oil particle is assumed to remain at low temperature because the water is agitated rapidly by tsunamis.

The effective heat transfer coefficient k_o is determined as follows so that $\dot{m}''_{b,i} = 0$ holds when the thickness of the oil particle is the minimum thickness before the fire is extinguished:

$$k_o = \left(\frac{\dot{m}''_{b,pool} L_v}{T_b - T_w}\right) d_{min} \tag{14}$$

where $d_{min}(= 0.001)$ is the minimum thickness before the fire is extinguished.

Thus, the prediction formula of the volume of the burning oil particle $V_{o,i}$ is expressed as follows:

$$\frac{\partial V_{o,i}}{\partial t} = -\left(\frac{\dot{m}''_{b,i} \pi r_{b,i}^2}{\rho_{o,i}}\right) \tag{15}$$

The heat release rate of the burning oil particle $\dot{Q}_{o,i}$ is expressed as follows:

$$\dot{Q}_{o,i} = \Delta H_o \dot{m}''_{b,i} \pi r_{b,i}^2 \tag{16}$$

where ΔH_o is the heat of combustion of oil.

3 Oil spill fire simulation at Kesennuma Bay in the 2011 Japan Tsunami [2]

The tsunami following the 2011 Great East Japan Earthquake was numerically reconstructed until 6 hours after the earthquake. Initial vertical displacement of water surface due to the deformation of sea bottom, which was an initial condition necessary for the tsunami simulation, was calculated using the method of Mansinha and Smylie [12] from the fault parameters of the 2011 Great East Japan Earthquake and Tsunami proposed by Fujii et al. [13]. The fault parameters had been identified to accurately reproduce the tsunami waveforms recorded at tide and wave gauges, GPS wave gauges and ocean-bottom tsunami sensors. The computational domains from fault to land consisted of five nested domains with the grid size of 1350 m, 450 m, 150 m, 50 m, and 10 m, respectively. The time increment was set to 0.2 s as a result of the consideration of the CFL condition. The boundary conditions at offshore and run-up front were given as free transmission and moving boundary, respectively, in the same manner as the method of Goto et al. [3]. The value of the Manning's roughness was selected depending on land use patterns based on the method of Kotani et al. [14].

As the majority of the spilled oil was marine diesel oil according to the Fire Departments of Kesennuma and Motoyoshi [1], which has a density of 820 kg/m^3, we focused on this type of oil in the simulation. The friction coefficient C_f was set to 0.006 on the basis of previous hydraulic experiments [15]. The diffusion coefficient $\boldsymbol{\kappa}_i$ was calculated as follows by reference to previous hydraulic experiments [16]:

$$\boldsymbol{\kappa}_i = 0.2\mathbf{u}_{*i}D_i \tag{17}$$

where $\mathbf{u}_{*i}$ is the friction velocity, and D_i is the total water depth. The oil spill locations were set to the locations where the fuel tanks had been found later, assuming that the amount of oil spill had been negligible while the tanks had been drifting. The amount of oil stored in each fuel tank was set on the basis of the information provided by the fire department. The oil spill was assumed to start at 50 min after the earthquake when the tsunami had arrived at any of the oil spill locations, and the spill rate was set to 0.2 m^3/s assuming that there had been a small crack on the surface of each fuel tank. The initial volume of the oil particle was uniformly set to 0.001 m^3. As a result, 7,530,000 oil particles were tracked in this simulation.

The time of fire outbreak on the sea was set to 5:30 pm, and the fire origin was set to the location reported by the Fire Departments of Kesennuma and Motoyoshi [1]. In the calculation, the oil particles that floated within a range of 5 m from the fire origin at 5:30 pm were set to the initial burning oil particles. The mass loss rate per unit area of oil obtained from pool experiments $\dot{m}''_{b,pool}$ was set to 0.045 kg/s/m^2 on the basis of previous combustion experiments [11]. The boiling point of oil T_b and the water temperature T_w were set to 603 K and 283 K, respectively. The heat of vaporization of oil L_v and the heat of combustion of oil ΔH_o were set to 250 kJ/kg and 42000 kJ/kg, respectively.

Figure 4 shows the dynamic states of the oil spill fires at Kesennuma Bay calculated by the proposed model. The oil particles spilled from fuel tanks were already distributed throughout the bay at 5:30 pm when the fire broke out on the sea. After the outbreak, the fire spread to surrounding oil particles one after another and immediately developed to large fires at the southern bay. Then, some of the burning oil particles were carried in the north direction by the tsunami, and transformed the northern bay into a sea of flames by also igniting the oil particles

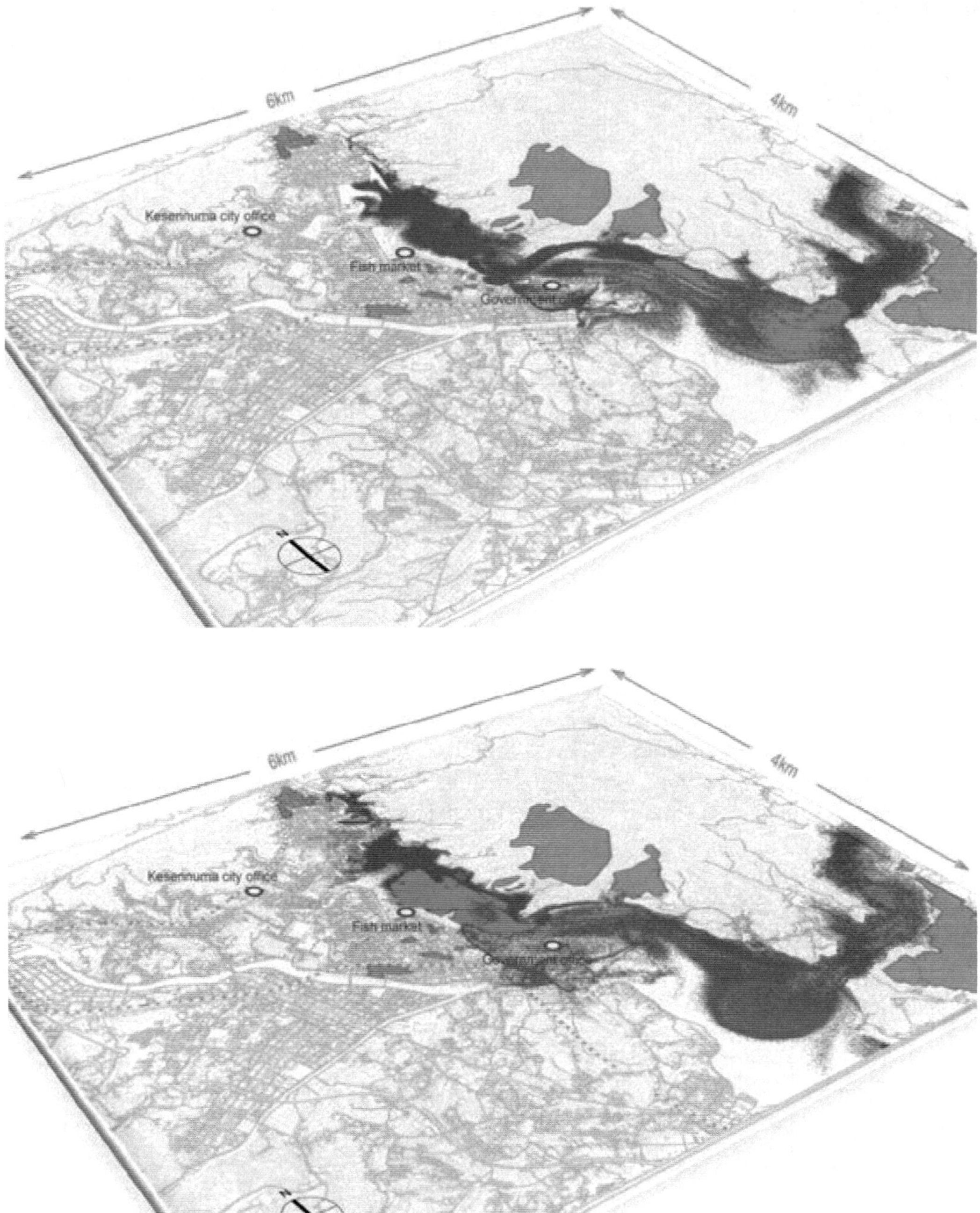

Figure 4. Dynamic states of the oil spill fires at Kesennuma Bay: the upper and lower figures are the calculated results at 5:40 pm and 5:50 pm, respectively. Unburned oil particles and burning oil particles are shown in black and red, respectively [2].

floating at the northern bay. The trend of fire spread calculated here is in good agreement with that reported by the fire departments. After that, the burning oil particles were carried in the south and east direction because of the backwash and were extinguished gradually.

4 Oil spill fire simulation at Osaka Bay in the future tsunami

Figure 5 shows the calculation examples of oil spill fires in the vicinity of petroleum industries at Osaka Bay in the tsunami following the future plate-boundary earthquake that will occur along the Nankai megathrust in the south of Japan. In this simulation, the tsunami flow that had been calculated by Kyaw et al. [17] was used. The oil spill locations were set to 197 points where the fuel tanks are built. The amount of oil stored in each fuel tank was uniformly assigned on the basis of the total amount of oil spill that has been predicted by the local government for this earthquake scenario [18]. The oil spill was assumed to start at 126 min after the earthquake when the inundation depth at the fuel tanks reaches the maximum. The oil spill rate was set to 0.2 m^3/s assuming that there had been a small crack on the surface of each fuel tank. The initial volume of the oil particle was uniformly set to 0.05 m^3. As a result, 449,160 oil particles were tracked in this simulation. The time of fire outbreak on the sea was set to 180 min after the earthquake, and the fire origin was set to the location near the installation area of fuel tanks.

After the outbreak, the fire spread to surrounding oil particles one after another and immediately developed to large fires on the sea near the installation area of fuel tanks. Then, some of the burning oil particles were carried in the north direction by the tsunami, and transformed the northern bay into a sea of flames. The oil spill fires were dominated by the transport of burning oil by the tsunami.

5 Conclusion

We summarized the development and validation of the numerical simulation model for tsunami-induced oil spill fires [2] and described the calculation examples of oil spill fires in the vicinity of petroleum industries at Osaka Bay, Japan in the tsunami following the future plate-boundary earthquake. However, we did not formulate the thermal impact of oil spill fires on surrounding environment, such as radiant and convective heat transfer to tsunami vertical evacuation buildings. Moreover, we did not discuss the variability of the behavior of tsunami-induced oil spill fires due to uncertain factors, such as location and time of ignition. Therefore, further model development and Monte Carlo simulation are needed in order to assess the thermal impact of tsunami-induced oil spill fires in various fire scenarios appropriately.

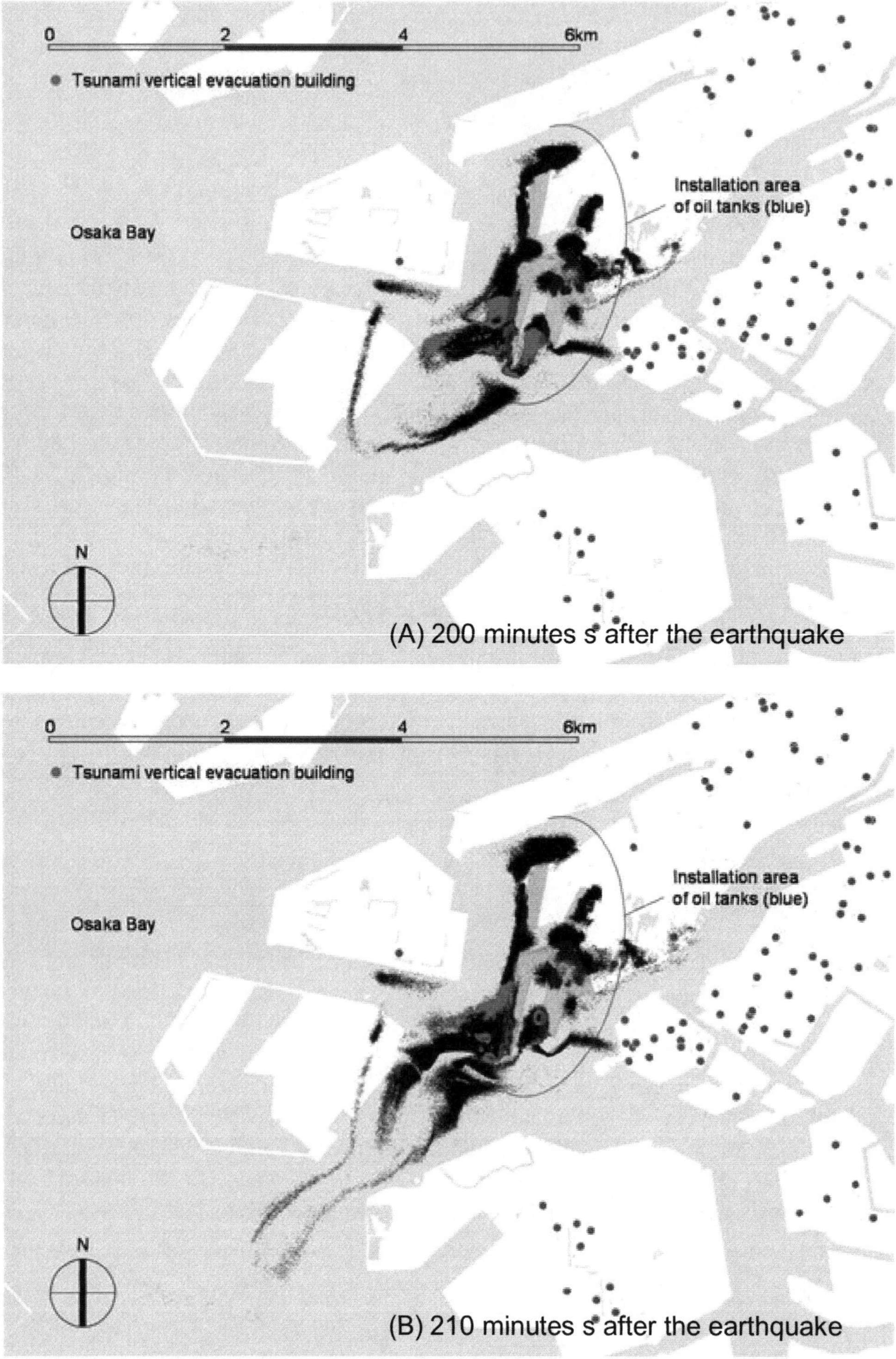

Figure 5. Calculation examples of oil spill fires in the vicinity of petroleum industries at Osaka Bay in the future tsunami.

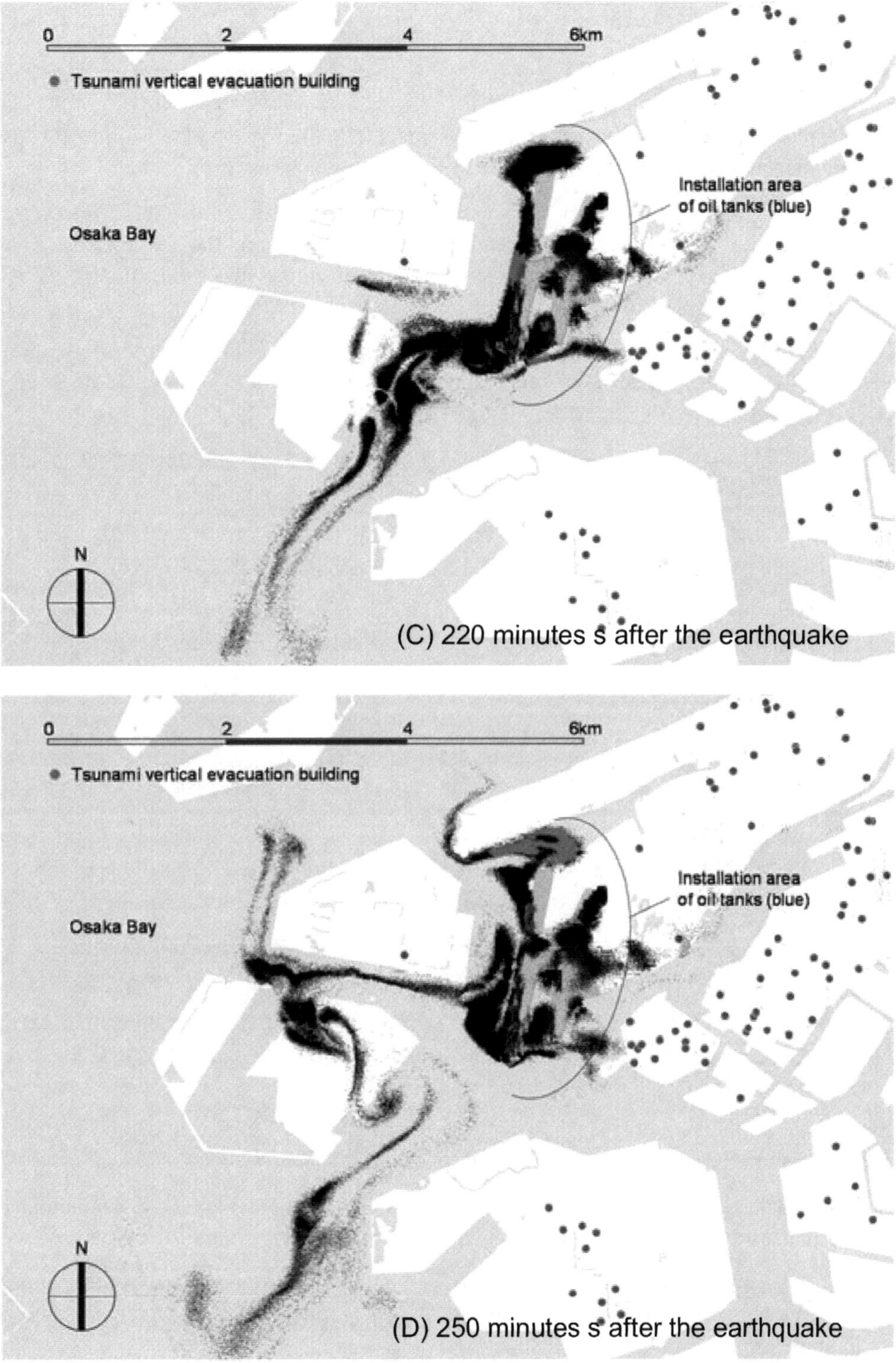

Figure 5. Continued.

6 References

[1] Fire Departments of Kesennuma and Motoyosi, 2012. Report on Fire Fighting Activities Following the 2011 Great East-Japan Earthquake (in Japanese).

[2] Nishino, T., Imazu, Y., A computational model for large-scale oil spill fires on water in tsunamis: Simulation of oil spill fires at Kesennuma Bay in the 2011 Great East Japan Earthquake and Tsunami, Journal of Loss Prevention in the Process Industries (2018), doi: 10.1016/ j.jlp.2018.02.009.

[3] Goto, C., Ogawa, Y., Shuto, N., and Imamura, F., 1997. Numerical method of tsunami simulation with the leap-frog scheme, IUGG/IOC Time Project, Intergovernmental Oceanographic Commission Manuals and Guides 35, UNESCO.

[4] Goto, C., 1985. A simulation model of oil spread due to tsunamis, Journal of Japan Society of Civil Engineers 357, 217-223 (in Japanese).

[5] Iwabuchi, Y., Koshimura, S., and Imamura, F., 2006. Study on oil spread caused by the 1964 Niigata earthquake tsunami, Journal of Disaster Research 1, 157-168.

[6] Arai, M., Saito, K., and Altenkirck, R. A., 1993. Flame propagation over a layer of crude oil floating on water, International Journal of Chemical Engineering 33.

[7] Guo, J., Lu, S. X., Zhou, J. B., Li, M. H., and Wang, C. J., 2012. Experimental study of flame spread over oil floating on water, Chinese Science Bulletin 57, 1083-1087.

[8] Kuwana, S., Tamizu, H., Ito, A., and Torikai, H., 2017. Flame spread over liquid fuel on a water layer – basic research on tsunami fire, Open Journal of Safety Science and Technology 7, 11-21.

[9] Gottuk, D. T., and White, D. A., 2016. "Liquid fuel fires," in The SFPE Handbook of Fire Protection Engineering, Fifth Edition, ed. M. J. Hurley (New York: Springer), 2552-2590.

[10] Babrausks, V., 1983. Estimating large pool fire burning rates, Fire Technology 19, 251-261.

[11] Blinov, V. I., and Khudyakov, G. N., 1961. Diffusion burning of liquids, Izdatel'stvo Akademii Nauk SSSR, Moscow. English Translation, U.S. Army Engineering Research and Development Laboratories, Information Resources Branch, Translation Analysis Section, Fort Belvoir, VI, No. T-1490a-c.

[12] Mansinha, L., and Smylie, D. E., 1971. The displacement fields of inclined faults, Bulletin of the Seismological Society of America 61, 1433-1440.

[13] Fujii, Y., Satake, K., Sakai, S., Shinohara, M., and Kanazawa, T., 2011. Tsunami source of the 2011 earthquake off the Pacific Coast of Tohoku, Earth, Planets and Space 63, 815-820.

[14] Kotani, M., Imamura, F., and Shuto, N., 1998. Tsunami run-up simulation and damage estimation by using GIS, Proceedings of Coastal Engineering, Japan Society of Civil Engineers 45, 356-360 (in Japanese).

[15] Lau, Y. L., and Moir, J. R., 1979. Booms used for oil slick control, Journal of the Environmental Engineering Division - ASCE 105, 369-382.

[16] Elder, J. W., 1959. The dispersion of marked fluid in turbulent shear flow, Journal of Fluid Mechanics 5, 544-560.

[17] Kyaw, W. P., Sugiyama, M., Takagi, Y., Suzuki, H., and Kato, N., 2017. Numerical analysis of tsunami-triggered oil spill from industrial parks in Osaka Bay, Journal of Loss Prevention in the Process Industries 50, 325-336.

[18] Disaster prevention plan for petroleum industrial complexes of Osaka Prefecture, Osaka Prefecture, 2017 (in Japanese).

18 AIChE Spring Meeting
14TH GLOBAL CONGRESS ON PROCESS SAFETY
ORLANDO FLORIDA 2018

Business Continuity Plan for the next large scale earthquake "NANKAI- Trough" at the refinery in JAPAN.

Hayato Abe
Idemitsu Kosan Co., Ltd.(Manufacturing&Technology Dept.)
Safety&Environmental Section in Aichi refinery
11, Minamihama-machi, Chita,Aichi, 478-8555, Japan
hayato.abe@idemitsu.com

Prepared for Presentation at
American Institute of Chemical Engineers
2018 Spring Meeting and 14th Global Congress on Process Safety
Orlando, Florida
April 22 – 25, 2018

Business Continuity Plan for the next large scale earthquake "NANKAI- Trough" at the refinery in JAPAN.

Hayato Abe
Idemitsu Kosan Co., Ltd.(Manufacturing&Technology Dept.)
Safety&Environmental Section in Aichi refinery
11, Minamihama-machi, Chita,Aichi, 478-8555, Japan
hayato.abe@idemitsu.com

Keywords: Natural Calamities, Business Continuity Plan

Abstract

On March 11th, 2011, Japan suffered the Great East Japan Earthquake of magnitude 9.0. It was the most powerful known earthquake to have hit Japan in recorded history. The earthquake and the tsunami left extensive damages to Japanese oil industry, the nuclear power plants, life lines such as electricity, gas, water service, communication net-work and traffic facilities. Total amount of loss reached US$150 billion.

The Japanese Government reviewed the estimate of probability and damage suffered from the next large scale earthquake "NANKAI-Trough" same Plate type earthquake, based on the experience of the Great East Japan Earthquake. The consequence of this review are so severe, the probability to break out is 70%／30years, the total amount of loss will be US$2,000 billion.

In this case, Japanese Oil Industry shall recover in early period to supply energies for reconstructing activities from the damage. Petroleum product is one of the most important energy source, which can transport to and storage at any place in such disaster case compared with many other sources.

IDEMITSU KOSAN Co.,Ltd. is operating The Aichi Refinery located in these area, estimated severe damage suffered from the next large scale Earthquake "NANKAI-Trough". Therefore we are trying to improve the countermeasure based on the Business Continuity Plan for this Earthquake.

For Example, reinforced basements of facility, structures under pipeline and oil tanks for earthquake-proof, disposition of big canon fire extinguish system （60,000L/min） with training based on fire pre-plan in order to protection for disasters such as vapor cloud explosion, full surface tank fire.

The goal of this BCP require re-shipping oils half of daily land carriage from Aichi Refinery within 24 hours after the earthquake. Now we are making effort to improve related facilities with first priority. We have to cooperate with public organization in control of public utilities such as the road traffic and the route of marine transportation, repairing and re-opening after the earthquake.

This presentation refers to work out the countermeasure of the BCP for large scale earthquake such as the "NANKAI-Trough" at the refinery.

1 Main section 1 heading (Introduction)

Almost Japanese refineries are located in the seashore-line, because Crude oils are imported from abroad on the sea-borne supply route. Japanese Crude oil products are under 1% of total amount of National consumption.

Japanese refineries have experienced so many earthquakes around in JAPAN .The Japanese lands are located on the dividing line between four crust Plates which have been moving and stored up strains.

It is the mechanism of periodical earthquakes that the energy is released from these strains over the sustainable limit. Japanese refineries near-by seashore-line will be suffered from Tsunami when the earthquake will break out around sea aria.

Japanese refineries have experienced great earthquakes around in JAPAN eight times in recent 25years. We learned from these earthquakes for improve our activity every time.

2.1 Lessons from massive earthquakes

In the quarter century, there have been a lot of massive earhquakes as you can see figuer 1. Among them, 2 earthquakes, 1st Tokachi-oki and 2nd Pacific coast of Tohoku brought heavy damage to our industry.

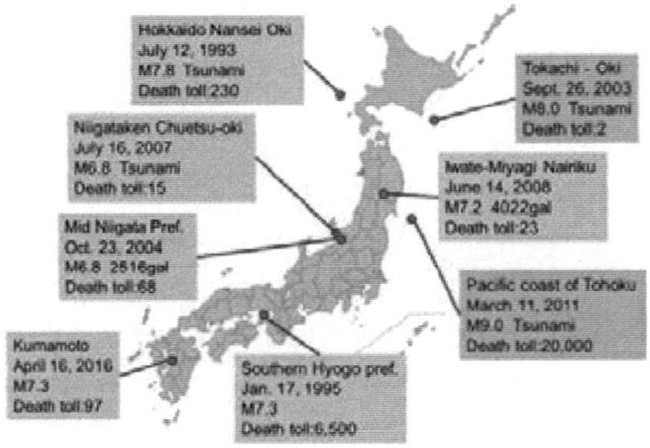

Fig.1 Massive Earthquakes in Japan within 25 years

2.1.1 Damages by The Tokachi-Oki Earthquake in Hokkaidou prefecture

The Tokachi-Oki Earthquake occurred before dawn on September 26, 2003. Its magnitude was 8.0, and the intensity at IDEMITSU Hokkaido Refinery was 86 gal. It wasn't so great earthquake, but it was also not a massive one at the site of Hokkaido refinery. But the velocity response spectrum was higher than our prediction. And the swing period and length were very long. The long swing was near with tanks' natural periods. It made oil in tanks swing in resonance with ground. Then sloshing phenomenon was occurred. A lot of tanks in Hokkaido refinery were damaged with roof deformations, oil spills on roofs and sunken roofs into the oil. And there were 2 fire, a rim seal fire and a full surface fire. The rim seal fire was distinguished in a relatively short time. But full surface fire was huge like Figure 2 and lasted 44 hours.

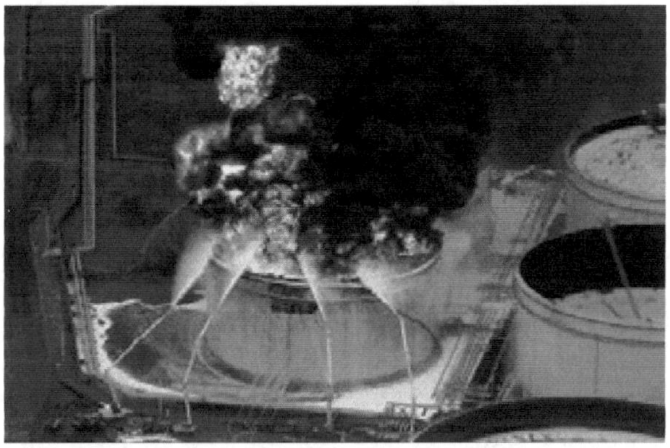

Fig.2 Full Surface Fire on Naphtha Tank in Hokkaido refinery

And a lot of tanks' floating roofs were sunk. Some of them were sunk keeping horizontal, but some others were sunk slanted. We had much difficulty draining oil from them safely, in order to prevent ignition by static charge, collision spark, and so on. A lot of parts of the tanks such as rolling ladder, instruments housing, gage pole, and rotation prevention pole were broken by hitting the swinging roof. And pontoons and deck plates of floating roofs were ruptured.

2.1.2 Lessons Learned from The Tokachi-oki earthquake

From the experience of heavy damages from The Tokachi-Oki earthquake, IDEMITSU learned much about natural calamities. Before that, we had not paid attention to prevention strategy against natural calamities. Therefore, we had not considered local specific ground characteristics nor the possibility of such a high velocity response spectrum. It triggered IDEMITSU re-examination of provisions for natural calamities gaining extreme severity. And, IDEMITSU re-examination and actions according to it triggered re-examination of Japanese fire prevention regulations. IDEMITSU conducted FEM analysis of the tank floating roof damage by sloshing.

Then we found the weak points in the case of long period earthquake. One of some weak points is pontoons and deck plates of floating roofs. It was found that the stress concentrates at the joint between Outer Lim Plate and Pontoon Bottom Plate in Figure 3.

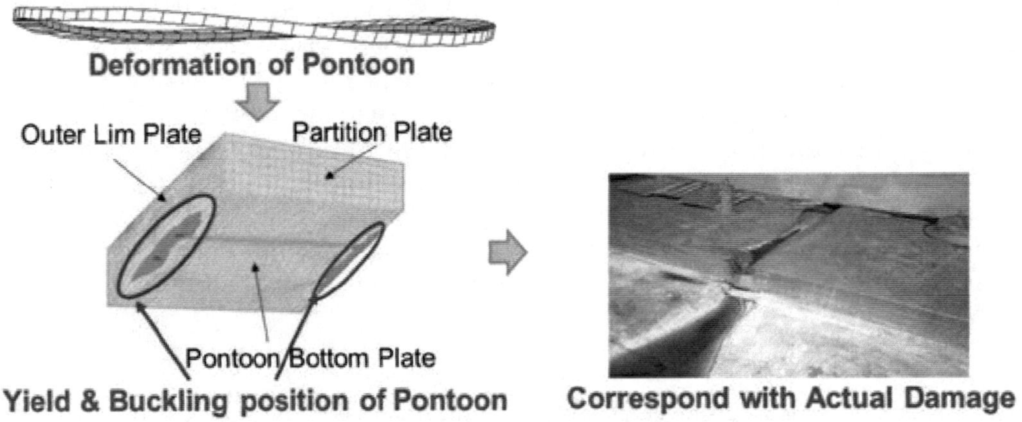

Fig.3 Cause & Improvement Analysis by FEM

And with many other studies, we improved our tank design and have been re-vamping our tanks. For example, the replacement of single deck floating roof with double deck, reinforcement of floating roof, removal of tank accessories, lowering the tank level and removal of at roof-drain piping. It will finally be completed in a few years.

IDEMITSU have built the Tank damage estimation system. We had experienced the difficulty to check all of 100 tanks' damages just after the Tokachi-oki earthquake 2003. Therefore the new system can figure out which tanks are damaged at once after an earthquake. The system provides with damage estimation by analyzing oil levels and seismic signals.

One of the estimation is whether Bulging or Sloshing the damage mode is.

The other estimation is stress distribution. It can figure out the damaged point, for example, shell or plate, and so on.

Another estimation is sloshing height. It can figure out in which tank the spill over occurred.

With this system, it can quickly be aware the trigger of disaster. Adding that, we disposition of Big canon fire extinguish system （60,000L/min） with training based on fire pre-plan in order to protection for disasters such as Full surface Tank fire.

2.1.3 Accelerative seismic assessment

Besides the unexpected long period earthquake threat, we also considered an accelerative earthquake would also be threat. IDEMITSU re-assessed this type of earthquake with concept of recurrence interval. We set the performance needed according to 3 levels of recurrence intervals in table1. (Set the small acceleration in case of short interval, on the other hand set the large acceleration in case of long interval.)

Then IDEMITSU researched the estimated seismic acceleration corresponding to each interval. IDEMITSU assessed own facilities resistance to earthquake using seismic acceleration assessment in table 2 and reinforced them.

Table 1. Adopted the Uniform Building Code standard

Recurrence Interval	Column Title
970 Years	- Prevention of human disaster - Prevention of large-scale disaster
475 Years	- Prevention of environmental pollution - Prevention of long-term stoppage of the refinery
100 Years	-Continue to operate

Table 2. Resistance to earthquake using seismic acceleration assessment

Recurrence Interval	Hokkaido	Chiba	Aichi	Tokuyama
970 Years	136 gal	311 gal	490 gal	161 gal
475 Years	88 gal	222 gal	363 gal	151 gal

For example, IDEMITSU reinforced LPG sphere as shown in figure 4. As provision to liquefaction, the tank basement and the piping support basement were jointed. We reinforce sphere's legs by adding and welding plates not to collapse. IDEMITSU also reinforced anchor of chimney by bouldering concrete, leg of electric precipitator by adding braces.

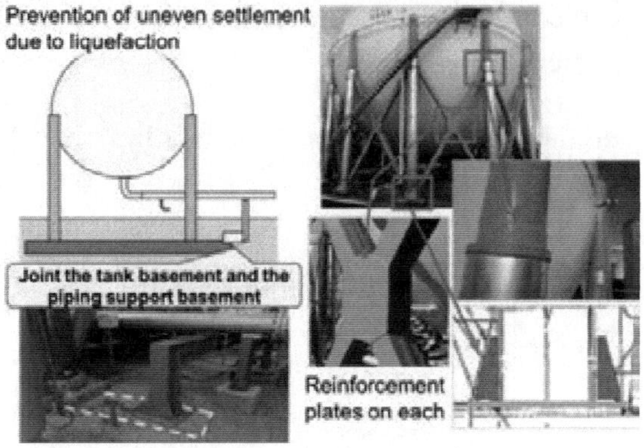

Fig.4 **LPG Sphere Reinforcement**

2.2.1 Damages by the Pacific Coast of Tohoku Earthquake

Half way through our re-examination of provision for natural calamities, The Pacific Coast of Tohoku Earthquake occurred.

As you know, it was the worst disaster in Japan's history. It was magnitude 9.0 and Tsunami waves came up to 38.9 meters (128 ft), and in some cases traveled up to 10 km (6 mi) inland. This earthquake shook over a large region of Japan including many refineries sites not only ours. A lot of refinery facilities were severely damaged. There was a lot of damage at refineries as I showed. But, fortunately, our sites weren't damaged at all. Of course, it mainly owes to our sites' location where the intensities of shock were relatively low. But, let me say, it partly owes to our provisions. LPG spheres in IDEMITSU Chiba were not damaged at all. And the sea water pump in IDEMITSU Hokkaido was not stopped by tsunami flooding. From this experience, we learned much again. Our effort to assess and make improvements after the 2003 tank fire was proven effective. IDEMITSU re-acknowledged the importance of plant reliability against natural calamities. We'd like to continue to research and apply Provisions to Natural Calamities. After the March 11th Earthquake, the government ordered us to re-assess & improve LPG tanks with the newest earthquake design regulations retroactively. We're now doing that with FEM simulations. This analysis technique is also a fruit of our effort after the 2003 tank fire.

Japanese recognized the serious Damage as loss of functions from this earthquake with look over these all aria. For example, the atomic power station, residents and public life support functions such as electricity, gas, water, means of communication, highway, liver, harbor and drainage facilities. It is the problem to be solved that is establishment of the BCP for keeping sufferer life, and restoration of industries in early period after the earthquake.

2.2.2 Lessons Learned from The Pacific Coast of Tohoku Earthquake

The Japanese Government reviewed the estimate of probability and damage suffered from the next large scale earthquake "NANKAI-Trough" same Plate type earthquake, based on the experience of the Great East Japan Earthquake. The consequence of this review are so severe, the probability to break out is 70%／30years, the total amount of loss will be US$2,000 billion.

This amount of loss is twice of the Japanese National annual budget. Therefore oil Industry shall recover in early period to supply energies for reconstructing activities from the damage. Petroleum product is one of the most important energy source, which can transport to and storage at any place in such disaster case compared with many other sources.

IDEMITSU KOSAN Co.,Ltd. is operating The Aichi Refinery located in these area, estimated severe damage suffered from the next large scale Earthquake "NANKAI-Trough". Therefore we are trying to improve the countermeasure based on the Business Continuity Plan for this Earthquake. The goal of this BCP require re-shipping oils half of daily land carriage from Aichi Refinery within 24 hours after the earthquake（Figure-5）.

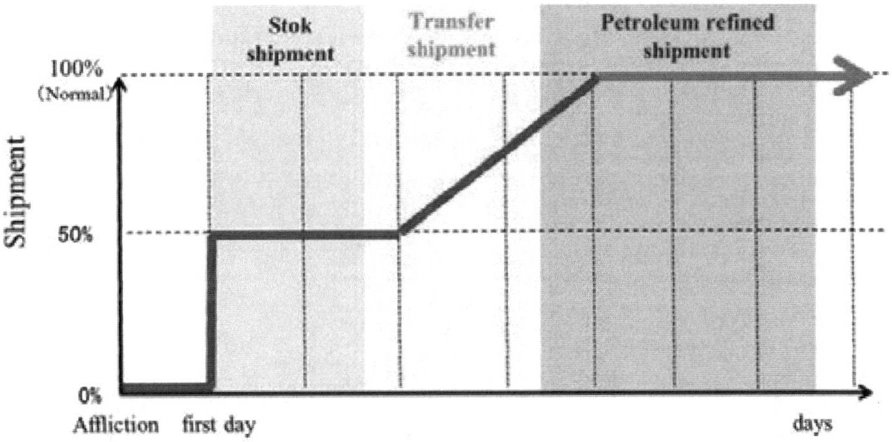

Fig.5 BCP Petroleum product supply target

Now we are making effort to improve related facilities with first priority. For Example, reinforced basements of facility, structures under pipeline and oil tanks for earthquake-proof, disposition of power generator system trailer provide for power failure and big canon fire extinguish system in order to protection for disasters as mentioned above.

In addition, IDEMITSU have to cooperate with public organization in control of public utilities such as the road traffic and the route of marine transportation, repairing and re-opening after the earthquake. Therefor We requested to the public organization adopt their BCP for restoration of traffics around the refinery in early stage. (Figure 6)

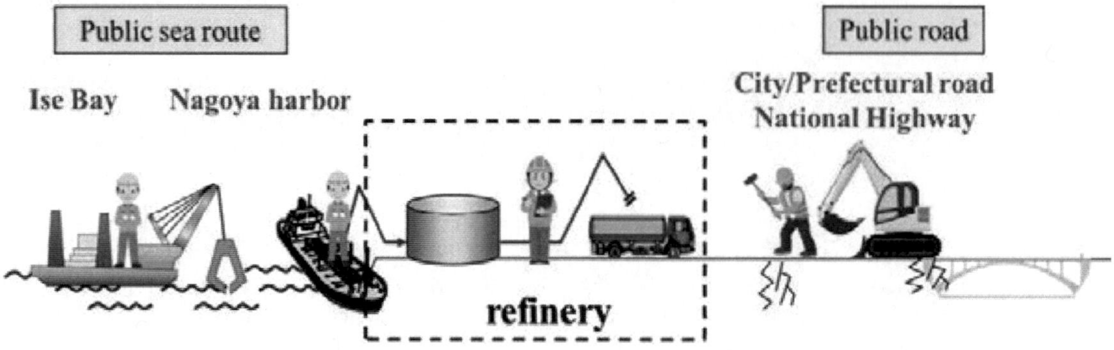

Fig.6 **Collaboration between government and refinery**

3 Conclusion

In industrial activity, we utilize Natural Phenomenon, such as Chemical Reaction, in human society as a production process taking advantage of Process Technology. On the other hand, Natural Phenomenon has negative aspects, typical one is Natural Calamity. In human society, needless to say, we need Safety & Continuity. Therefore, we need Provision Technology to help minimize the damage from Natural Calamity. In addition to that we have to be organized for re-shipping oil products on the time schedule of the BCP, including the cooperation with public organizations. IDEMITSU have to keep moving forward in both aspects.

4 References

[1] Kikuchi,T., Miura,M., "Damage to Floating Roof Tank by Sloshing" Proceeding of the ASME Pressure Vessel and Piping Conference,PVP2005-71244.

[2] Miura,M.,Kikuchi,T., "The Sloshing Simulation of Floating Roof Tank", Proceeding of the ASME Pressure Vessel and Piping Conference,PVP2005-71439.

[3] Kikuchi,T.," Earthquake countermeasure at refinery / factory - preparation for natural disaster", A lecture at Japan Society of Mainte-nology 18 January 2010

[4] Kikuchi,T.," Lessons learned from the damage of the Tokachi-oki earthquake and practice of natural disaster countermeasures", at Japan Society for Promotion of Science July 2012

[5] Kikuchi,T., " Idemitsu earthquake and tsunami measures at refineries and factories - About prevention of serious damage due to natural disaster", A lecture at High Pressure Institute of Japan 27 October 2013

Pollution, Control Status and Strategy of Volatile Organic Compounds in China's Petrochemical and Refining Industries

Jia Ke, China University of Petroleum (East China), Qingdao, P R China
Dongfeng Zhao, China University of Petroleum (East China), Qingdao, P R China; Center for Safety, Environmental, and Energy Conservation Technology, Qingdao, P R China
Yi Liu, China University of Petroleum (East China), Qingdao, P R China; Center for Safety, Environmental, and Energy Conservation Technology, Qingdao, P R China

Abstract

With rapid industrialization and urbanization in China, haze and photochemical smog, caused by fine particulate matters ($PM_{2.5}$, PM_{10}) and ozone (O_3), have become one of the most severe environmental pollution issues in China, which were seriously affecting the daily life and physical health of residents. Thus, as an important precursor of fine particulate matters and ozone, volatile organic compounds (VOCs) are now included in the national control of key pollutants and total control factors. Because of its strategic importance in industrial system and its huge VOCs emissions (accounts for 34.3% of the total emissions), petrochemical and refining industries have recently elicited public concern and given new impetus to scientific interest.

This paper reviews the VOCs emissions, and summarizes the policies, regulations, emission standards and the existing pollutant permit system for controlling VOCs in China's petrochemical and refining industries. This paper also analysis the main problems in current petrochemical and refining industries VOCs control status and management. Some countermeasures and suggestions, such as developing and improving VOCs monitoring methods, establishing VOCs emission list database, researching the best feasible control technology in petrochemical and refining industries and so on, are proposed to strengthen the control and management of VOCs emissions. Besides, the prospects and recommendations are put forward for China's future open, whole-process VOCs management model. The findings not only can provide a scientific basis for enhancing and implementation of existing VOC control strategies, but also can provide technical support for governments and enterprises to improve environmental quality and alleviate regional air pollution.

Keywords: Petrochemical and Refining Industries; Volatile Organic Compounds; Policies; Environmental Management

Introduction

With the rapid development of China's industrialization process, air pollution type has shifted from traditional coal smoke pollution, which dominated by dust and SO_2 emissions, to a composited air pollution, which dominated by fine particulate matters ($PM_{2.5}$, PM_{10}) and ozone (O_3) pollution. Air pollution is increasingly intensifying. The continuous phenomena of acid rain, haze and photochemical smog have become one of the most serious environmental issues in many cities and regions in China, seriously affecting people' daily life and physical health.

Many articles show that volatile organic compounds (VOCs) are important precursors of atmospheric ozone (O_3) and secondary organic aerosol (SOA). SOA is considered to be one of main contribution sources to $PM_{2.5}$ in the ambient air, which has contribution to the formation of the composited air pollution. Among various VOCs anthropogenic emission sources, industrial sources are the largest source of VOCs emissions. The petrochemical and refining industries, which has significant VOCs emissions among industrial sources, is undoubtedly placed on the front line of VOCs control. Therefore, VOCs research and effective control in the petrochemical and refining industries has become a hot topic in the field of atmospheric environmental management and control in China.

The status of VOC emissions in the petrochemical and refining industries, including policies, regulations, standards of VOCs control and management, were analyzed and studied. Some main problems in VOCs control current status and control management were proposed. In addition, solutions and recommendations were provided. The results can will provide technical supports to improve environmental quality.

1 China Air Quality Status

Nowadays, China has established a comprehensive national environmental air quality monitoring network and has set up more than 5,000 monitoring stations, which included 1,436 air quality monitoring stations in 338 cities at or above prefecture level (contained 135 clean control stations), 96 regional (rural) ambient air quality monitoring stations, 16 background ambient air quality monitoring stations, and 38 atmospheric particulate components/photochemical monitoring stations in 2+26 cities (i.e. Beijing, Tianjin, and 26 cities in the Beijing-Tianjin-Hebei and surrounding area).

According to the Ambient Air Quality Standards (GB 3095-2012) and the Technical Regulation for Ambient Air Quality Assessment (HJ 663-2013), six ambient air quality pollutants, i.e. SO_2, NO_2, $PM_{2.5}$, PM_{10}, O_3 and CO, were evaluated based on the ambient air quality monitoring data. The ratio of different air quality levels was assessed in different cities from 2013 to 2016. And its results were showed in Table 1. The average annual concentrations of ambient air quality pollutants from 2013 to 2016 were shown in Table 2.

Table 1 Air quality conditions from 2013 to 2016

Year	**percentage of air quality standards**				
	excellent and good	**slightly polluted**	**intermediately polluted**	**heavily polluted**	**very heavily polluted**
2013	60.5	22.9	8	6.2	2.4
2014	66.3	21.6	6.5	4.4	1.2
2015	76.7	15.9	4.2	2.5	0.7
2016	78.8	14.8	3.7	2	0.6

Note: Days with AQI = 0~100 indicates perfect, AQI =101~150 indicates slight pollution, AQI =151~200 indicates intermediate pollution, AQI =201~300 indicates heavy pollution and AQI >300 very serious pollution.

Table 2 Average annual concentrations of ambient air quality pollutants from 2013 to 2016

Year	**$PM_{2.5}$**	**PM_{10}**	**90th percentile concentration of O_3 daily maximum 8-hour average**	**SO_2**	**NO_2**	**95th percentile concentration of daily CO average (mg/m^3)**
2013	72	118	139	40	44	2.5
2014	62	105	140	35	38	2.2

2015	50	87	134	25	30	2.1
2016	47	82	138	22	30	1.9
Standard valve	35	70	160	60	40	4

The percentage of cities met national air quality standard was increased from 4.1% in 2013 to 24.9% in 2016, and the percentage of cities failed to meet national air quality standard was decreased from 95.9% in 2013 to 75.1% in 2016. By comparing the ambient air qualities data from 2013 to 2016, it can be found that the urban air quality has improved as a whole. It indicated that China's air pollution prevention and control has made significant progress under control and management. The annual average concentrations of $PM_{2.5}$, PM_{10}, NO_2, SO_2, and CO decreased year by year. The number of heavy pollution days in most cities has decreased. However, pollution issues, such as high-pollution loads of $PM_{2.5}$ and PM_{10} in key areas, composited pollution and heavy air pollution days which the primary pollutants were $PM_{2.5}$, O_3 and PM_{10} during these days, are still serious. The overall situation of air pollution control is still severe.

As an important precursor of SOA and ozone, VOCs have become the fourth most important atmospheric pollutants in China after sulfur dioxide, nitrogen oxides, and dust, and have been listed in national key prevention and control pollutants and total control factors. Therefore, the control and reduction of VOCs has also become one of the essential issues in the prevention and control of atmospheric pollution. In addition, petrochemical industry has attracted much attention due to its huge VOCs emission, and has become one of the key industries for VOCs management and control.

2. Volatile Organic Compounds

Volatile Organic Compounds (VOCs) are composed of hundreds of organic compounds that are directly emitted into the atmosphere. They are a class of organic compounds. For different industries, the chemical composition varies, but mainly includes organic compounds such as alkanes, unsaturated hydrocarbons, halogenated hydrocarbons, benzene series, organic acids, alcohols, and esters. For volatile organic compounds, the United States, the European Union, Japan, and other countries have been research and control VOCs decades ago. However, China starts relatively late to research and control VOCs, and is still at exploratory stage.

2.1 Definition of VOCs

There is no formal and standard definition of VOCs in the world nowadays. In china, many standards also defined VOCs, but it existed difference. The specific definitions were described in Table 3 below.

Table 3 Definition of VOCs in different standards

No.	Standard	Definition
1	Emission standard of pollutants for synthetic leather and artificial leather industry GB 21902-2008	Any organic compound (excluding methane) that has a boiling point of less than 250°C or can be discharged into the air as a gaseous molecule at atmospheric pressure
2	Emission standards of air pollutants for petroleum refining and petrochemicals manufacturing industry DB 11/447-2007	Organic compounds with a vapor pressure higher than or equal to 0.01 kPa at 20°C
3	The discharge standard of pollutants	Volatile organic compounds that saturated vapor

	for bio-pharmaceutical industry DB 31/373-2010	pressure at 0.1mmHg or above at 20°C, or melting point below room temperature and boiling point below 260°C, excluding methane
4	Indoor air quality standard GB/T 18883-2002	Volatile organic compounds that retention time between hexane and n-hexadecane analyzed by non-polar columns (polarity less than 10), which sampled by Tenax GC or Tenax TA
5	Integrated emission standard of air pollutants GB 3095-2012	Organic compounds with a vapor pressure higher than or equal to 0.01 kPa at 20°C or has corresponding volatility under specified conditions
6	Emission standard of pollutants for petroleum refining industry GB 31570-2015	Organic compounds involved in atmospheric photochemical reactions, or determined or calculated by specific methods
7	Emission standard of pollutants for petroleum chemistry industry GB 31571-2015	Organic compounds involved in atmospheric photochemical reactions, or determined or calculated by specific methods

As can be seen from Table 1, there are different definitions of VOCs in different standards. These definitions were defined in terms of volatility (such as vapor pressure, boiling point, etc.). Under this definition, VOCs basically cover all organic compounds. However, it is not clear that which substances have O_3 and $PM_{2.5}$ formation potential. They have limitations. In addition, there are some definitions that are defined based on monitoring methods. Under this definition, some of the higher melting VOCs are excluded from the definition, which cannot fully reflect the definition.

For petrochemical and refining industries, the definition is mainly based on photochemical reactivity. Under this definition, the factors can be found to control VOCs, so as to achieve the control of the formation of ozone in ambient air. It is also possible to find out specific substances with high photochemical activity at the source stage. The formation of ozone and $PM_{2.5}$ in the ambient air can be achieved by not using or instead of using such substances, or developing new processing technologies.

2.2 Sources of VOCs

The sources of VOCs include natural sources and human sources. Natural sources mainly come from plant emissions, such as plant releases, volcanic eruptions, and forest fires. They are uncontrolled. The most important emissions sources are forests and shrubs. The most important natural emissions are isoprene and monoterpenes. The human sources mainly come from the use of automobile exhaust, fuel evaporation, petrochemicals, and paints. It includes various industrial sources, agricultural sources, transportation sources, and life sources.

In China, human sources mainly come from industrial sources, transportation sources (motor vehicle emissions) and biomass combustion. Due to the rapid industrialization process, VOCs emissions from industrial sources increased year by year and accounted for more than 50% of the total VOCs emissions. Industrial emission sources are complex, involving production, usage, storage, and transportation processes. The petrochemical and refining, coatings, inks, pesticides, and automotive industries accounted for more than 80% of total industrial VOCs emissions. Research showed the VOCs emissions from petrochemical and refining industry represented about 34.3% of the total VOCs emissions.

2.3 Photochemical Reaction of VOCs

Under UV light, VOCs have a series of photochemical interactions with nitrogen oxides (NOx), which formed O_3, peroxyacetyl nitrate (PAN), highly reactive free radicals (ON, RO_2, HO_2), aldehydes and so on. These pollutants can further form secondary aerosols (SOA) through oxidation, nucleation, and condensation processes, which promotes the formation of $PM_{2.5}$.

When there is no VOCs in the air, the interactions between NO_2, NO, and O_3 in the atmosphere can reach equilibrium:

$$NO_2 \rightarrow NO+O$$

$$O+O_2 \rightarrow O_3$$

$$O_3+NO \rightarrow NO_2+O_2$$

When there is no VOCs in the air, the following reactions occur:

$$VOCs+OH \rightarrow HO_2$$

$$HO_2+NO \rightarrow NO_2+OH \text{ (reaction quickly)}$$

NO, produced by NO_2, is consumed by HO_2, resulting in NO being unable to react with ozone and breaking the equilibrium of NO_2, NO, and O_3 interactions. It eventually leading to the accumulation of O_3.

2.4 Damages of VOCs

Due to the diversity of VOCs' physical and chemical properties, the scope and impact of VOCs are also relatively widespread. VOCs are not only harmful to environmental, but also harmful to human health.

1) Impact on the environment

Some of the substances of VOCs have strong photochemical reactivity, can cause photochemical smog, and can indirectly affect the radiation balance on the Earth's surface, which in turn affects the climate. Some VOCs can destroy stratospheric ozone characteristics and reduce stratospheric ozone concentrations, such as chlorofluorocarbons (CFCs). With the diffusion and migration of VOCs pollution, the environmental problems such as acid rain, ozone layer destruction and atmospheric warming are no longer belonged to one city or country, but spread to a global scale.

2) Impact on human health

Most VOCs are toxicologically and harmful to human health. At present, more than 50 pollutants are contained in priority control pollutants list, which announced by the US EPA. VOCs can cause maladjustment of multiple systems, such as the respiratory system, nervous system and cardiovascular system. VOCs also have carcinogenicity and genetic toxicity. When the concentration of VOCs in the environment reaches a certain concentration, people may feel headaches, nausea, and vomiting. In severe cases, convulsions, coma, and even life-threatening conditions may occur.

3. Status of VOCs Emissions in Petrochemical and Refining Industries

As a pillar industry for the economy development of China, petrochemical and refining industries arises a series of environmental pollution problems, especially VOCs pollution problems during the acceleration of industry economic growth. Researchers used different estimation methods to estimate the VOCs emissions. The VOCs emissions from human-made source in China are about 20 Mt/a, and the data rapidly rising. Among the human-made source, industrial sources accounted for about 55%, in which petroleum refining and storage, organic chemicals, and fine chemicals were key industries for VOC emissions.

In November 2015, the Ministry of Ecology Environment (MEE) issued the "Guidelines for

VOCs Pollution Source Inspection Work for petrochemical and refining industries" in accordance with the theoretical basis for the classification and analysis of pollution sources in the process chain, and divided VOCs sources into 12 categories. These are the leakage of equipment static and dynamic sealing points; the volatilization loss of organic liquids storage and blending process; the volatilization loss of organic liquids loading and unloading process; the volatilization loss of wastewater in collection, storage, treatment and disposal process; organized emissions; the release from circulating cooling water system; abnormal conditions (including startup, shutdown and maintenance) emissions; flares emissions; combustion emissions; unorganized emissions; sampling process emissions and accident emissions. The document requires that all petrochemical and refining industries should basically complete the VOCs comprehensive control and establish VOCs monitoring system by 2017. The total VOCs emissions will be reduced by more than 30% from 2014. By the end of 2017, the petrochemical and refining industries have comprehensively carried out comprehensive VOCs control work.

The VOC emissions of each source from petrochemical and refining industries with different capacities are shown in Table 4.

Table 4 VOC emissions from different petrochemical and refining industries (t/a)

Source	**Sinopec Qingdao Refining Corp., Ltd.**	**Haerbin Petrochemical**	**Guangxi Petrochemical**	**Jiujiang Petrochemical**
Capacity (million t/a)	1000	420	1000	800
Leakage of equipment static and dynamic sealing points	90.2	212.9	126.3	199.4
Volatilization loss of organic liquids storage and blending process	319.1	997.3	2087.8	2027.8
Volatilization loss of organic liquids loading and unloading process	26.1	511.9	1613.9	1084.4
Volatilization loss of wastewater in collection, storage, treatment and disposal process	228.4	130.5	159.4	1061.5
Release from circulating cooling water system	225.6	124.3	121.4	853.2
Combustion emissions	7.5	19.4	221.3	13.8
Organized emissions	48.6	2.6	370.2	87.97
Unorganized emissions	472.5	0.0	0.0	167.5
Sampling process emissions	0.0	0.0	19.9	0
Flares emissions	7.2	143.6	81.7	0.02
Abnormal conditions emissions	0.0	0.0	3.0	0.4
Accident emissions	0.0	0.0	0.0	0
Total	1425.1	2142.4	4805.0	5495.9

4 VOCs Control Status

Facing the rising industrialization and deteriorating ambient air quality, China has begun to pay attention to VOCs pollution, and has gradually established a management idea, taking whole process and refinement management and control as center. VOCs management and control measures will be conducted from the source, process, and ends aspects. In an attempt to reduce air pollution and control VOCs pollution, the Chinese government has made the decision to enforce stricter regulations, including laws, standards, regulation, action plans, VOCs pollution prevention technologies and others.

4.1 Policy System

For air pollution, the two key laws are Environmental Protection Law (EPL) and Law of the People's Republic of China on the Prevention and Control of Atmospheric Pollution. As the air pollution has been shifting from typical coal smoke pollution to composited air pollution, the two laws are amended on April 2104, August 2015 respectively and implemented the next year. The current version of Prevention and Control of Atmospheric Pollution Law has expanded into the scope of legal supervision, and defined legal responsibilities of VOCs prevention and control.

Since 2010, in order to govern VOCs and improve environmental air quality, the State has successively issued various regulations and action plans, such as 'Guiding Opinions on Promoting Air Pollution Joint Prevention and Control Work to Improve Regional Air Quality', '12th Five-year Plan' on Air Pollution Prevention and Control in Key Regions', 'the State Council Air Pollution Prevention and Control Action Plan', 'Control Pollutant Release Permit System Implementation Plan'. These regulations and action plans listed VOCs as key pollutants for key air pollution prevention and control, together with particulate matter, SO_2, and NO_2. They are also placed $PM_{2.5}$ and O_3 and its precursors in a core position in the prevention and control of air pollution in China. It is proposed that by 2020, the state will establish and improve the VOCs pollution prevention and management system, and implement VOCs pollution reduction in key areas and key industries.

As one key industry, petrochemical and refining industries, the MEE issued the 'Comprehensive Plan for the Control of Volatile Organic Compounds in the Petrochemical and Refining Industries', the 'Guidelines for VOCs Pollution Source Investigation in the Petrochemical and Refining Industries' and the 'Guidelines for Petrochemical and Refining Industries Leak Detection and Restoration Work' and other documents to comprehensively control VOCs in the petrochemical and refining industries, and to greatly reduce VOCs emissions from the petrochemical and refining industries.

4.2 Emission Standards

To strengthen the control of atmospheric pollution, the National Ambient Air Quality Standard (NAAQS) (GB 3095-2012), which newly limits the ambient levels of $PM_{2.5}$ and ozone and sets clear reduction goals for air quality planners and regulators, was revised on 29 February 2012 and enacted on 1 January 2016. As for petrochemical and refining industries, some industrial emission standards have issued, like 'Emission Standard of Pollutants for Petroleum Refining Industry' (GB 31570-2015), 'Emission Standard of Pollutants for Petroleum Chemistry Industry' (GB 31571-2015) and so on.

In addition to the national standards, various local standards have been established. For example, Beijing has established the 'Integrated Emission Standard for Air Pollutants' (DB 11/501-2007) and the 'Emission Standard of Air Pollutants for Refining and Petrochemical Industry' (DB 11/447-2007). Shanghai has newly established the 'Integrated Air Pollution Emission Standard' (DB 31/933-2015). The standards and documents established by various provinces have further strengthened the control of VOCs pollution.

Furthermore, the state has also put forward more stringent emission standards for vehicles. Since 2017, the national vehicle emission pollutants will fully implement the national V emission standards. For petrochemical and refining industries, the more stringent gasoline and diesel standards enables petrochemical enterprise to further optimize the oil processing technology and reduce the sulfur and benzene concentration in petroleum products. To a certain extent, it also can reduce the VOCs emissions from petrochemical and refining industries.

4.3 Regional Joint Prevention and Control

The complex air pollution in China is not just a local issue but a domestic regional cross-boundary problem. In order to solve this problem, regional coordination to control cross-boundary air pollutants is necessary to reduce the impact of atmospheric pollution on downwind locations, and improve the regional air quality.

Until now, nine regions, including the Beijing-Tianjin-Hebei region, the PRD region, the YRD region, central Liaoning region, Shandong Peninsula, Western Taiwan Straits Metropolitan region, Cheng-Yu Metropolitan region, Wuhan Metropolitan region, and Cheng-Zhu-Tan region, have been designated as pilot regions for the joint prevention and control of regional air pollution.

The effectiveness of regional joint prevention and control is gradually emerging. For example, on 20 August 2015, ahead of the 70th-anniversary celebrations of the end of World War II, the Beijing and surrounding areas shut down industrial facilities and reduced car emissions. This action resulted in $PM_{2.5}$ concentration lower than the national air quality standard. The restrictions resulted in an average Beijing $PM_{2.5}$ concentration of 19.5 $\mu g/m^3$, the lowest that had ever been on record in the capital, together with the Parade Blue phenomena. At the same time, various local governments (such as Beijing, Shanghai, Guangdong, etc.) are also developing different policies and plans to conduct VOCs prevention and control.

Besides, in order to improve the air quality in autumn and winter, the state has new requirements for new and existing enterprises in "2+26" cities (including Beijing, Tianjin, and 26 cities in the Beijing-Tianjin-Hebei and surrounding area). The pollutants in these enterprises will implement specific emission limits from March 1, 2018 and October 1, 2018 respectively.

4.4 VOCs Prevention Technologies

VOCs pollution prevention and control follows the principle of comprehensive prevention and control combined with source control, process control and end treatment.

Source control mainly refers to develop and use less contaminated raw materials, eliminate materials which have many types of VOCs and volatility, and use alternatives, alternative technologies. For petrochemical and refining industries, adopting advanced clean technologies, improving the conversion and utilization efficiency of crude oil are encouraged. Reducing or avoiding the emission of VOCs at the source is the primary way to control VOCs pollution. It is easy for VOCs to violate and leak during usage, transportation, and storage. Process control can be achieved by controlling leakage, collecting VOCs and so on. In addition, it also can be achieved by taking measures, such as improving the automation and sealing of equipment components, optimizing process, improving equipment, and using clean process, to reduce VOCs emissions.

End treatment is the final step to control VOCs emissions, which is the research focus and hotspots. Based on the technology principles, the end treatment techniques of VOCs can be broadly divided into two categories: decomposition and destruction technologies and recycling technologies. On the basis of investigation of VOCs pollution control technologies status, the MEE issued the 'National Advanced Pollution Prevention and Control Technology Catalogue (VOCs Prevention and Control Technology)', listing various proposed advanced VOCs end treatment technologies and its application scope (below table 5 only lists processing technologies involving the petrochemical industry). These technologies are innovative. The effects are stable and the cost is reasonable. The technologies should be encouraged for enterprises to promote applications

Table 5 Advanced VOCs end treatment technologies and its application scope

No.	Technology	Application scope
1	Activated carbon adsorption-nitrogen desorption condensation solvent recovery technology	Petrochemical and other industries
2	Oil and Gas Active Carbon Adsorption and Recovery Technology during Oil Storage and Transportation	Oil and gas recovery during product oil loading or storage
3	Oil and Gas Membrane Separation and Adsorption Recovery Technology for Oil Storage and Transportation	Petrochemical industry oil and gas recovery
4	Stationary organic waste heat storage combustion technology	Medium or high-concentration VOCs exhaust gas purification in petrochemical and other industries
5	Rotary heat storage combustion purification technology	Medium or high-concentration VOCs exhaust gas purification in petrochemical and other industries
6	Regenerative catalytic combustion (RCO) technology	Medium and high concentration VOCs exhaust gas treatment
7	Low concentration organic waste gas biological purification technology	Low concentration organic waste gas treatment
8	Advanced oxidation-biological purification coupled processing technology	Purification of low-concentration VOCs and malodorous gases from the production processes and sewage treatment plants (stations) in industries such as petroleum refining

4.5 Pollutant Permit System

China's pollutants permit system began in the 1980s, but it did not play a central role at that time. Until November 2016, the new pollutant permit system began to be implemented nationwide. It was clearly proposed that by 2020, the pollutants permit covering all stationary pollution sources should be issued, and a permit emission control system should be established, and impelling the "one-signature" management. In recent years, the state has issued 'Provisional Regulations on the Management of Emission Permits', 'Administrative Measures on Emission Permits (Trial)' and other documents to present detail provisions on the procedures for the application, review, issuance, and management of pollutant permit. At the same time, regulations on the allowable emission concentrations, permitted emissions, and environmental management for air pollutants were also provided.

For the petrochemical and refining industries, all the enterprises should be issued the pollutant permit before the end of 2018. In order to guide and regulate the application and issuance of emission permits for enterprises, the MEE issued the 'Technical Specification for Application and Issuance of Pollutant Permit in Petroleum Industry' (HJ 853-2017). At present, 34.46% of China's petrochemical and refining enterprises have completed the application of pollutant permit in the application platform. The VOC annual permit emission limit has implemented in three VOCs sources, i.e. Leakage of equipment static and dynamic sealing points, Volatilization loss of organic liquids storage and blending process and Volatilization loss of organic liquids loading and unloading process.

No pollutant emissions will be allowed without air pollution permits. One enterprise can held one permit. By implementing the pollutant permit system, it will change the phenomenon of disorderly discharge of enterprises in China, and promote the continuous improvement of environmental quality. The content of the permit license is not only to control the concentration of permitted emissions and the amount of permitted emissions, but more importantly, to control pollution prevention and control measures, finally to reduce pollutants.

Overall, from the review of the policies on the control of VOCs and/or photochemical smog, the stricter and aggressive measures help to alleviate air pollutants. Even though the atmospheric pollution is likely to be persistent because of on-going urbanization and industrialization, it is possible that the air quality can be gradually improved if appropriate measures are undertaken.

5 Problems and Suggestions

The Chinese government began to attach great importance to controlling VOCs emissions, regarding VOCs as a top priority among atmospheric pollutants for effectively improving air quality. Even though laws, regulations, standards and measures to control VOCs have been introduced in recent years, they still have problems and need to be further strengthened and standardized.

5.1 VOCs Monitoring Methods

Some standards put forward regulations and requirements to VOCs monitoring. However, the VOCs monitoring indicators and the monitoring technology methods are not clear or uniform. The characterization of VOCs is not uniform, some are based on the sum of characteristic pollutants, and some are based on NMHC. While in practice, VOCs are generally collected by solid adsorption or sampling bags, and analyzed by gas chromatography, gas chromatography/mass spectrometry, high performance liquid chromatography (HPLC) analysis, and a series of portable devices developed on this basis, such as TVA2020C, PHX-21. In addition, the pollutant permit system requires enterprises to implement self-monitoring. Now the self-monitoring of the pollutant permit has designed nearly 610,000 monitoring devices, while there are only 41077 related to the petrochemical and refining industries. Besides, the self-monitoring programs in enterprises are still based on manual monitoring, and the proportion of automatic monitoring is only 12.80%.

Therefore, it is necessary to formula and improve the VOCs monitoring standard system, establish scientific VOCs monitoring methods, and carry out research on petrochemical and industries industry monitoring method standards. In addition to monitor the conventional chimneys, leak detection standards for the devices (production devices, non-organized gas collection devices, etc.) should also be considered. Secondly, to improve monitoring capabilities, establish on-line monitoring networks for VOCs (using infrared instruments), and increase efforts to research portable monitoring devices and on-line monitoring devices. Finally, VOCs monitoring and big data platform are built to achieve standardized, precise, stable and reliable VOCs monitoring data.

5.2 VOCs Emission Inventory

The VOCs emission inventory is an indispensable parameter for the formulation of VOCs control measures. Compared with foreign countries, China's VOCs emission inventory research started late and is still on the beginning stage. There is no uniform and standardized emission source classification system, and there are few special studies on various types of emission sources. The petrochemical and refining industries have a poor level of refinement of emission source inventory, sluggish data updates, and non-comprehensive coverage. Besides, it lacks localized VOCs emission factors. The currently used emission factors are generally revised empirical emission factors that do not reflect China's actual situation. And the statistics of the activity level of the emission source sector is incomplete and lacks continuity.

Therefore, it is extremely necessary to establish an updated emission inventory that covers the spectrum of each pollution source in petrochemical and refining industries as much as possible. In addition, based on the actual situation in China, China's exemptions lists and the corresponding industry exemptions lists should be formulated. On the basis of the definition of photochemical reactivity, poisonous and harmful atmospheric pollutants shall be set up synchronously to control toxic and hazardous characteristics of VOCs. It is recommended to quote the list of HAPs formulated by the US EPA, and gradually improve the list of toxic and harmful air pollutants in China according to national conditions.

5.3 VOCs Control Technologies

The components and volumes of VOCs emitted from different processes in petrochemical and refining industries are different, and the adapted technologies are different. At present, the effect of VOCs treatment technology is different, and the non-methane hydrocarbon (NMHC) treatment efficiency is quite low. Moreover, the economic cost of some technologies is high, and many VOCs processing facilities cannot operate stably, continuously, and reliably.

Therefore, when selecting VOCs control technologies and facilities, enterprises should consider the characteristics of VOCs, as well as comprehensive consideration of various technologies' performance indicators, construction and operating costs, and implementation of emission standards. If necessary, two or more processes combination and multi-stage treatment can be used to achieve compliance with the emission standards. Special research should be conducted on VOCs in petrochemical and refining industries. Measures like continuous research and innovate treatment technologies, development relevant alternatives and improve existing equipment or processes are need to take effort to reduce VOCs emissions.

6 Conclusion

With the rapid progress of urbanization and industrialization in China, atmospheric pollution has become more and more serious. Currently, the type of atmospheric pollution is no longer a simple pollution, but a regional and complex pollution with extensive secondary pollutants. As an important precursor to cause secondary pollution, VOCs have attracted attention from countries in the world. Because of the huge VOCs emissions, wide range of sources in petrochemical and refining industries, accurately understanding the current status of VOCs emissions in the petrochemical and refining industries in China and the status of VOCs management and control are important prerequisites to improve China’s ambient air quality. It is important for China’s future comprehensive air pollution control decisions. Therefore, this article mainly discusses the status of China's atmospheric environment, the current status of VOCs emissions from petrochemical and refining industries, and the VOCs control measures.

China's management and control of VOCs in petrochemical and refining industries is still in its infancy. The future prevention and control of VOCs in petrochemical and refining industries remains an arduous task. In response to the current problems in the control of VOCs, some countermeasures and suggestions are proposed:

1) Formula and improve the VOCs monitoring standard system as soon as possible, and determine the characterization methods that comply with the VOCs in petrochemical and refining industries. At the same time improve the monitoring capabilities and the accuracy of online monitoring devices.

2) Accelerate the research on emission inventory and localization emission factors of VOCs in petrochemical and refining industries, establish an exemption list and a list of toxic and hazardous

pollutants, and provide technical support for the active implementation of emission inventory and source resolution of VOCs in the petrochemical industry;

3) Research and development best VOCs pollution prevention technologies for petrochemical and refining industries. Comprehensive consideration should be given when selecting VOCs treatment technologies. When necessary, a variety of technologies can be used together to achieve the pollutants meet the standards.

In the long-term perspective, the VOCs management and control model need to carry out work from the policy level, technical level, and management level. There are requirements and guidance in terms of policies, and they are technically supported and guaranteed in terms of technologies. There are breakthroughs and improvements in terms of management. Only by that, the VOCs pollution problem can effectively solved and the ambient air quality can be improved.

Acknowledgements

The work is supported by the International Cooperation Program for Innovative Talents of the China Scholarship Council (liujinwei [2016] 7988). Thanks for the funding.

References

1. Derwent R G, Jenkin M E, Saunders S M, et al. Photochemical ozone formation in north west Europe and its control[J]. Atmospheric Environment, 2003, 37(14):1983-1991.
2. Wu R, Xie S. Spatial Distribution of Ozone Formation in China Derived from Emissions of Speciated Volatile Organic Compounds[J]. Environmental Science & Technology, 2017, 51(5):2574.
3. Middleton P, Stockwell W R, Carter W P L. Aggregation and analysis of volatile organic compound emissions for regional modeling[J]. Atmospheric Environment Part A General Topics, 1990, 24(5):1107-1133.
4. Crutzen P J, Lawrence M G, Pöschl U. On the background photochemistry of tropospheric ozone[J]. Tellus Series B-chemical & Physical Meteorology, 1999, 51(1):123-146.
5. Klimont Z, Streets D G, Gupta S, et al. Anthropogenic emissions of non-methane volatile organic compounds in China[J]. Atmospheric Environment, 2002, 36(8):1309-1322.
6. Wei W, Wang S, Hao J, et al. Projection of anthropogenic volatile organic compounds (VOCs) emissions in China for the period 2010–2020[J]. Atmospheric Environment, 2011, 45(38): 6863-6871.
7. Saikawa E, Kim H, Zhong M, et al. Comparison of emissions inventories of anthropogenic air pollutants and greenhouse gases in China[J]. Atmospheric Chemistry & Physics, 2017, 17(10):1-41.
8. Huang G, Brook R, Crippa M, et al. Speciation of anthropogenic emissions of non-methane volatile organic compounds: a global gridded data set for 1970-2012[J]. Atmospheric Chemistry & Physics, 2017, 17(12):1-36.
9. Zheng J, Zheng Z, Yu Y, et al. Temporal, spatial characteristics and uncertainty of biogenic VOC emissions in the Pearl River Delta region, China[J]. Atmospheric Environment, 2010, 44(16):1960-1969.
10. Ohara T, Akimoto H, Kurokawa J, et al. An Asian emission inventory of anthropogenic emission sources for the period 1980–2020[J]. Atmospheric Chemistry & Physics, 2007, 7(16):6843-6902.
11. Xia L, Cai C, Zhu B, et al. Source apportionment of VOCs in a suburb of Nanjing, China, in autumn and winter[J]. Journal of Atmospheric Chemistry, 2014, 71(3):175-193.

12. Ying L, Min S, Lu S H, et al. Source apportionment of ambient volatile organic compounds in the Pearl River Delta, China: Part II.[J]. Atmospheric Environment, 2008, 42(25):6261-6274.
13. Rumei L I, Yuanyuan W U, Peng L, et al. Characteristics and sources apportionment of ambient volatile organic compounds(VOCs) in summer in Shuozhou[J]. Environmental Chemistry, 2017.
15. Han L, Wang X R, He M, et al. Inventory and environmental impact of VOCs emission from the typical anthropogenic sources in Sichuan province[J]. Environmental Science, 2013, 34(12):4535-4542.
16. Wallace L A, Pellizzari E D, Hartwell T D, et al. The influence of personal activities on exposure to volatile organic compounds.[J]. Environmental Research, 1989, 50(1):37.
17. Zhou J, You Y, Bai Z, et al. Health risk assessment of personal inhalation exposure to volatile organic compounds in Tianjin, China.[J]. Science of the Total Environment, 2011, 409(3):452.
18. Lin L, Jun L U, Ying-Ge M A, et al. Reviews of VOCs Control at Home and Abroad[J]. Administration & Technique of Environmental Monitoring, 2011.
19. Wang H L, Zhang G N, Nei L, et al. Study on control and management for industrial volatile organic compounds (VOCs) in China.[J]. Environmental Science, 2011, 32(12):3462.
21. Guo S, Tong L, Zhou X, et al. VOCs Emission Control in Petrochemical Industry[J]. Environmental Protection of Chemical Industry, 2014.
22. Zhang G N, Hao Z P, Jiang M, et al. Study on foreign regulations and standards of stationary sources VOCs emission control.[J]. Environmental Science, 2011, 32(12):3501.
23. Louie P K K, Ho J W K, Tsang R C W, et al. VOCs and OVOCs distribution and control policy implications in Pearl River Delta region, China[J]. Atmospheric Environment, 2013, 76(S1):125-135.
24. Zhang X, Xue Z, Sun X, et al. Review on Control of Ambient Volatile Organic Compounds (VOCs) and Suggestions in China[J]. Environmental Science & Management, 2014.
25. Khan F I, Ghoshal A K. Removal of Volatile Organic Compounds from polluted air[J]. Journal of Loss Prevention in the Process Industries, 2000, 13(6):527-545.
26. Guo H, Ling Z H, Cheng H R, et al. Tropospheric volatile organic compounds in China[J]. Science of the Total Environment, 2017, 574: 1021-1043.

513535 A Look inside a Chemical Plant Waste Water Treatment Unit

Abstract:

There are a wide variety of pieces of equipment that compose waste water treatment inside of chemical plants. In this presentation, I will discuss the process flow of the waste water treatment plant for the Syngenta St. Gabriel chemical plant to help young engineers understand the complexities of such a system and the current commercially available systems we use to treat waste water to meet government regulations. We have a pre-treatment step, carbon adsorption, and then a biological unit that help to treat the water before it can be sent out to the Mississippi River.

Using Connected Systems to Improve Process Safety

Scott M. Wozniak
Honeywell UOP
25 East Algonquin, Des Plaines, Illinois, 60017 USA
Scott.Wozniak@honeywell.com

Ajay Kakkanattu
Honeywell UOP
Unitech Trade Center, Sector 43, Gurgaon 122 002, Haryana, India
Ajay.Kakkanattu@honeywell.com

Prepared for Presentation at
American Institute of Chemical Engineers
2018 Spring Meeting and 14th Global Congress on Process Safety
Orlando, Florida
April 22 – 25, 2018

Using Connected Systems to Improve Process Safety

Scott M. Wozniak
Honeywell UOP
25 East Algonquin, Des Plaines, Illinois, 60017 USA
Scott.Wozniak@honeywell.com

Ajay Kakkanattu
Honeywell UOP

Keywords: Connected Systems, Process Safety, Leading Indicators, API 754

Abstract

A performance indicators program provides useful information for driving improvement and when acted upon, contributes to reducing risks of major hazards by identifying the underlying causes and taking action to prevent recurrence. To address this issue, API published the second edition of API Recommended Practice (RP) 754, Process Safety Performance Indicators for the Refining and Petrochemical Industries. This Recommended Practice (RP) identifies leading and lagging process safety indicators that are useful for driving performance improvement. The current evolution of industry with the introduction of connectivity based solutions to improve the various performance aspects of the Process units has opened many opportunities in the area of process safety. There are a lot of challenges to analyze the Process safety lead and lag indicators in a connected environment platform. This paper discusses the potential for increasing the process safety performance of the Unit with the use of connected systems by recording the data, analyzing data in terms of API RP 754 and bench marking the performance. By having this data from API RP 754 and in real-time makes this a true process safety indicator leading indicators. Leveraging IIoT to drive process safety excellence. Zero is good...

Introduction:

In 2010, the American Petroleum Institute (API) issued "Recommended Practice (RP) 754: Process Safety Performance Indicators for the Refining and Petrochemical Industries" [1]. This standard has gained broad acceptance and use not only by the refining and petrochemical sectors in the United States but also by other sectors of the energy industry and by other process industries in the U.S. and around the globe. Industry regulation related to process safety is at an all-time high. Every operator is committed to safety and risk avoidance. So why do the incidents and accidents still happen? An essential element of any improvement program is the measure of existing and future performance. Therefore, to continuously improve upon process safety performance, it is essential that companies measure these indicators with high priority.

Acquiring a clear and realistic picture of process safety performance in industry is highly important in the current environment. In order to achieve that, proper data capture and thorough analysis is warranted with a higher degree of accuracy than is currently done with manual input. Since many process safety metrics are updated manually, the collection of the data without any systemic delay because this requires a manual step and can be error prone. This paper stresses the importance of analyzing both lagging and leading indicators to enable real-time monitoring of process safety performance.

Connected Systems: The Industry Trend

More and more customers today are looking for highly customized products. At the same time, manufacturers are under pressure to produce efficiently at mass-production costs. One way organizations can meet this challenge is by integrating business and manufacturing systems with automation technology. It can help manufacturers operate more efficiently using a variety of data sources from both operations and enterprise systems. More specifically, companies are able to turn massive amounts of data into action by using advanced analytics to identify bottlenecks, troubleshoot issues, understand asset interdependencies, and reduce costs. Typically, many of the traditional industry management processes are manual and time intensive. To understand the true potential of integrating business and manufacturing systems from the table to industry, it is helpful to review the situation many companies operate in today. Getting manufacturing performance data to the top executives is a slow process, subject to human error and interpretation. Often it begins with plant managers creating spreadsheets from system-generated reports. Executives review these spreadsheets at the end of each week and use the data to create visual dashboards for higher-level management. Unfortunately, complicated graphs and spreadsheets do not give enough actionable items to help until after the fact. But imagine what could be accomplished if a company could use technology to create real-time connections among the automated manufacturing floor and all other business systems, both internal and external. Not only that, but what if the data was instantly converted into visuals, so all employees could easily analyze and understand the information?

With information at their fingertips, managers can identify and resolve issues up and down the supply chain to increase efficiencies, improve profitability, and raise customer satisfaction. In fact, the seamless integration of plant information with business systems presents so many opportunities, companies are just scratching the surface of what is possible.

Figure 1: Connected Plant

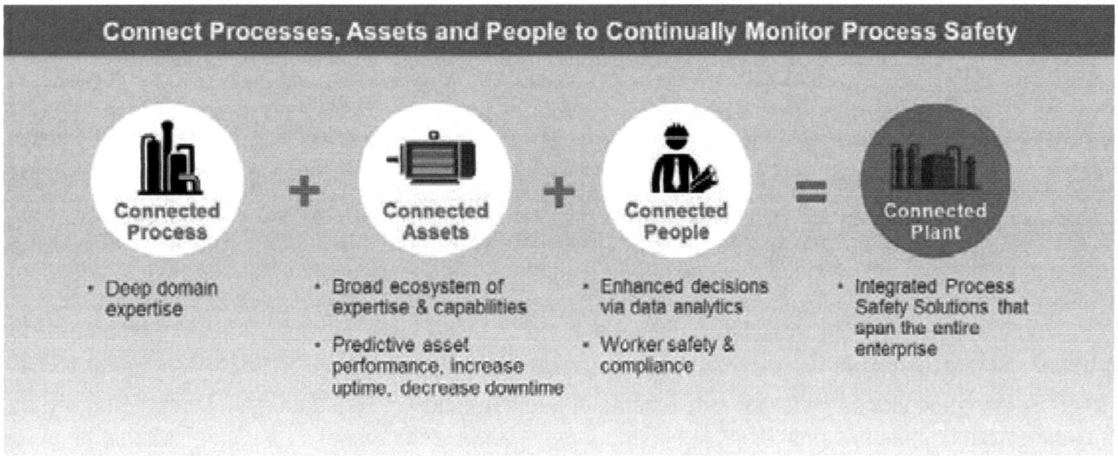

Real-time, connected manufacturers enjoy many benefits not available to companies still working within a more traditional environment, which typically has misaligned departmental silos, manual processes, and disconnected systems. Putting the right data in the hands of decision makers when they need it allows companies to take immediate action in response to changing market conditions. As in the modern industry, the Process safety is analyzed more dynamically as a company performance indicator to win trust of the stake holders. This reiterates the importance digitalizing of process safety to incorporate the dynamic nature of the industry safety performance. Figure-1 illustrates how a highly hazardous industrial environment can be connected with the assets, processes and people. All three elements can be assessed based upon the leading and lagging indicators for it real time performance to analyze safety as outlines in API RP 754 guidelines.

In Need of Connected Process Safety Update?

The connection between process safety and operational excellence is increasingly recognized. The industry is in pursuit of world-class performance that is delivered by enabling everyone within the organization to consistently make the most effective operational decisions, considering a fully integrated picture of the risk, cost and productivity of operational reality. The focus is on major accident hazards and how to eliminate the low-frequency, high-consequence events that can have devastating impacts to assets. Advancing safety performance and risk reduction is almost always a stated goal in company literature and annual reports, evidence from the 2017 survey indicates that companies do not have readily accessible and carefully structured plans in place to achieve that goal. In fact, many believe that companies do not have well-defined safety performance measures, including leading and lagging safety indicators. Although reducing MAH risk is considered a major driver for improving safety performance, companies often fails to always maintain a sense of vulnerability about exposure to MAH risk. It is important to make the process safety as digital and independent as possible to reflect the correct process safety performance of the unit due to the following reasons.

- Process safety is specialized knowledge, not typically understood by operations and maintenance, leading to implementation gaps.
- Production takes priority over safety, which often leads to shortcuts and safety incidents, despite corporate safety policies.
- Corporate lip service to PSM policies are not backed up with effective and efficient planned preventative maintenance.

There is a well-established body of mature engineering science around process safety that defines how to design and operate plants to keep assets and people secure, to keep process fluids in pipes and to stop loss of primary containment, which could escalate into a major accident. However, despite the specifications arising from this body of knowledge, once the facility is built and operating, safety barriers begin to degrade over time. Additionally, many of the process safety metrics are often updated manually and these activity is periodical. A large quantity of data needs to be analyzed and verified to reflect the correct process safety performance of the unit. This also contributes to the degradation of safety performance over time which often goes unidentified.

It is necessary to develop a single, shared view of the operational reality, ensuring that the right information about hazards and risk are available in a format that can be accessed, consumed, understood and used to support decision-making from boardroom to frontline, is essential to reducing exposure to MAH risk.

How Connected Systems Can Measure Process Safety

We can categorize all the performance data in three layers such as Plant Incident data, Real time process data and Operational Performance data. We can deploy process safety indicators for capturing these data in a reliable and continuous way. We need well defined process safety indicators and their tracking to get real Process safety picture of the industry. There are a variety of resources that can help to achieve the required data digitally to update the process safety performance. Since every company has different ways of managing their process safety it is essential to come up with a standardized format by which they can manage the process safety in a better way than currently doing.

Figure 2: Process Safety Performance Update

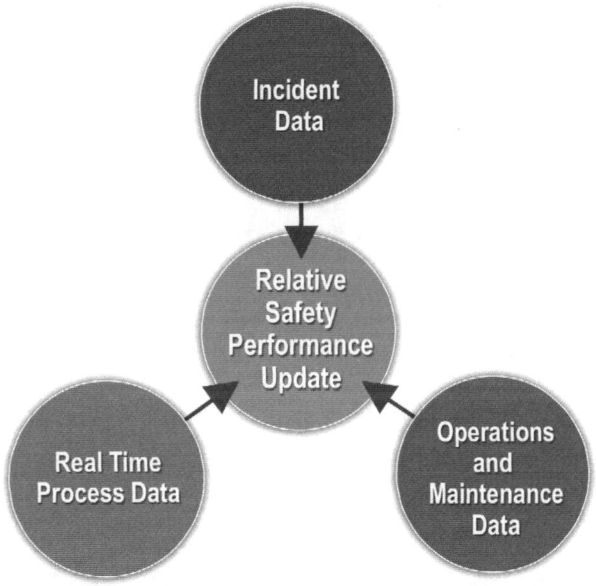

A connected Process Safety Management (PSM) management system can be developed based on PSM related regulatory requirements and best practices across industries. This will help companies to identify the potential risks due to the hazards in the system, enable companies to stay in compliance with regulations, protect the workers, the communities and their business.

Categorizing the Incident data and operation and maintenance data based on the requirements defined by API-754 is a very important step to digitalize the process safety performance. There are many operational management tools available to capture the data digitally which can be used to develop a Continuous Process Safety update. Also use of enterprise management tools which are widely used to manage the operations which will help to generate data for the leading Indicators especially for the process safety elements like MOC, PHA, MIQA, etc.

Process Safety Hierarchy

Most process systems are protected by layers of safeguards sometimes referred to as Levels or Layers of Protection, as shown in Figure 2. They are an integral part of any Process Safety Management program [1] the lowest layer consists of normal operational controls, which are not safeguards even though they control the great majority of deviations.

Figure 3: Process Safety Hierarchy

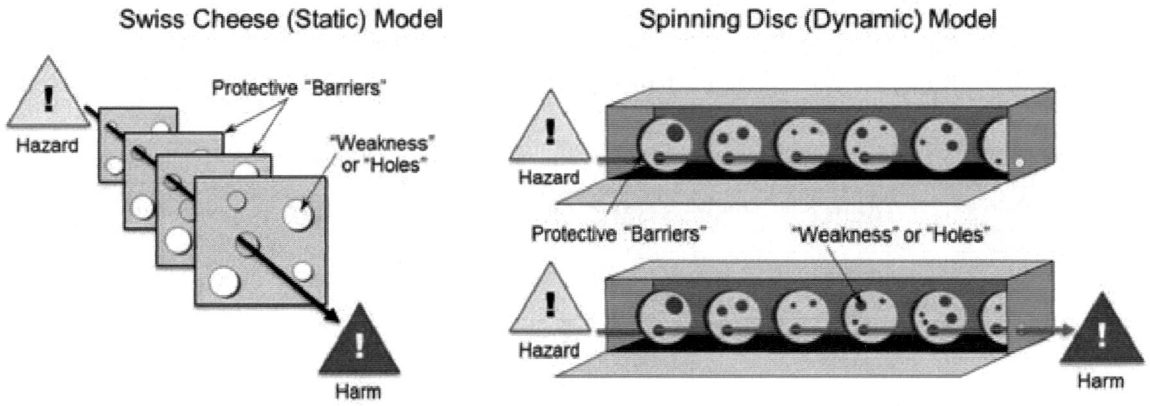

- Hazards are contained by multiple protective barriers
- Barriers may have weaknesses or holes'
- When holes align, the hazard passes through the barriers resulting in the potential for adverse consequences.
- Barriers may be physical engineered containment or behavioral controls dependent on people
- Holes can be latent/incipient, or actively opened by people

Theoretically, any breach of the safety layers can be identified as potential event leading to a safety consequence. But in practice, it is very hard to identify and record each breach of any of the safety layers as it happens real time, which is highly manual intensive and less reliable. Generally events associated with no visible consequences, like near misses, do often get overlooked in the industry and not investigated with the same importance as it should be. This makes the some of the safety layers weakened and the probability of a safety layer breach simultaneously increases. With the periodic assessment of many of these manually updated matrices may not provide the actual picture of the safety performance.

The potential consequences occurred by weakened preventive and mitigation safe guards is demonstrated in figure 3. It is to be noted that every breach of the safe guards may not land up in the potential worst case consequences. But it is wise idea to identify the safeguards and their performance by their merit in real time to make better decision to protect the assets and human life by preventing the high consequence events.

Figure 4: Process Safety Metrics: Tier 1/2/3/4 Indicators

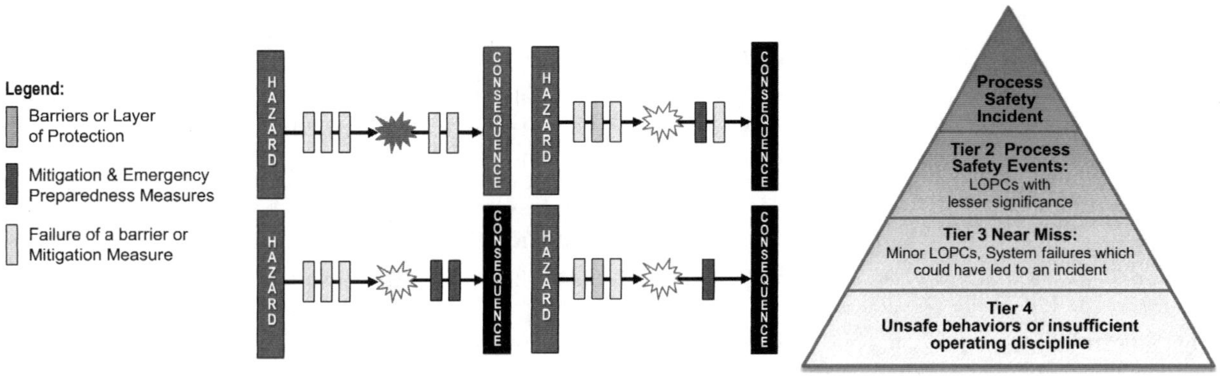

An essential element of any improvement program is the measure of existing and future performance. Therefore, to continuously improve upon process safety performance, it is essential that companies in the chemical and petroleum industries implement effective leading and lagging process safety metrics.

This paper describes the recommendations assembled by the Center for Chemical Process Safety (CCPS) Process Safety Metric committee for a common set of company and industry leading and lagging metrics.

Within this document is a description of three types of metrics:

"Lagging" Metrics – a retrospective set of metrics that are based on incidents that meet the threshold of severity that should be reported as part of the industry-wide process safety metric

"Leading" Metrics – a forward looking set of metrics which indicate the performance of the key work processes, operating discipline, or layers of protection that prevent incidents

"Near Miss" and other internal Lagging Metrics – the description of less severe incidents (i.e., below the threshold for inclusion in the industry lagging metric), or unsafe conditions which activated one or more layers of protection. Although these events are actual events (i.e., a "lagging" metric), they are generally considered to be a good indicator of conditions which could ultimately lead to a more severe incident.

These three types of metrics can be considered as measurements at different levels of the "safety pyramid" illustrated in Figure 1. Although Figure 1 is divided into four separate layers (Process safety incidents, other incidents, Near miss, and Unsafe behaviors/Insufficient operating discipline), it is easier to describe metrics in terms of the categories shown above.

Digitalizing the Performance Indicators:

The Performance indicators demonstrate the actual picture of the plant's safety performance. Digitalizing the key indicators will help the companies to illustrate the process safety performance real time thus helping them make sound decisions to prevent any major hazard incidents. Digitalizing the process safety indicators can be done in different ways depending upon the capability of the facility owner to produce accurate data in the digital platform. The most advanced method we can envisage is the cloud based approach where all the live data is captured and analyzed in a cloud based platform and calculated to yield a process safety score. Another method can be data capture and analysis in the site based servers where the details are analyzed for yielding a process safety performance update.

Digitalization for process involves the integrated engineering and operation of process plants across the whole plant life-cycle, as well as simulation through the use of digital twins and utilizing big data to inform better strategic decision-making. This method can considerably help address the process safety key indicators evaluation requirements, along with training programs for operation and maintenance. In order to produce a clear picture, it is highly recommended to take inputs from various management systems and systematically examine these input data to produce a meaningful output.

Figure 5: Digital Picture of Process Safety

The questions often arises are "which type of data we would need to be updated for a real time process safety picture"? And "how those data can be collected"? The adoption of process safety performance indicators have solved the first part of that question, the second question poses the bigger challenge. Proper identification and selection of the leading indicators can resolve the issue. These data can be collected from various sources such as DCS, Operation and maintenance management tools etc.

As a huge volume of data is collected from different sources it is important to analyze, compare these data to produce a more meaningful input to the process safety update. For example a DCS system can generate data for breach of a safety layer resulted demanded an action of Safety instrumented System (SIS). Similarly a separate input from maintenance management system can give you the test frequency and design and manufacturing details of the same SIS system. Combining these two data and analyzing them together in the digital platform will give you a clearer picture of the status of the SIS system. This activity would otherwise consume time and effort to get to the similar conclusion without the digital platforms.

Figure 6: Connected Plant – Architecture

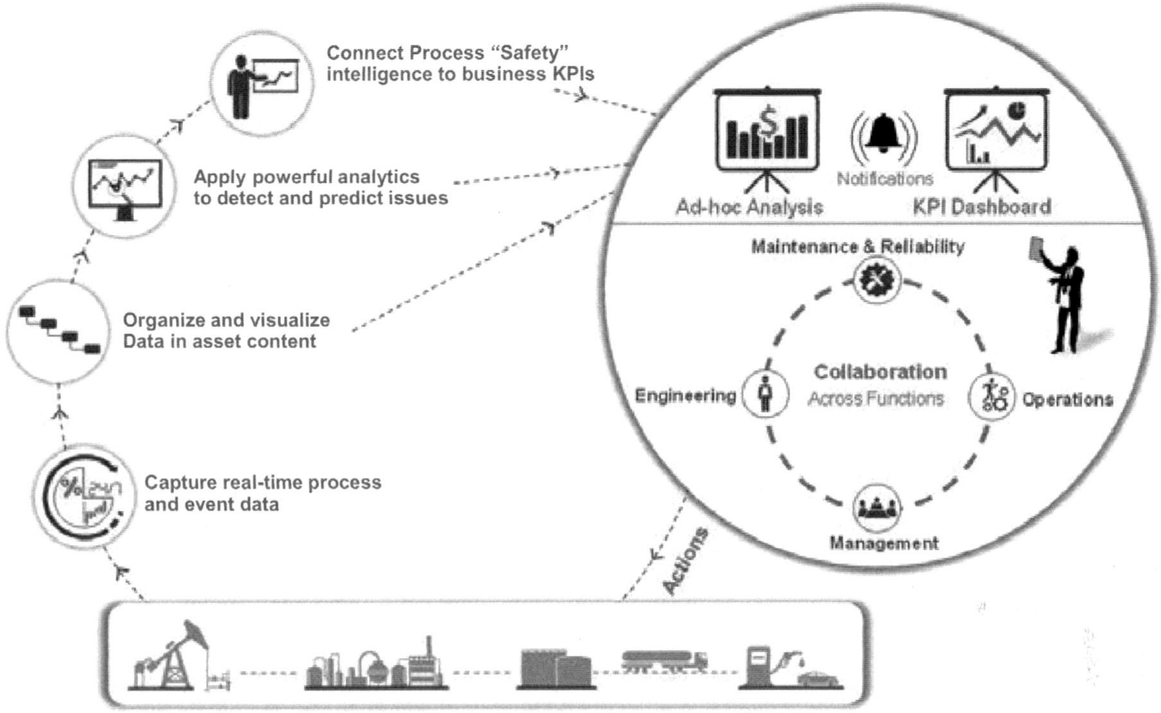

Challenges of Digitalizing the Process Safety Indicators:

Digitalizing the process safety indicators is an exciting idea to get the process safety performance in the newest of industrial trends. Since the novel thought process is to get all the process safety data in the digital platform and analyze them on its merit to evaluate the safety performance of the unit, then we need to overcome some constraints to convert some manual data into digital platform. It is comparatively easier to capture the process data relating to the process safety performance from the DCS or the plant data historian with the current technology available in the market even though some data capture may require additional detection and processing mechanism which are not generally followed in the industry. This more tends to be a leading indicator based evaluation for the plant performance. In order to capture the Tier 1 and Tier 2 Indicator data, extensive use the Operational Management tools in the digital space are required. In the current industrial environment these type of Operational management tools are not uncommon, but it might be difficult to find the similar operational platform used industry wide. The major challenge is to establish an operational management data base tool with sufficient amount of information to get updated data with minimum possible lag to capture data to evaluate the process safety performance.

Another challenge would be reducing the manual intervention in the data capture in real time. As we discussed above it will be near impossible to evaluate the consequences of a potential Tier 1 and 2 LOPC event as it happened, the digital platform may require a manual intervention to upkeep the evaluation. Since Tier 1 LOPC can make a serious impact on the process safety performance card, it is quite possible that the event may get under reported. It may be worthwhile for the process safety evaluation team to experiment a slab system where a user can select the pre-defined consequences to analyze the process safety impact.

The Tier 3 indicators/ events are the leading indicators with which the safety performance can be realistically estimated in a digital platform. Many of the indicators are easily traceable and the quantification of their activations are easily measurable. For example, the activation of Safety interlocks and safety shut down systems are easier to track as it is digitally logged in the existing control systems like DCS. These details can be captured into a platform outside of the DCS environment for the further analysis. Existing operational management tools are capable of such digital identification of the information to further process it. However digitally capturing some of the Tier 3 indicators are found to be challenge as in the case of Pressure Relief Systems activation and releases. The pressure relief valve activation is considered to be Tier 3 Indicator event, but the same is not digitally recorded in the system as in the case of SIS activation. This poses a challenge to the safety management system to identify these events and predict the safety performance accurately based on the actual performance of the unit. Another of such challenge is posed by the activation of mechanical trip systems.

In order to develop a comprehensive safety performance analysis mechanism, proper identification and live analysis of all events are important. The computation of all these details can fetch us a more reliable safety performance score for the unit. The measurability of all the events shall be increased to achieve more reliability in the safety score. The installation of external digital flow meters at the safety critical pressure relief devises can be a good idea to increase the measurability of the potential releases. In an industry point of these flow meters can be redundant in nature in most of the operational phase of the unit, but the data from the same can be digitally registered in the safety management system as a Tier 3 event. This gives the system an independent nature to analyze the safety removing the operator dependency to log the malfunction details and the safety score will become more reliable and active.

For example a utility failure which results in a Refinery or Petrochemical complex may result in many of Tier-3 events. But this may be captured as a single safe operating limit (SOL) excursion by the operator instead of capturing all the excursions resulted due to the same initiating cause. It may be worthwhile to mention that by automating the Process Safety matrix capturing all these SOL's can be captured as Tier 3 events and record and analyze the data with more accuracy.

Digitalizing Tier 4 indicators poses a higher degree of challenge given the sheer number of occurrences and the difficulty of all the potential Tier 4 events get reported and recorded in much meaningful manner. The owner may well have scattered mechanisms to capture many of the Tier 4 indicators and several of them may be done in a manual mode. However compliance check with the Tier-4 indicators are difficult as a large amount of qualitative information attached with those. Selection of right type of indicators which possesses high attributes of detectability, measurability of data with accuracy is greatly recommended for establishing the Tier 4 indicators for a company. For example, training and competency is an important indicator and the details of training and competency is maintained, or expected to be maintained, in all hazardous process industries. But it is extremely hard to avail the record in the digital back ground for this indicator in many part of the world. Defining a minimum required competency for each role and identify the trainings requirements against the target manning requirement will establish the basic frame work. A manual interaction through a digitally proven enterprise training management system can initiate the tracking activity in terms of planning, scheduling and assigning the appropriate training requirements. Digitally tracking the records against the training plan already keyed in will keep the data live and comparable with other standard tier-3 process indicators. This would keep everybody ready and up to date to complete the activities on the stipulated time period removing the potential lag associated with the otherwise periodically tracked mechanisms. Also the indicator will have a strong visibility in front of higher management, since it can affect the overall process safety score in more active manner.

Conclusion:

We have shown that a performance indicators program provides useful information for driving improvement and when acted upon, contributes to reducing risks of major hazards by identifying the underlying causes and taking action to prevent recurrence. API Recommended Practice (RP) 754, Process Safety Performance Indicators for the Refining and Petrochemical Industries is the gold standard when identifying when a process safety incident has occurred (i.e. Tier 1/2/3/4). We have shown how a connected system can improve the safety performance by establishing a framework linking many different elements together with the objective of managing process hazards to acceptable levels. The process safety lead and lag indicator program is a highly effective program and operators of highly hazardous facilities worldwide do follow this program rigorously to achieve and maintain the high performance in Process Safety. It is important that the process safety performance indicators needs to be updated actively with accurate data on a continuous basis. Digitalizing the process safety indicators program can utilize good quality data to produce a much reliable process safety performance analysis with minimum or less frequent human inputs.

REFERENCES

1. "ANSI/API Recommended Practice 754: Process Safety Performance Indicators for the Refining and Petrochemical Industries," 1st ed., Amer. Petrol. Inst., Washington, D.C. (2010).

2. "ANSI/API Recommended Practice 754: Process Safety Performance Indicators for the Refining and Petrochemical Industries," 2nd ed., Amer. Petrol. Inst., Washington, D.C. (2016).

3. "Process Safety Leading and Lagging Metrics …You Don't Improve What You Don't Measure," Ctr. for Chem. Proc. Safety, Amer. Inst. of Chem. Engrs, New York City (2011).

4. Process Safety: Understand the Changes in API RP 754

   The second edition of this standard on process safety performance indicators contains some important revisions. By Kelly K. Keim, ExxonMobil Research and Engineering, on behalf of the American Petroleum Institute (Mar 08, 2017)

5. White paper "What Are the Benefits of Connected Manufacturing?" By David Parrish, global senior director of industrial machinery and components marketing for SAP, Published in ISA Interchange

6. Safety: Process safety and risk management— An industry view- By Jones, S. - Petrotechnics, Aberdeen, UK, Published in Hydrocarbon Processing (November 2017)

18 AIChE Spring Meeting
14TH GLOBAL CONGRESS ON PROCESS SAFETY
ORLANDO FLORIDA 2018

Technology Impact on Process Safety through the Cloud

Annant Srivastava
Siemens
4615 Southwest Freeway Suite 900, Houston 77027 TX
annant.srivastava@siemens.com

Anil Patel
Siemens
4615 Southwest Freeway Suite 900, Houston 77027 TX
anilkumar.patel@siemens.com

Prepared for Presentation at
American Institute of Chemical Engineers
2018 Spring Meeting and 14th Global Congress on Process Safety
Orlando, Florida
April 22 – 25, 2018

Technology Impact on Process Safety through the Cloud

Annant Srivastava
Siemens
4615 Southwest Freeway Suite 900, Houston 77027 TX
annant.srivastava@siemens.com

Anil Patel
Siemens
4615 Southwest Freeway Suite 900, Houston 77027 TX
anilkumar.patel@siemens.com

Keywords: cloud computing, technology impact, Internet of Things, IoT, process safety management, mechanical integrity, pressure relief analysis, management of change

Abstract

For oil and gas operators, process safety continues to be a major concern and whenever encountered with challenges, the solutions demanded prompt and well defined process and dedicated resource pool to mitigate the risk to their people and their assets. Over the past decade, advancements in integrated technology have flowed at an accelerated pace. The emerging cloud technological advancements provide oil and gas industry with solutions to some of its most critical processes.

Database driven asset management systems, pressure relief analysis and a well-defined management of change process in a holistic environment with access to all the right stakeholders, have contributed in reducing the risk management challenges for the industry. This paper will discuss the technology trends in the area of asset integrity, pressure relief analysis and management of change that facilitates to utilize data pertaining to process safety management (PSM) elements in a more meaningful way. The paper will present the process life cycle around the cloud technologies and will briefly talk about the positive impacts with some case studies

1 Technology Trends

The oil and gas industry is operating in an environment of opportunity coupled with high risk and diminishing return on investments.[4] The current conventional way of managing assets and information in traditional client server local infrastructure not only presents risks and inefficiencies but also leads to complex work processes in the traditional way of accessing information also exists in major part of process safety area.[6] Moreover the continued downward trend of oil prices has presented an urgent need of information technology transformation within this industry. [5]

The technological advancements in the area of cloud computing and Internet of Things have the potential to improve business functions and work processes within the oil and gas industry.[9]

1.1 Cloud Technology

Cloud computing provides seamless uninterrupted availability of data across organizations in a most robust and secured platforms.[6] Cloud enables companies to run high performance servers and allows multiple applications to run leveraging the vast infrastructure that can grow on demand without disrupting the ongoing operations in the plant. Cloud Computing is "the practice of using a network of remote servers hosted on the Internet to store, manage, and process data, rather than a local server or a personal computer". [7]

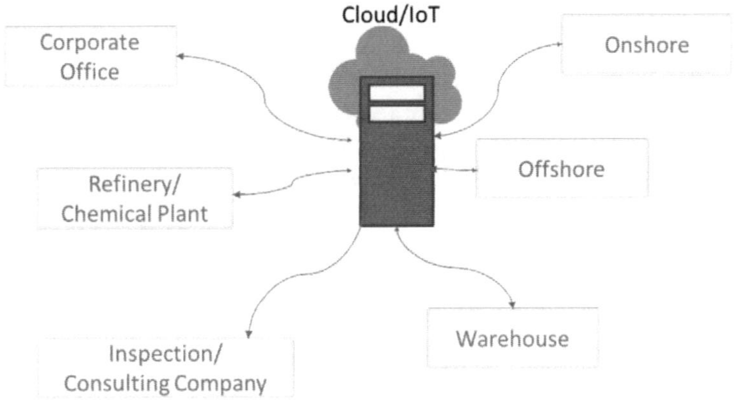

The data can be exchanged securely in encrypted format from any remote locations to central servers residing in an entirely different geographical location.

The data related to process safety is critical and is sensitive in nature, cloud computing coupled with highly available secured networks are more reliable and safer than the on premises data centers where the central servers are more vulnerable to attacks. The data stored in cloud resides in highly secured infrastructure completely unknown to end users.

Cloud computing is offered in variety of ways, that can be publicly or privately served depending upon the nature of the business the organization is involved. Software as a service model is best suited in process safety area where the companies enable their employees and their customers to view the data real time and mitigate risks and make decisions in a timely manner.

1.2 Internet of Things

Internet of Things is the interconnection via the Internet of computing devices embedded in everyday objects, enabling them to send and receive data.[8] The Internet of Things (IoT) integrates sensing, communications, and analytics capabilities enabling companies to directly manage their existing assets and keeping their critical information live. The IoT is a suite of technologies and applications that equip devices and locations to generate all kinds of information—and to connect those devices and locations for instant data analysis and, ideally, "smart" action. Conceptually, the IoT implies physical objects being able to utilize the Internet backbone to communicate data about their condition, position, or other attributes.[11] The businesses can deploy an IoT solution using sensors in different locations to generate simulated data. This data is then moved to a cloud-based analytics platform for high availability and accessibility. For an efficient flow of data exchange, IoT provides wide arrays of wireless sensing devices that can connect to assets namely pumps, motors, compressors, turbines etc. The data can then be monitored remotely real time enabling companies to establish an efficient process safety work processes.

2 *Process Safety Life Cycle*

In the petrochemical industry, several applications like asset integrity management system, inspection data management systems, relief system design and management of change run simultaneously collecting and managing data at various stages within the asset lifecycle management. [3]

Using cloud technology for Oil and gas, data management also makes it easy for the project manager to share pertinent information with all project stakeholders. Additionally, data can be easily cross analyzed and interpreted, leading to decisions being made and conveyed faster, thus reducing adverse impacts to the community, company, and environment. This efficiency also translates into minimization of costs associated with delays, slowdowns, catastrophes, or other unplanned interruptions. [2]

The cloud computing enables organizations to centrally operate these applications without having to install them on individual computers.

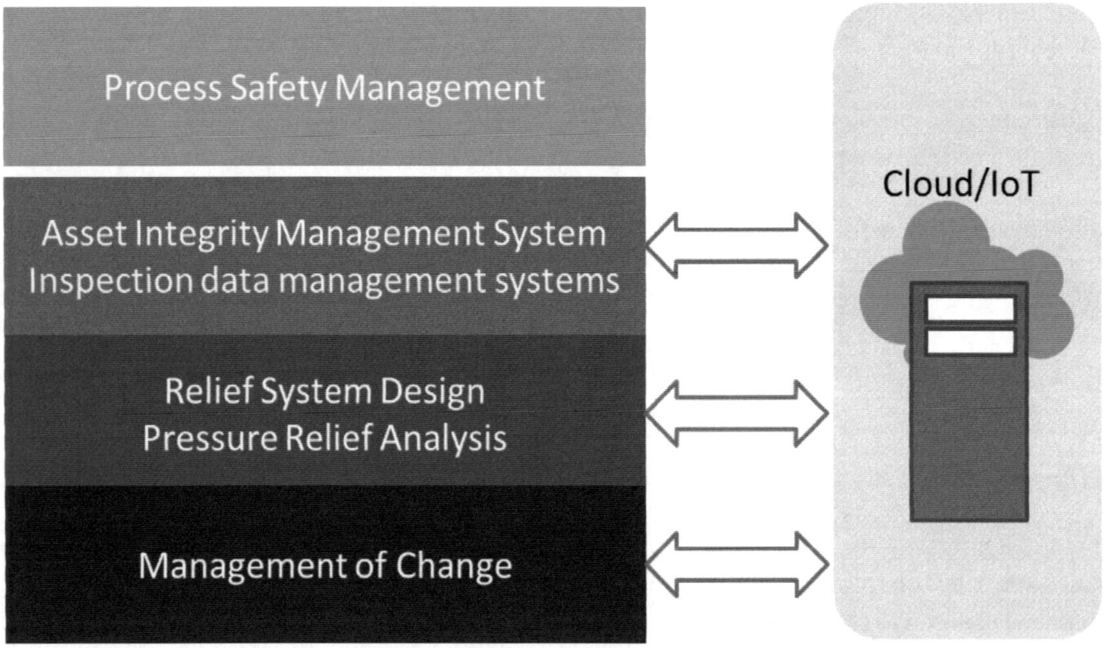

In the process safety element of mechanical integrity, having a centralized, robust and on demand scalable asset tracking system plays a vital role. Ability to analyze the thickness data, minimum thickness & corrosion rate calculations in a holistic environment using reliable and robust infrastructure over cloud greatly enhances the work processes within the process safety lifecycle.

As part of mechanical integrity and asset reliability work process for safety critical equipment, owner operators need to schedule the inspection activities for an asset and track it through completion and synchronize it with ERP systems to timely close the work orders. Data availability for the critical equipment whenever demanded enables stakeholders to proactively take decisions and track the status of the inspection activities from a centralized database management system over cloud. As cloud computing has inherent capability of scaling its infrastructure on demand, stakeholders can store complete inspection documentation and history and process safety data of the entire facility.[1]

Similarly, in the area of relief system design, seamless accessibility of comprehensive safety reporting of all relief and disposal system devices over secured cloud infrastructure enables stakeholders to validate and keep the relief system design documentation evergreen throughout the facility lifecycle.

An effective management of change process is critical to process safety program in order for the owner operator to anticipate changes from the outcome of inspection results and/or relief system design study to accommodate the necessary changes in their process. A highly available and shared management of change system over cloud with these processes enables owner operators to effectively manage the documentation and review of mandatory workflow steps giving the stakeholders ability to sign off MOCs and make effective decisions leading to change.

3 Case Study

3.1 Company Profile

Business in Process Safety Consulting offers pioneering methodology and assisting customers with design, rating, documentation and management of pressure relief and vapor/liquid disposal systems barriers for Oil and Gas Production, Refining &Petrochemicals.

3.2 Challenge

Large amount of data that was generated in silos and data was archived with lack of contextual information and inefficient data discovery.

Growing data demanded a scalable infrastructure with minimum downtime.

Process Safety consultants wanted to perform comprehensive pressure relief analysis and ensure process equipment has adequate overpressure protection in a holistic environment that can help seamless exchange of data across geographically dispersed team.

3.3 Solution

The process safety applications and process safety data information were seamlessly migrated to Cloud in an encrypted and secured platform.

Using the latest cloud technologies, users were accessing the process safety applications along with process safety data through web browser in a secured network.

The cloudwatch feature of the selected cloud vendor continuously monitored systems and servers for performance and availability measurements.[10]

3.4 Benefit

The entire team of process safety consultants from different locations is now working in a centralized environment that is scalable on demand thereby minimizing the downtime required to scale up the infrastructure in the past.

The flare study was conducted as the pressure relief analysis was completed by process safety consultants in a virtual private cloud environment. This helped team achieve a quick turnaround time.

4 Conclusions

The pace with which process safety area adapts to cloud computing will ensure the overall change in the work processes thereby enabling a much safer environment by eliminating much of the hassle associated with using an IT network, unplanned breakdowns and damages associated with it.

Much like other verticals in the oil and gas industry, process safety is striving hard to keep the cost controlled in such a volatile phase. Advancements in technologies like Cloud computing and IoT are helping these organizations to improve efficiencies and save cost. These digital disruptions in the long term can become a turning point for process safety industry in how they assess and manage risks in the overall safety life cycle.

References

[1] Guidelines for Mechanical Integrity Systems CCPS, pp. 183-202, 2006.

Web resources:

[2] How Cloud Computing is Revolutionizing Oil and Gas Projects, Available at http://www.landpoint.net/how-cloud-computing-is-revolutionizing-the-oil-and-gas-industry/ Accessed on Feb 12, 2018.

[3] Occupational Safety and Health Standards https://www.osha.gov/pls/oshaweb/owadisp.show_document?p_table=STANDARDS&p_id=9760_ Accessed on Feb 12, 2018.

[4] Oil Prices in Crisis: Considerations and Implications for the Oil and Gas Industry https://www2.deloitte.com/content/dam/Deloitte/us/Documents/energy-resources/us-oil-prices-in-crisis-considerations-and-implications-for-the-oil-and-gas-industry-02042015.pdf

[5] The digital transformation in upstream oil and gas https://www2.deloitte.com/insights/us/en/industry/oil-and-gas/digital-transformation-upstream-oil-and-gas.html

[6] The Great IT Myth: Is Cloud Really Less Secure Than On-Premise? Ben Rossi, Information Age
http://www.information-age.com/great-it-myth-cloud-really-less-secure-premise-123459135/

[7] Cloud Computing
https://www.google.com/#q=cloud+computing

[8] Internet of things
https://www.google.com/search?q=internet+of+things&rlz=1C1GGRV_enUS786US786&oq=internet&aqs=chrome.0.69i59j69i57j69i61l3.1930j0j7&sourceid=chrome&ie=UTF-8

[9] The Next Big Innovation In Oil & Gas: Cloud Computing
https://oilprice.com/Energy/Energy-General/The-Next-Big-Innovation-In-Oil-Gas-Cloud-Computing.html

[10] Amazon CloudWatch
https://aws.amazon.com/cloudwatch/

[11] Smart buildings: How IoT technology aims to add value for real estate companies, Deloitte University Press, April 19, 2016.
https://www2.deloitte.com/insights/us/en/focus/internet-of-things/iot-commercial-real-estate-intelligent-building-systems.html

Siemens Energy, Inc.
Process Safety Consulting
4615 Southwest Freeway, Suite 900
Houston, TX 77027 USA
www.siemens.com/energy/oilandgasconsulting

The information contained in this paper represents the current view of the authors at the time of publication. Process safety management is complex and this document cannot embody all possible scenarios or solutions related to compliance. This document contains examples for illustration and is for informational purposes only. Siemens makes no warranties, express or implied, in this paper.

Data Mining and Analysis for Alarm Management

Pankaj Goel[a, b], Aniruddha Datta[a] and M. Sam Mannan[b]
[a] Department of Electrical and Computer Engineering,
[b] Mary Kay O'Connor Process Safety Center, Artie McFerrin Department of Chemical Engineering
Texas A&M University, College Station, Texas 77843, USA
pankaj.goel@tamu.edu, datta@ece.tamu.edu, mannan@tamu.edu

Prepared for Presentation at
American Institute of Chemical Engineers
2018 Spring Meeting and 14th Global Congress on Process Safety
Orlando, Florida
April 22 – 25, 2018

Data Mining and Analysis for Alarm Management

Pankaj Goel[a,b], Aniruddha Datta[a] and M. Sam Mannan[b]
[a] Department of Electrical and Computer Engineering,
[b] Mary Kay O'Connor Process Safety Center, Artie McFerrin Department of Chemical Engineering
Texas A&M University, College Station, Texas 77843, USA

Keywords: alarm management, data mining and analytics, alarm flooding, safety

Abstract

With the advancement in automation technology and connected devices, modern industrial systems routinely have thousands of sensors deployed in the field that are connected to control system units. In addition, most of these sensors are configured as alarms in the control system. This has resulted in a higher number of alarms, poor system performance, additional workload on operators, and in some cases has led to abnormal situations. These situations can further escalate to a catastrophic incident, if not managed properly. Although there are several solutions available to address these challenges related to the alarms and their management, most of them are complex, costly and time-consuming. To overcome these challenges, we propose an alarm management framework which is relatively simple and efficient, and involves four distinct levels - *design, rationalized, advanced, and intelligent*. This paper provides the details of a generic method developed with the use of data mining and analytics including pre-processing, similarity analysis, and correlation mapping. The method benchmarks the system for the defined Key Performance Indicators (KPI's), identifies the bad actors, provides insights about the alarm system which can be used as an input to the rationalization process to manage the alarm flooding, and hence improve the overall efficiency of the system. An example historical alarm and event log dataset is used as a case study to demonstrate the proposed generic method, assess, benchmark the alarm system, provide details about the bad actors and assist in the overall decision support. From the example, it is clear that the new method results in reliable decision making, contributing to better alarm management and safer operations.

1 Introduction

Alarm systems play a crucial role in operating a process in the safe region and serve as a layer of protection in preventing the escalation of an abnormal situation. *"An alarm is an audible and/or visible means of indicating to the operator an equipment malfunction, process deviation, or abnormal condition requiring a response" [1]*. Alarm systems acts as an early detection indicator that alerts the operator to an abnormal process condition [2, 3] or malfunction of an equipment in service. The process of designing, implementing, understanding and operating systems of alarms is known as alarm management. According to the International Society of Automation [1] (ISA), *"Alarm management is the set of processes that ensures an effective alarm system."* With the advancement in automation technology and connected devices, modern industrial systems routinely have thousands of sensors deployed in the field that are connected to control system units. Additionally, due to the ease in configuration techniques available at the software level, these sensors can be configured as alarms in the control system. This has resulted in a higher number of alarms, poor system performance, additional workload on operators, and in some cases abnormal situations. These situations can further escalate to catastrophic incidents, if not managed properly [4, 5]. Although there are several solutions available to address these challenges related to the alarms and their management, most of them are complex, costly, and time-consuming. To overcome these challenges, the authors propose an alarm management framework which is relatively simple and efficient, and involves four distinct levels - *design, rationalize, advance*, and *intelligent* as described in Section 2.

A generic method developed with the use of data mining and analytics theory including pre-processing, similarity analysis, and correlation mapping is explained in Section 3. The method benchmarks the system for the defined KPI's, identifies the bad actors, provides insights about the alarm system which can be used as an input to the rationalization process to manage the alarm flooding, and hence improves the overall efficiency of the system. An example of anonymized historical alarm and event log dataset is used in Section 4 to demonstrate the effectiveness of the proposed generic method, assess, benchmark the alarm system, and identify the bad actors and their details. From the results of the case study example, it is clear that the new offline method results in reliable decision making, contributing to better alarm management and safer operations.

2 Alarm management framework

One of the biggest challenges in alarm management for industrial facilities is alarm floods. *"Alarm flooding is a condition during which the alarm rate is greater than what the operator can effectively manage (e.g. more than 10 alarms per 10 minutes)"* [1, 6]. Alarm floods result in abnormal situations that may lead to a process upset and finally a potential catastrophic incident. Figure 1 shows the evolution of alarm management and some of the major incidents related to alarm management issues as one of the root cause. There have been several standards and guidelines developed in the past few years which assist in the designing and management of alarm systems. It is pertinent to note that these documents provide a set of guidelines or instructions to the user on the design of alarm systems with certain KPI's against which all alarm systems are required to be benchmarked. However, these documents do not provide any detailed technical methodologies or steps to be followed to achieve these requirements.
To address the issues of alarm management, the authors propose an integrated alarm management framework as shown in Figure 2, which consists of four stages: Design, Rationalize, Advance (alarming techniques) and Intelligent alarm systems.

The design stage is the most critical part of an effective alarm management. During the design stage, it is recommended that users design their alarm systems with the help of available standards, guidelines and best practices used in the industry, and their own organization. It includes developing the philosophy document and the design requirements. While designing such systems some of the important characteristics that should be built into the design are [3]:

a. Every designed alarm should require an operator action,
b. Priority selection for the alarms according to the rule of 85/15/5 for (Low/Medium/High) priorities
c. Uniqueness
d. Timeliness
e. Suitable description to understand the alarm
f. The relevance for the operational value

The rationalize part includes tasks of alarm classification, prioritization, rationalization and documentation [7]. The rationalization process includes inputs such as the alarm philosophy document and a list of all potential alarms. The rationalization step provides the result in the form of a master alarm database with design requirements. There may be a significant reduction in the number of configured alarms and nuisance alarms after this process is completed correctly. The rationalization process can be implemented both during the design of a new system or on an existing running system.

Advanced alarming techniques are required to manage the alarm floods in case of process changeover conditions such as start-up and shut-down. In such situations, suppression techniques are used [8], which work on the principle of hiding or masking the alarms which are not relevant to the operator after a particular process event. These techniques include: alarm shelving (manual suppression) initiated by the operator, and designed suppression (automatic suppression) based on operating states or plant conditions and out of service alarms.

With advances in digitization, more data is being collected and stored daily by the operating companies. The use of advanced analytics and expert knowledge in these cases will lead to more informed decisions and business intelligence. In the case of an intelligent alarm management, the use of advanced analytic tools for both offline and online analysis will provide a platform to operate the systems at optimized levels. A method is developed and explained for offline analysis for alarm and event logs and explained in the subsequent sections.

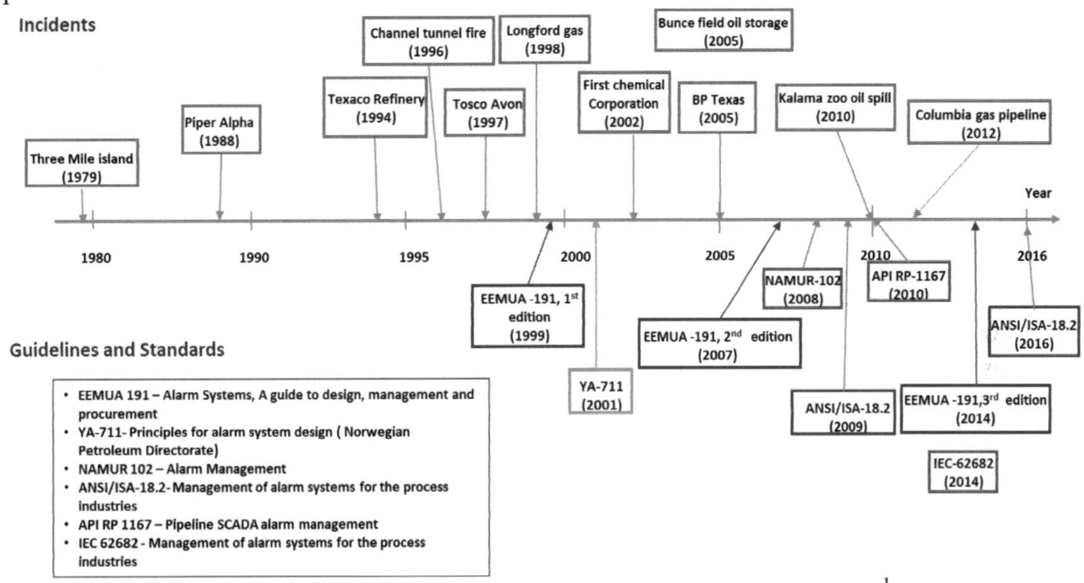

Figure 1: Advancements in alarm management [3] [1]

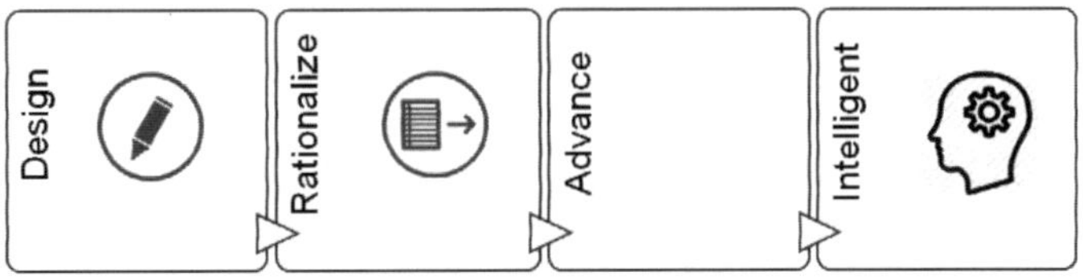

Figure 2: Alarm management framework

3 Methodology

In process plants, the data is generated continuously by different sources which include process variable readings, production data, alarm and event log data, machine system monitoring, system fault records and more[9]. The alarm and event log data captured can provide very relevant and useful information regarding bad actors and flood sequences that can be used for better alarm management. A bad actor is defined as "*an alarm that is suspect and cannot be relied upon to deliver accurate information to the operator, such as stale, chattering, duplicate or suppressed alarms* " [10], To demonstrate this we have developed a methodology with anonymized alarm and event log and process operation knowledge using

[1] *Reprinted from Journal of Loss Prevention in the Process Industries, 50, Goel, P., Datta, A., & Mannan, M. S. (2017), Industrial alarm systems: Challenges and opportunities, 23-36, Copyright (2017) with permission from Elsevier.*

'Python' (an open source language) [11]. This method includes two steps: (1) Pre-processing of the dataset, (2) Analysis and evaluation. The flow diagram of the complete process is illustrated in Figure 3.

For the purpose of this study, an alarm event (X_i) is defined as $X_i(t) = 0, if\ O\ (t) \in Op\ or\ 1, otherwise.$ Here, the alarm is inactive whenever the value of the operating condition O(t) lies within the normal operating range (Op) and active in the case of a deviation. Clearly, $X_i(t) = 0$ and $X_i(t)=1$ correspond to inactive and active alarms respectively.

A flood event, in this case, is defined as; $A= [a_1, a_{2,...,} a_N]$, where the symbol *[.]* indicates a sequence, N is the total number of alarms that occurred in an alarm sequence A, and a_i is the alarm tag occurring in a chronological order with a time stamp *(t)*. a_i can be represented by a tuple with two attributes $a_i=(e_i, t_i)$, such that e_i is the alarm tag and t_i is the time stamp of the alarm that occurred.

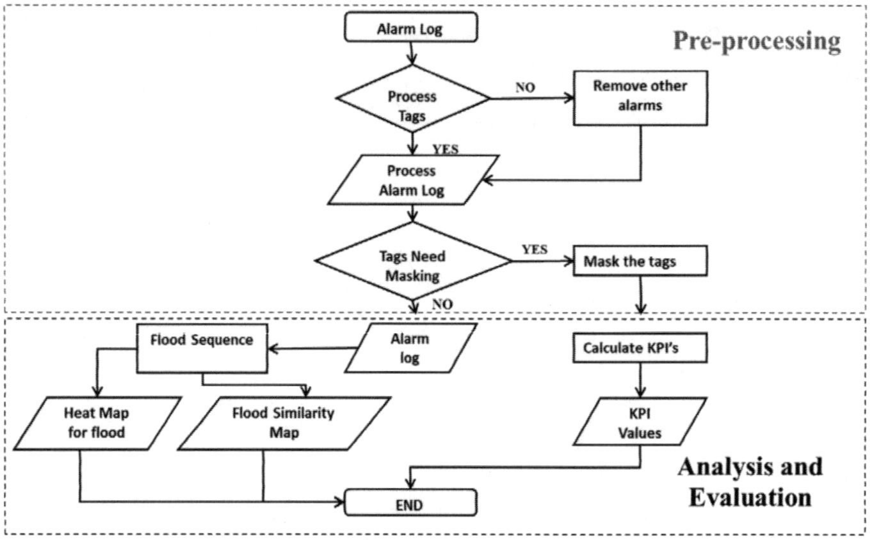

Figure 3: Flowchart of the methodology

Algorithm: Analysis of alarm flood patterns in alarm and event logs
Input: Alarm and event messages logged in Distributed Control System (DCS) extracted in the form of an excel file from the Asset Management System (AMS) system
Output: Mined KPI's, Visual plots identifying floods, similarity correlation maps between flood sequences

Read $A= [a_1, a_2, \ldots a_N]$ where $a_n=(e_n, t_n)$, $n=1,2\ldots N$ $\forall$ $e_n \in$ alarm type, t_n is time stamp
Remove a_n if $a_n \in$ 'SYS' or 'Non- Process tag'
Do $e_n \leftarrow m$ where $m=1, 2, 3, \ldots M$ (*)
Count $\{e_n \mid t_n \in (0, 24)\}$
Identify $(e_n, t_n) \mid \{count(e_n \mid t) > 10\}$ $\forall$ $t \in (t_n, (t_n+10))$
Count Priority (P); where for Priority $\exists$ P: (low, medium, high, and emergency)
Count (a_n) $\forall$ MSGTYPE $\neq$ 'RTN' $\mid e_n$
Count $(a_n \mid t)$ $\forall$ $t \in (t_n, (t_n+1))$
Generate heat-maps for top 45 alarm tags each day
Generate correlation map using 'Gestalt pattern matching' to identify similar alarm flood sequences
Return heat-maps and correlation maps for the given alarm floods

'Gestalt pattern matching' is used decide how similar two one-dimensional patterns are. This algorithm returns a value that can be used as a confidence factor, or percentage, showing how alike any two strings are [12].

When an alarm and event log is captured from a control system, it includes alarms other than process variables, such as system alarms, hardware alarms *etc.* It is necessary to remove such alarms to obtain the appropriate picture of the system. Hence, these alarms are removed as a part of the pre-processing process of the database. To make data anonymized, the tag e_i is converted into a numerical form such that all unique instrument tags are mapped to a unique numerical symbol.

Once the pre-processing is complete, an alarm log with only process variables is available for the analysis. This log includes both the normal sequence and the flood sequence of the alarms. With the help of data mining and analysis methods, KPI's are calculated and the results are provided to the user. The flood sequences are used in this case to find out the similarity and correlations. A pseudo-code for the methodology is provided above.

4 Case Study and results

This section illustrates the implementation of the developed method. An anonymized industrial alarm and event log is used to demonstrate the application. The details of the used dataset are given in Table 1. The data for three days of a plant is used for the analysis with 8793 alarms and events in three days for 192 unique instrument tags. After the preprocessing *i.e.* removing the system alarms and other undesired events from the log, the dataset is reduced to 4513 items.

Table 1: Dataset details

Detail	Value
No. of total items (alarms + events)	8793
No. of total tags	192
Total number of days	3
Items after preprocessing	4513

The pre-processed items were analyzed to find the following Key Performance Indicators - average alarms per day, peak alarms/10 minutes, average alarms/ 10 minutes and priority distribution. These values were compared against the values prescribed in EEMUA 191 guideline [6] and ISA 18.2 standard [1]. The results are summarized in Table 2. It is observed that there are a significant number of average alarms on Day 2 and Day 3 operations (almost more than 3 times as specified in the standard). The number of peak alarms/ 10 minutes is considerably higher on these days suggesting possible alarm floods during these days. The average alarms/ 10 minutes is also high, which is indicative of a poor alarm management strategy being in place. The prioritization captured in this case is observed as 63/30/5 for low/medium/high priority settings.

After the pre-processing and KPI's calculation, the heat map ("*a graphical representation of data in the form of a map or diagram in which data values are represented as colors*") for the top 45 alarms tags with the highest count appearing during the 24 hour time period is plotted. The data values are shown as colors in Figures 4, 5, and 6; color bar on the right side of the heat map provides the number of times the alarm occurred. The rectangular block shows the count of alarms per 10 minutes for each tag. The python seaborn library is used to produce the heat maps. It is observed from the heat maps that even for Day 1 there are some bad actors which appear as alarms several times during a 10-minute frame. For *e.g.* tag 142 appeared during 1040-1050 minutes time frame 6 times alone. During the same time frame, there are a few other tags that appeared multiple times resulting in a total number of alarms per 10 minutes to be

>10 alarms. Hence, such maps can be used to perform the tag-wise detailed analysis and observe the patterns and bad actors during an alarm flood or normal sequences. A similar analysis is shown for Day 2, and Day 3 which shows a significant number of flood patterns due to several tags activating alarms in a 10-minute time frame. The similar pattern was observed during the KPI calculation for Days 2 and 3.

A flood sequence may contain several tags. To understand the relationship between alarm flood sequences; the correlation plots are developed as shown in Figure 7. In this case, we are using correlation test to find an association or related alarms between different alarm flood sequences. As observed from the figure, sequence 203 and 189 and 204 and 190 are highly correlated, which indicates that serval alarm tags are common in both the flood sequences. Similarly, the analysis is performed on different alarm flood sequences to find the similarity between the sequences. The overall accuracy of this method is 93%.

Table 2: Benchmarking results of alarm and event log against the KPI

KPI	EEMUA 191[6]	ISA 18.2[1]	Analysis Result
Average alarms per day	<144 (up to 288 may be manageable)	~ 150 (~300 may be manageable)	Day 1: 142 Day 2: 504 Day 3: 526
Peak alarms /10 minutes	<10	<=10	Day 1: 11 Day 2 : 39 Day 3: 41
Average alarms / 10-minutes interval	1	~1(~2may be manageable)	Day 1: ~1 Day 2 : 2.1 Day 3: 2.2
Priority Distribution % (low/med/high)	80/15/5	80/15/5	63/30/5 (L/H/E)

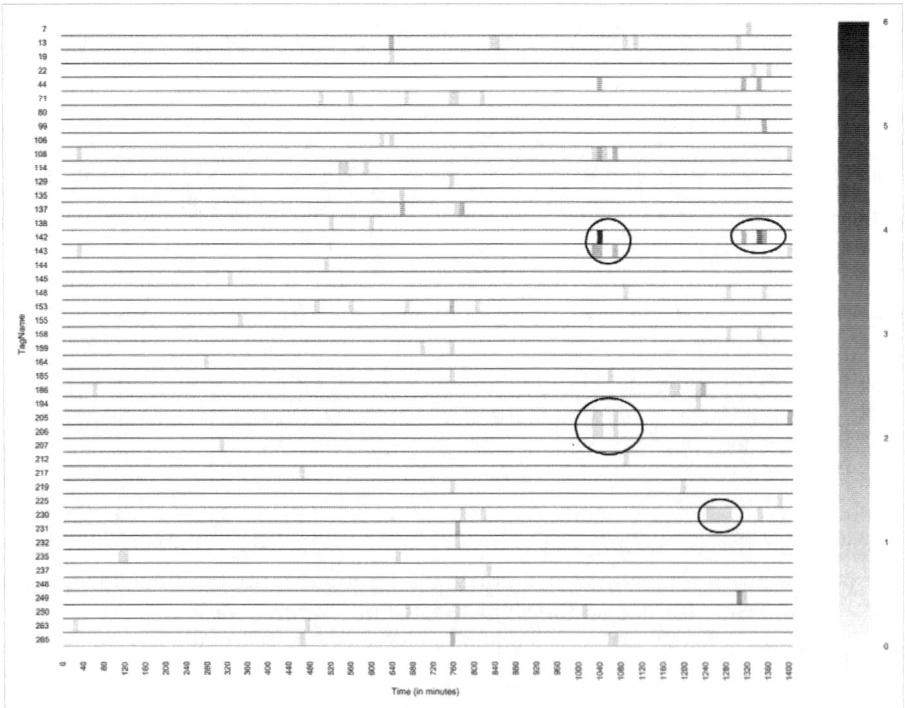

Figure 4: Log analysis Day 1

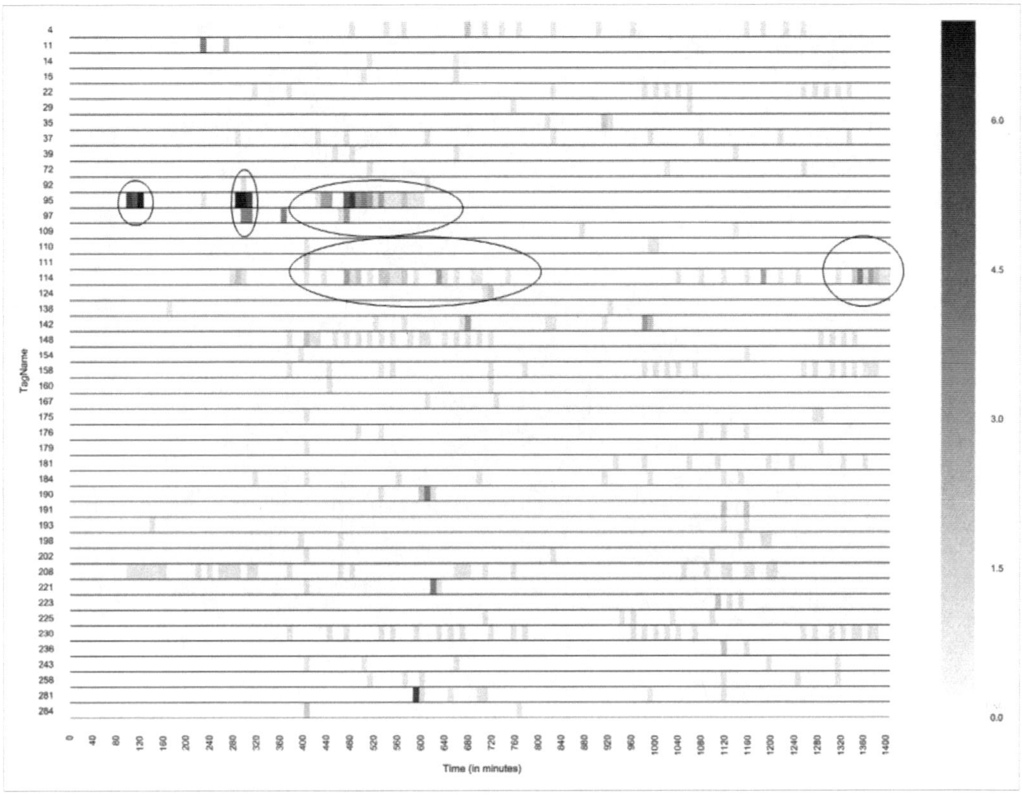

Figure 5: Log analysis Day 2

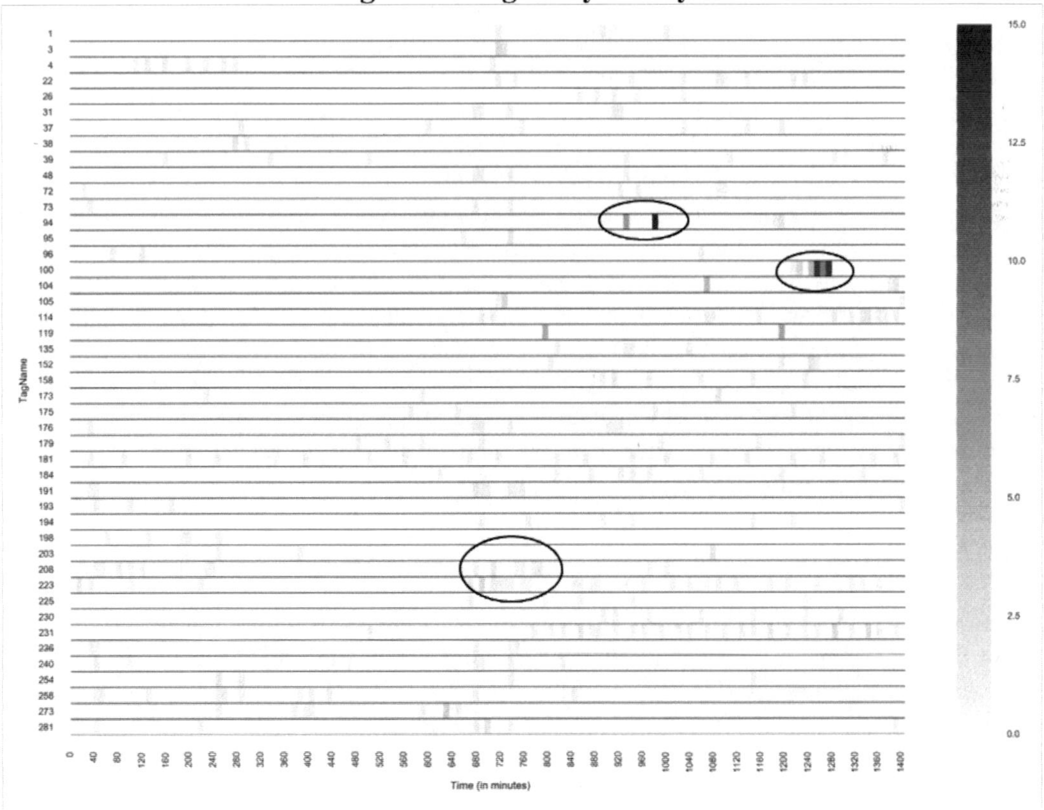

Figure 6: Log analysis Day 3

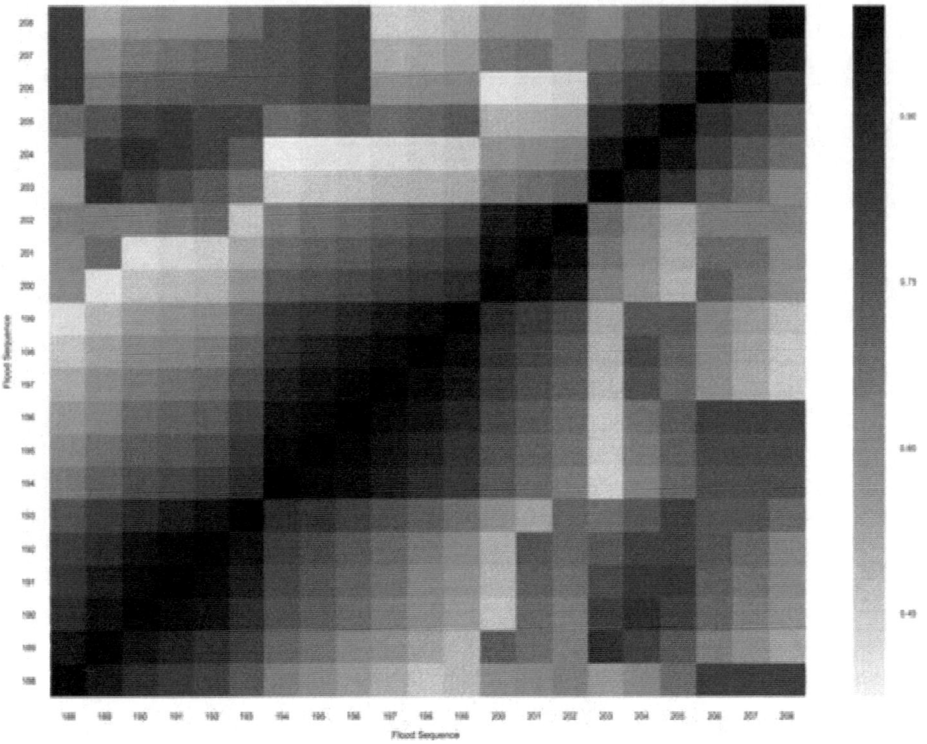

Figure 7: Flood sequence analysis Day 3

5 Conclusions

Industrial alarm system management has improved over the past few years. However, there are still some critical challenges related to alarm management, which need to be addressed. The advances in automation technology and the increase in connected devices have resulted in a higher number of alarms, poor system performance, additional workload on operators, and in some cases has led to abnormal situations. To provide a solution to these challenges, we conceptualized an alarm management framework with four distinct levels - *design, rationalized, advanced*, and *intelligent*. In addition, a generic method developed with the use of data mining and analytics theory is used to perform the offline analysis of alarm and event logs. An anonymized industrial dataset is used to demonstrate the proposed method. The KPI's were calculated and benchmarked against the available guidelines and standards. The heat map and correlation maps were plotted for the dataset to infer the data with ease and provide meaningful information about the bad actors and assist in the overall decision making. The results from these methods can be used as an input to the alarm rationalization process.

6 Acknowledgments

The authors are thankful for the valuable suggestions of the anonymous reviewers, which have improved the quality of this work. The authors acknowledge receiving useful suggestions from Mr. Aniket Bonde a student at Texas A&M University.

References

1. ANSI/ISA-18.2, *Management of Alarm Systems for the Process Industries*. 2016.
2. Jain, P., et al., *A Resilience-Based Integrated Process Systems Hazard Analysis (RIPSHA) Approach: Part I Plant System Layer.* Process Safety and Environmental Protection, 2018.
3. Goel, P., A. Datta, and M.S. Mannan, *Industrial alarm systems: Challenges and opportunities.* Journal of Loss Prevention in the Process Industries, 2017. **50**: p. 23-36.
4. MANNAN, M. and H. WEST, *1.6 Process Alarm Management.* 2006.
5. Jain, P., et al., *Did we learn about risk control since Seveso? Yes, we surely did, but is it enough? An historical brief and problem analysis.* Journal of Loss Prevention in the Process Industries, 2017. **49**: p. 5-17.
6. EEMUA, *191-Alarm Systems: A Guide to Design, Management and Procurement Edition 3*. 2013.
7. Stauffer, T., N. Sands, and D. Dunn. *Alarm Management and ISA-18–a Journey, not a Destination*. in *Texas A&M Instrumentation Symposium*. 2010.
8. Rothenberg, D.H., *Alarm management for process control.* Momentum, 2009.
9. Goel, P., A. Datta, and M.S. Mannan. *Application of big data analytics in process safety and risk management*. in *2017 IEEE International Conference on Big Data (Big Data)*. 2017.
10. Grosdidier, P., et al., *A path forward for DCS alarm management.* Hydrocarbon processing, 2003. **82**(11): p. 59-68.
11. Van Rossum, G. *Python Programming Language*. in *USENIX Annual Technical Conference*. 2007.
12. John W. Ratcliff, D.E.M. *Pattern Matching: The Gestalt approach*. 1987 [cited 2018; Available from: http://collaboration.cmc.ec.gc.ca/science/rpn/biblio/ddj/Website/articles/DDJ/1988/8807/8807c/8807c.htm.

514059 Image Classification in Manufacturing Analytics: Improving a Pellet Classification System with Deep Neural Networks

Abstract:

Manufacturing analytics is of paramount importance in many plants today, and its significance increases in the current context of big data analytics and industry 4.0 initiatives[1]. The fields of statistics, chemometrics, and machine learning are expected to provide tools that effectively handle data characteristics, such as process measurements (e.g. temperature, pressure), spectra from online analyzers and quality control labs, pictures taken with standard or hyper-spectral cameras, and all other data sources that provide insight into the state of product processes and quality.

The task of image classification[2] is considered in this paper. A supervised learning problem is presented whereby an image is the input and the output is a unique label attributed to the image from a set of available classes. Image classification is one of the main tools used for quality assurance and control at Dow Chemical, and developing a suitable classifier is a relevant industrial challenge due to accuracy and robustness requirements. In this context, recent developments in deep learning[3] that have proven successful in increasing image classification accuracy and providing state-of-the-art results in computer vision. Traditional approaches utilized for image classification are based on prior knowledge and pre-defined features. However, in this work, we leverage deep neural networksâ€™ (DNN) ability to automatically learn features from images[4] and test their performance in a real industrial context of pellet shape prediction. Moreover, other less complex techniques such as partial least squares discriminant analysis (PLS-DA) and random forests (RF) are explored in order to assess the benefits of adopting DNN as opposed to a more traditional classifier.

PLS-DA, RF, and DNN models were developed for two classification tasks: pellet shape classification (distinguishing good and bad pellets), and detecting tails in a pellet (distinguishing whether a pellet contains tails or not). After developing these models, the results were consistent for both classification objectives. Compared to the in situ classification system currently used at Dow Chemical, RF were able to better utilize the same pre-defined features and improve prediction accuracy significantly, while PLS-DA had the worst performance. DNN obtained the highest accuracy overall (higher than 96%), and the main advantage is that there is no need to specify a priori features because they are extracted from the raw image itself. Furthermore, visualizing the output of some layers of the network showed that activations occurred in regions that are meaningful for the classification tasks, further supporting that DNN effectively modeled the relevant features of the pellet.

References

1. Reis, M. and G. Gins, *Industrial Process Monitoring in the Big Data/Industry 4.0 Era: from Detection, to Diagnosis, to Prognosis.* Processes, 2017. **5**(3): p. 35.
2. Duchesne, C., J.J. Liu, and J.F. MacGregor, *Multivariate image analysis in the process industries: A review.* Chemometrics and Intelligent Laboratory Systems, 2012. **117**(Supplement C): p. 116-128.
3. LeCun, Y., Y. Bengio, and G. Hinton, *Deep learning.* Nature, 2015. **521**(7553): p. 436-444.
4. Simonyan, K. and A. Zisserman, *Very deep convolutional networks for large-scale image recognition.* arXiv preprint arXiv:1409.1556, 2014.

513668 Immersive Introduction to Iot

Abstract:

ABSTRACT for AIChE 2018 Spring Meeting

Immersive Introduction to IoT

Joel Berg

The MathWorks, Inc.

joel.berg@mathworks.com

508-647-4415

You've heard of the "Internet of Things" (IoT), and have a notion of it, but what is it really? And how many presentations do you attend where the speaker actually asks you to turn on and use your phone?!

This talk will begin with a short overview of some common building blocks in IoT architecture, including sensing and measurement devices, wireless communications, cloud-based data aggregation, real-time model updating, and decision/action rules taken by cyber or physical actors.

Now the fun starts, with the presentation becoming more of a group "event." We will describe a process-oriented scenario, and ask audience members with smartphones to participate as actors within a living, cyber-physical model. Dynamic information will be shown and continuously updated on the main screen, and participants will play active roles according to signals they are seeing displayed and a brief set of instructions, by way of their devices (smartphones). The aggregated inputs of the audience will govern the system state evolution, which will also be displayed. The event will end with a short discussion and inventory of observations and insights deriving from the demonstration.

The objectives of this talk/event are to educate attendees about IoT, introduce the ThingSpeak® platform, provoke discussions about IoT use cases, and, last but not least, have some fun.

512435 Comparing Several Methods for Product Properties Prediction

Abstract:

Comparing several methods for product properties prediction

Jean Jérôme Da Costa[1], Benoit Celse[1], Fabien Chainet[1], Marion Lacoue Nègre[1], Cyril Ruckebush[2], Didier Espinat[1],*

1. *IFP Energies Nouvelles, Rond-Point de l'Echangeur de Solaize, 69360 Solaize, France*
2. *LASIR CNRS, Université Lille Nord de France, Sciences et Technologies, Cité scientifique, bât C5, 59665 Villeneuve d'Ascq Cedex - France*

**E-mail: benoit.celse@ifpen.fr*

HIGHLIGHTS

- Comparison between linear regression models and interpolation methods on toy examples
- Interpolation methods are much more efficient especially for non linear data
- Comparison between kriging and spline interpolation
- Kriging models can provide a posteriori analysis of the results
- Application on cloud point and viscosity index prediction

INTRODUCTION

In order to respect specification, petroleum products properties prediction is an increasingly need for refining industry. Machine learning is then more and more used to develop predictive models from process and physicochemical analytical data. The most classical models are based on classical (multilinear and nonlinear) regression referring to some equations which describe physicochemical phenomena or derived from empirical observations (Riazi 2005). However, regression models require to identify the adapted analytical expression for the regression function. That may be very complex in some situations.

Many chemometrics applications in modeling petroleum products properties were also proposed using partial least squares (PLS) regression (Braga, Jez Willian B., Junior, Araci Araújo Dos Santos et Martins 2014). Although these models provide relatively good performances, they use spectroscopic data such as NMR (nuclear magnetic resonance) or NIR (Near Infrared spectroscopy). This type of data are not currently available for refiners.

Some authors (Piloto-Rodríguez et al. 2013) proposed to use artificial neural networks (ANNs) for modeling complex properties. The results are good. However, the optimization of ANN requires to perform very stringent statistical steps to insure the model robustness and a lot of effort to prevent from over-fitting. This might be very tricky.

The aim of this work is to compare interpolation and linear regression methods on toy and real examples (could point, viscosity index ...) in order to suggest the best methodology to use for product properties prediction.

MATERIALS & METHODS

Kriging is an interpolation method based on stochastic multiGaussian assumptions (Roustant, Ginsbourger et Deville 2012)). Its goal is to provide the best linear unbiased estimate (B.L.U.E.) and was first used in applied geostatistics for modeling properties related to natural resources. More recently, other applications of kriging were developed such that metamodeling, which is an increasingly need in computing sciences (DiceKriging, DiceOptim: Two R Packages for the Analysis of Computer Experiments by Kriging-Based Metamodeling and Optimization 2012). One major point of kriging is the ability to provide a measure of the prediction uncertainty that depends on the data configuration (Roustant, Ginsbourger et Deville 2012).

The aim of splines interpolation is to draw contour lines "as smooth as possible", that is a map which looks like what a draftsman would obtain manually (Rooij, P. L. J. van et Schurer 1973). The splines consist of polynomials of degree m being local. The polynomials describe pieces of a line or surface. For degree m=1,2 or 3, a spline is called linear, quadratic or cubic respectively. This method is currently used in non-parametric statistical learning (Schumaker 2015).

RESULTS & DISCUSSIONS

Results on toy examples point out that interpolation methods have a high capacity of adaptability to linear or nonlinear situations without requiring previous analysis. Although, kriging and splines provide very close performance in some situations, kriging is more valuable since it provides a measure of uncertainties related to the predicted value based on stochastic assumptions. It is based on a specific distance between the new sample to predict and the calibration data base. This point is very important in modeling physicochemical properties of petroleum products. It is even particularly essential when the number of descriptors increases significantly (higher than 3).

Kriging and MLR models are developed and compared in real examples: cloud points and viscosity index. Although MLR provides good performances that are in accordance with the uncertainties of the standard measure, kriging enables to improve accuracy. In our knowledge, this is the first kriging model for the prediction of these two properties.

Globally, interpolation methods are well adapted to the modeling of petroleum products properties since they provide good performances when the number of samples is limited. Use of kriging methods to high dimensional study remains always challenging.

Bibliography

Braga, Jez Willian B.; Junior, Araci Araújo Dos Santos; Martins, Ingrid S. (2014) Determination of viscosity index in lubricant oils by infrared spectroscopy and PLSR. In : Fuel, vol. 120, p. 171–178. DOI: 10.1016/j.fuel.2013.12.017.

DiceKriging, DiceOptim: Two R Packages for the Analysis of Computer Experiments by Kriging-Based Metamodeling and Optimization (2012) : University of California, Los Angeles.

Piloto-Rodríguez, Ramón; Sánchez-Borroto, Yisel; Lapuerta, Magin; Goyos-Pérez, Leonardo; Verhelst, Sebastian (2013) Prediction of the cetane number of biodiesel using artificial neural networks and multiple linear regression. In : Global Conference on Renewable energy and Energy Efficiency for Desert Regions 2011 "GCREEDER 2011", vol. 65, n° Supplement C, p. 255–261. DOI: 10.1016/j.enconman.2012.07.023.

Riazi, M. R. (2005) Characterization and properties of petroleum fractions. 1st ed (Online-Ausg.). West Conshohocken, Pa : ASTM International (ASTM manual series MNL 50).

Rooij, P. L. J. van; Schurer, F. (1973) A bibliography on spline functions. Eindhoven : Technische Hogeschool (TH report / Technische Hogeschool Eindhoven Nederland, 73-WSK-01).

Roustant, Olivier; Ginsbourger, David; Deville, Yves (2012) DiceKriging, DiceOptim: Two R Packages for the Analysis of Computer Experiments by Kriging-Based Metamodeling and Optimization. In : Journal of statstical software, vol. 51, n° 1.

Schumaker, Larry L. (2015) Spline functions. Computational methods. Philadelphia : Society for Industrial and Applied Mathematics (Other titles in applied mathematics, 142).

513857 Model-Based Development of High-Purity Hydrogen Production Unit for Hydrogen Refueling Stations

Abstract:

A conceptual high-purity hydrogen production unit (100 kg hydrogen per day), comprising methane-steam reforming, water-gas shift reactors and pressure-swing adsorption process, for hydrogen refueling stations was developed by the model-based methodology. The system should satisfy the main prerequisites for actual application such as 1) higher than 70 % hydrogen production efficiency on the basis of higher heating values and 2) lower than 10 ppm of CO concentration at the outlet of the CO removal process. The heat exchanging system between heat sources and sinks in the reforming process was optimized by the commercial process simulation software package and the heat and mass transport phenomena in the steam-methane were also intensively studied by the computational fluid dynamic models. In addition, the optimal vacuum pressure swing adsorption process was also developed to purify the hydrogen from the product gas of the water-gas shift reactor. Thus, the developed system was tested and the operating data was compared with the mathematically predicted values so that the fidelity of the model-based process development techniques is evaluated. The proposed model-based process development techniques will contribute to the optimization and scale-up of the process.

Designing for Innovation: Process and Technology Configurations for Oil Sands Production

Evar C. Umeozor, Ian D. Gates
Department of Chemical and Petroleum Engineering, University of Calgary

Abstract

Bitumen from the oil sands is an unconventional oil resource and the major source of Canadian oil production, representing about 70% of total daily volumes. However, the high viscosity of the oil and the nature of its deposition necessitate the use of special tools and recovery techniques that are currently energy, water and emissions intensive. Recent lows in the price of crude oil and evolving environmental conservation goals call for the development of processes and technologies to improve operational performance. This work presents a mixed-integer linear optimization model aimed to simultaneously improve three operational performance indicators representing the economic, emission, and energy intensities of the processes and technologies deployable in a Greenfield facility. The proposed model identifies Pareto-optimal facility configurations for in-situ heavy oil recovery from the oil sands. The optimization criterion is formulated as a weighted sum of the performance indicators of the facility. The performance attributes are captured as the greenhouse gas emissions, process heat and electricity requirements, and the capitalized costs of the facility. Facility operation is constrained to account for field conditions via governing physical equations and other policy-related requirements. By identifying individual unit operations in the bitumen recovery process, covering both the surface and sub-surface elements, existing challenges/limitations of current in-situ recovery methods are used to highlight process and technology improvements that lead to enhanced overall performance. We demonstrate that the boiler, separator, and reservoir segments of the typical in-situ oil sands operation are key areas for improving the performance indicators of the production operations. Overall, solvent-based recovery is observed to be very promising from environmental and efficiency perspectives but very sensitive to the economics of the solvent being used. Following solvent-based methods is steam with non-condensable gas (NCG) addition. Steam and NCG injection offers a relatively balanced alternative with regards to the process performance indicators. Our results highlight the importance of integrating sub-surface and surface decisions when assessing innovations in oil sands process designs.

514279 Model-Guided Optimization of Industrial Azeotrope Distillation Column

Abstract:

A highly non-ideal mixture is separated using azeotrope distillation by introducing an entraining agent. The key components and process impurities in the feed along with the entrainer form multiple azeotropes.

The column exhibited chronic underperformance leading to significant loss of valuable chemicals. A rigorous model was developed using literature and proprietary data to represent the highly non-ideal coupled system. Although the model was not able to exactly match plant data, it was successfully used as a tool to improve the performance of the industrial column. In this presentation, the authors will be sharing the applied methodology.

513180 Knowledge Transfer and Support of New Technology Acquisitions

Abstract:

Successful acquisition of new chemical process technologies requires many steps both from a commercial and technical due diligence standpoint. Two of the most important steps are the meticulous transfer the entire knowledge base to the acquiring company followed by the development of plans for continued technology support after the commercial deal is completed.

In many instances, the technology seller provides limited technology support for a finite time after the commercial deal is complete, making it imperative that every effort be made to efficiently transfer as much of the technology knowhow as possible before support ends.

The plan going forward may include the retention of key technical employees that were intimately involved in the development of the technology and/or sales force employees that know the market and customer base, but ultimately the acquiring company must include the development of these critical resources within its own operating footprint.

Those resources include research and development, engineering, marketing and licensing as well as many other services key to the commercialization, sales and support of the newly acquired technology.

The entire process is illustrated in the cases studies of recent acquisitions by TechnipFMC.

513334 Optimal Selection and Design of a Methanol Process with Enhanced CO_2 Utilization

Abstract:

The recent discoveries of significant reserves of shale gas have spurred various monetization pathways including the production of methanol. There are several routes to convert shale/natural gas to methanol. The decisions for technology selection and process design were typically based on techno-economic criterial. Because of the growing interest in reducing the greenhouse gas emissions of gas processing, carbon footprint is emerging as a primary criterion in the selection and design of the process. The purpose of this work is to carry out process synthesis, selection, and design of a shale-gas to methanol process with multiple objectives including profitability and carbon footprint. Several reforming pathways along with the associated downstream processing are considered. In addition to the screening of individual types of reforming, the study also considers combined reforming of methane which incorporates three types of reforming, namely steam reforming, partial oxidation, and dry reforming. It is desired to consider existing technologies and to also modify the design of an existing methanol technology to include combined reforming. Utilization of CO_2 is addressed while maintaining the suitable syngas ratio for methanol synthesis, and minimizing CO_2 emissions, waste water generation, and energy requirements of the overall process. A methanol from natural gas production plant with a capacity of 5000 metric tonnes per day is considered as the base case scenario. The plant employs conventional steam reforming to generate syngas for methanol synthesis. Aspen HYSYS along with optimization software (LINGO) are used to simulate the process scenarios and to determine the optimum reforming configuration, respectively. Mass and energy integration are performed to improve the sustainability and profitability of the process, through identifying water and heat sinks in the design. A techno-economic analysis of the proposed design is conducted, including the cost of retrofitting, carbon taxes, and utility requirements.

AUTHOR INDEX

AUTHOR INDEX